Lecture Notes in Networks and Systems

Volume 114

The series "Lecture Notes in Networks and Systems" publishes the latest developments in Networks and Systems—quickly, informally and with high quality. Original research reported in proceedings and post-proceedings represents the core of LNNS.

Volumes published in LNNS embrace all aspects and subfields of, as well as new challenges in, Networks and Systems.

The series contains proceedings and edited volumes in systems and networks, spanning the areas of Cyber-Physical Systems, Autonomous Systems, Sensor Networks, Control Systems, Energy Systems, Automotive Systems, Biological Systems, Vehicular Networking and Connected Vehicles, Aerospace Systems, Automation, Manufacturing, Smart Grids, Nonlinear Systems, Power Systems, Robotics, Social Systems, Economic Systems and other. Of particular value to both the contributors and the readership are the short publication timeframe and the world-wide distribution and exposure which enable both a wide and rapid dissemination of research output.

The series covers the theory, applications, and perspectives on the state of the art and future developments relevant to systems and networks, decision making, control, complex processes and related areas, as embedded in the fields of interdisciplinary and applied sciences, engineering, computer science, physics, economics, social, and life sciences, as well as the paradigms and methodologies behind them.

**** Indexing: The books of this series are submitted to ISI Proceedings, SCOPUS, Google Scholar and Springerlink ****

More information about this series at http://www.springer.com/series/15179

Atef Zaki Ghalwash · Nashaat El Khameesy ·
Dalia A. Magdi · Amit Joshi
Editors

Internet of Things—Applications and Future

Proceedings of ITAF 2019

Springer

Editors
Atef Zaki Ghalwash
Faculty of Computers
and Information
Helwan University
Helwan, Egypt

Nashaat El Khameesy
Information and Computer
Systems Department
Dean Sadat Academy
Cairo, Egypt

Dalia A. Magdi
French University in Egypt
Cairo, Egypt

Amit Joshi
Global Knowledge Research
Foundation
Ahmedabad, India

ISSN 2367-3370 ISSN 2367-3389 (electronic)
Lecture Notes in Networks and Systems
ISBN 978-981-15-3074-6 ISBN 978-981-15-3075-3 (eBook)
https://doi.org/10.1007/978-981-15-3075-3

This Springer imprint is published by the registered company Springer Nature Singapore Pte Ltd.
The registered company address is: 152 Beach Road, #21-01/04 Gateway East, Singapore 189721, Singapore

Preface

This volume contains the papers presented at ITAF 2019: The First World Conference on Internet of Things: Applications & Future, held at Cairo, Egypt, 14 & 15 October, 2019; collaborated by Global Knowledge Research Foundation. The associated partners were Springer, InterYIT IFIP. The 1st ITAF congress featured two days of focused networking and information sharing at the IOT cutting edge. This first edition brought together researchers, leading innovators, business executives, and industry professionals to examine the latest advances and applications for commercial and industrial end users across sectors within the emerging Internet of Things ecosphere. It targeted state-of-the-art, as well as emerging topics related to Internet of Things such as Big Data Research, Emerging Services and Analytics, Internet of Things (IOT) Fundamentals, Electronic Computation and Analysis, Big Data for Multi-discipline Services, Security, Privacy and Trust, IOT Technologies, and Open & Cloud technologies.

The main objective of the conference was to provide opportunities for the Researchers, Academicians, Industry persons, Students, and expertise from all over the world to interact and exchange ideas and experience in the field of Internet of Things. It also focuses on innovative issues at international level by bringing together the experts from different countries.

It introduced emerging technological options, platforms, and case studies of IOT implementation in areas by Researchers, leaders, engineers, executives, and developers who will present the IOT industry which are dramatically shifting business strategies and changing the way we live, work, and play.

The ITAF Conference incited keynotes, case-studies, and breakout sessions, focusing on smart solutions leading Egypt in IOT technologies into 2030 and beyond.

The conference started by the welcome speech of Assoc. Prof. Dalia A. Magdi conference chair, ITAF 2019, Head of Information Systems Department, French University in Egypt, followed by the speech of Prof. Taha Abdallah, acting President of the French University in Egypt; Dr. Amit Joshi, Organizing Secretary, ITAF 2019, Director of Global Knowledge Research Foundation; Mr. Aninda Bose, Sr. Publishing Editor, SpringerNature.

On behalf of ITAF2019 board we thank all respectable Keynote speakers Eng. Mohamed Safwat IoT Innovation Leader & Senior Consultant, Orange Business Services; Dr. Ahmed Samir Lead Data Scientist- Ph.D., Big Data Team, Vodafone; Eng. Omar Mahmoud Software Engineer, R&D Office of the CTO, Dell Technologies; Eng. Ahmed Abdel Bakey Software Engineer, Data and AI Expert Labs, IBM, Egypt; Prof. Mike Hinchey, Chair, IEEE, UK and Ireland section, Director of Lero, and Professor of Software Engineering at the University of Limerick, Ireland; Nilanjan Dey, Ph.D., Asst. Professor Department of Information Technology, Techno India College of Technology, Rajarhat, Kolkata, India; Mr. Mihir Chauhan, entrepreneur and researcher, Assistant Professor at Sabar Institutes of Technology for Girls and Venus International College of Technology; Prof. Alaa El Din M. El Ghazali Professor of Computer and Information Systems, Sadat Academy for Management Sciences, former President of Sadat Academy for Management Sciences; Prof. Galal Hassan, professor of information systems engineering at the Faculty of Computers and Information, Cairo University, certified usability analyst & internationally certified trainer; Prof. Nevine Makram Labib, Professor of Computer Science and Information Systems, head of Computer and Information Systems department, Sadat Academy for Management Sciences (SAMS); and Prof. Kamal ElDahshan, faculty of Science, Al-Azhar University.

A lot of researches were submitted in various advanced technology areas, 31researches were reviewed and accepted by the committee members to be presented and published. There were 6 technical sessions in total, and talks on academic and industrial sector were focused on both the days.

On behalf of conference chairs and editors, we owe a special thanks to Prof. Layla Abo Ismaeil Member of Parliament & Secretary of Education and Scientific Research Committee; Prof. Naglaa El Ahwani Chairman of Board of Trustees of French University in Egypt for their attendance. We also thank Orange Business Services, Vodafone, Dell Technologies, and IBM, Egypt for their support and participation, we also thank all keynote speakers, researchers, and attendees of this conference.

Cairo, Egypt

Assoc. Prof. Dalia A. Magdi
Conference Chair, and Editor

Contents

Editors and Contributors

About the Editors

Prof. Atef Zaki Ghalwash is a Professor and Dean of the Faculty of Computers & Information, Helwan University, Egypt. He holds an M.Sc. from Cairo University, Egypt, and a Ph.D. from the University of Maryland, USA. He established the first BLE & IoT lab at the Faculty of Computers & Information, HU. With research interests in AI, IoT, and security, he has published articles and technical reports in various international journals and conference proceedings, in addition to several books on computing and informatics.

Prof. Nashaat El Khameesy is a distinguished Professor of Computers and Informatics, Sadat Academy, Egypt. He has authored more than 250 publications and technical reports in many international journals and conferences. He is currently Dean of the New Cairo Academy for Management Sciences, Cairo, Egypt, vice-president of the AMSE (since 2002), and a co-founder of the Egyptian Society of Information Systems and Computer Technology (ESIS-ACT).

Assoc. Prof. Dalia A. Magdi is Chair of the ITAF Conference. She is currently Head of the Information Systems Department, French University in Egypt; and a member of the Editorial Board or reviewer for many international journals, such as the SCIREA Journal of Information Science, SCIREA Journal of Computer Sciences, Internet of Things and Cloud Computing (IOTCC) Journal, Horizon Journal of Library and Information Science, and JCSS in the areas of Computer Science and Security. She has published many books internationally, such as A Proposed Enhanced Model for Adaptive Multi-agent Negotiation Applied On E-commerce, LAP LAMBERT Academic Publishing.

Dr. Amit Joshi is Chair of InterYIT, IFIP, and Director of the Global Knowledge Research Foundation. He is an active member of the ACM, IEEE, CSI, AMIE, IACSIT Singapore, IDES, ACEEE, NPA, and many other professional societies.

Currently, he is a Chairman of the Computer Society of India (CSI) Udaipur Chapter and Secretary of the ACM Udaipur Professional Chapter. He has presented and published more than 50 papers in national and international journals/ conferences of the IEEE and ACM. He has also organized more than 40 national and international conferences and workshops through the ACM, Springer, and IEEE. He is currently the Director of Emanant TechMedia (P) Limited, which focuses on ICT and web-based services.

Contributors

Ahmed H. Abase Computer Science Department, Cairo University, Giza, Egypt

Sami A. Abd El-hafeez Faculty of Science, Port Said University, Port Said, Egypt

Hoda Abdelhafez Faculty of Computer & Informatics, Suez Canal University, Ismailia, Egypt;
College of Computer & Information Sciences, Princess Nourah University, Riyadh, Kingdom of Saudi Arabia

Noha MM. AbdElnapi Faculty Computer Science, Nahda University, Beni Suef, Egypt

Mona Sayed Abdul-Karim Electrical Engineering Department, Faculty of Engineering, Minia University, Minia, Egypt

Amira Hassan Abed Department of Information Systems Center, Egyptian Organization for Standardization and Quality, Cairo, Egypt

Fatmah Alamri College of Computer & Information Sciences, Princess Nourah University, Riyadh, Kingdom of Saudi Arabia

Meaad Alamri College of Computer & Information Sciences, Princess Nourah University, Riyadh, Kingdom of Saudi Arabia

Nourah Alharthi College of Computer & Information Sciences, Princess Nourah University, Riyadh, Kingdom of Saudi Arabia

Abdelmageid A. Ali Faculty of Computers and Information, Minia University, Minia, Egypt

Nabil A. Ali Suez Institute of Management Information Systems, Suez, Egypt

Mashael Al-Saud College of Computer & Information Sciences, Princess Nourah University, Riyadh, Kingdom of Saudi Arabia

Rabab Hamed M. Aly The Higher Institute for Management Technology and Information, Minia, Egypt

Shahad Alzamil College of Computer & Information Sciences, Princess Nourah University, Riyadh, Kingdom of Saudi Arabia

Abeer A. Amer Sadat Academy for Management Sciences, Cairo, Egypt

Wael A. Awad Faculty of Science, Port Said University, Port Said, Egypt

Ahmed Dahroug Technology & Maritime Transport, Arab Academy for Science, Cairo, Egypt

Hala M. Ebeid Faculty of Computer and Information Sciences, Ain Shams University, Cairo, Egypt

Gaber Abd El bake Faculty of Computer and Information Sciences, Ain Shams University, Cairo, Egypt

Mohamed B. El_Mashade Electrical Engineering Department, Faculty of Engineering, Al-Azhar University, Cairo, Egypt

Noha E. El-Attar Faculty of Computers and Artificial Intelligence, Benha University, Benha, Egypt

Mohamed Elawady Computer Engineering and Control Engineering Department, Faculty of Engineering, Tanta University, Tanta, Egypt;
Computer Engineering Department, Behera High Institute, Behera, Egypt

Khaled Elbehiery DeVry University, Denver, CO, USA

Hussam Elbehiery Ahram Canadian University (ACU), Cairo, Egypt

Omer El-emam Faculty of Computer and Information Sciences, Ain Shams University, Cairo, Egypt

Ahmed H. El-Gamal IT Department, Faculty of Computers and Information Sciences, Mansoura University, Mansoura, Egypt

M. M. Elgayar IT Dept., Faculty of Computers and Information Sciences, Mansoura University, Mansoura, Egypt

Sherif Elhennawy Information Systems Auditing Consultant, Cairo, Egypt

Menna M. Elkholy Faculty of Computer and Information Sciences, Ain Shams University, Cairo, Egypt

Shrouk A. Elmasry Faculty of Science, Port Said University, Port Said, Egypt

Doaa S. El-Morshedy Faculty of Science, Port Said University, Port Said, Egypt

Eslam Abo Elsoud Arabic Academy for Science Technology & Maritime Transport, Alexandria, Egypt

Gurjot Singh Gaba School of Electronics and Electrical Engineering, Lovely Professional University, Jalandhar, India

Menna Ibrahim Gabr Faculty of Commerce & Business Administration, BIS, Helwan University, Helwan, Egypt

Ahmad M. Gamal El-Din Sadat Academy for Management Sciences, Cairo, Egypt

Mariam Gawich French University in Egypt, Cairo, Egypt

A. Genina French University in Egypt, Cairo, Egypt

Jeanne Georges French University in Egypt, Elshorouk, Egypt

Hesham F. A. Hamed Electrical Engineering Department, Faculty of Engineering, Minia University, Minia, Egypt

Ibrahim M. Hanafy Faculty of Science, Port Said University, Port Said, Egypt

Ibrahim Hassan Faculty of Fine Arts University, Alexandria University, Alexandria, Egypt

Mohamed Samir Hassan Faculty of Computers and Information, Helwan University, Helwan, Egypt

Abdel Fatah Hegazy Arabic Academy for Science Technology & Maritime Transport, Alexandria, Egypt

Iman M. A. Helal Faculty of Computers and Artificial Intelligence, Cairo University, Giza, Egypt

Noha A. Hikal IT Department, Faculty of Computers and Information Sciences, Mansoura University, Mansoura, Egypt

Ismail A. Ismail Department of Computer Science, 6 October University, Cairo, Egypt

Sara Kamal Faculty of Computers and Artificial Intelligence, Cairo University, Giza, Egypt

Mohamed H. Khafagy Computer Science Department, Fayoum University, Faiyum, Egypt

Nevine Makram Labib Sadat Academy for Management Sciences, Cairo, Egypt

Dalia A. Magdi Information System Department, French University in Egypt, Elshorouk, Egypt;
Computer and Information System Department, Sadat Academy for Management Sciences, Cairo, Egypt

Nourane Mahdy Technology & Maritime Transport, Arab Academy for Science, Cairo, Egypt

Hesham Mahmoud Management Information System, Modern Academy for Computer Science and Management Technology in Maadi, Cairo, Egypt

Sherif A. Mazen Faculty of Computers and Artificial Intelligence, Cairo University, Giza, Egypt

Alshaimaa M. Mohammed Faculty of Science, Computer Science & Mathematics Department, Suez Canal University, Ismailia, Egypt

Marwa Mostafa Data Reception, Analysis and Receiving Station Affairs, National Authority for Remote Sensing and Space Science, Cairo, Egypt

Reham R. Mostafa IS Department, Faculty of Computers and Information Sciences, Mansoura University, Mansoura, Egypt

Mohamed Mounir Communication and Electronics Department, El-Gazeera High Institute for Engineering and Technology, Cairo, Egypt

Khalid Salih Nasr Electrical Engineering Department, Faculty of Engineering, Minia University, Minia, Egypt

Mona Nasr Faculty of Computers and Information Helwan University, Department of Information Systems, Helwan, Egypt

Fatma A. Omara Faculty of Computer Science and Information, Computer Science Department, Cairo University, Giza, Egypt

Nahla F. Omran Department of Mathematics, Faculty of Science, South Valley University, Qena, Egypt

Kamel H. Rahouma Electrical Engineering Department, Faculty of Engineering, Minia University, Minia, Egypt

Mohammed Abo Rizka Arab Academy for Science Technology & Maritime Transport, Cairo, Egypt

Heba M. Sabri Sadat Academy for Management Sciences, Cairo, Egypt

Rowayda A. Sadek Faculty of Computers and Information, Helwan University, Helwan, Egypt

Amany Sarhan Computer Engineering and Control Engineering Department, Faculty of Engineering, Tanta University, Tanta, Egypt

Ahmad M. Sayed Faculty of Computer and Information Sciences, Ain Shams University, Cairo, Egypt

Basant Sayed Department of Information Systems, Higher Institute of Qualitative Studies, Garden, Egypt

M. B. Senousy Sadat Academy for Management Sciences, Cairo, Egypt

Essam Shaaban Department of Information Systems, Faculty of Computers & Information, Beni Suef University, Beni Suef, Egypt;
Canadian International Collège CIC, Zayed, Egypt

Hossam-Eldeen M. Shamardan Faculty of Computers and Information, Helwan University, Helwan, Egypt

Amr Ehab Muhammed Shokry Arab Academy for Science Technology & Maritime Transport, Cairo, Egypt

Manal Tantawi Faculty of Computer and Information Sciences, Ain Shams University, Cairo, Egypt

Rezq Thabet Faculty of Computer and Information Sciences, Ain Shams University, Cairo, Egypt

Mohamed F. Tolba Faculty of Computer and Information Sciences, Ain Shams University, Cairo, Egypt

GeoLocalitySim: Geographical Cloud Simulator with Data Locality

Ahmed H. Abase, Mohamed H. Khafagy and Fatma A. Omara

Abstract Cloud simulator is a framework which supports cloud modelling, testing functionality (e.g. allocating, provisioning, scheduling, etc.), analysing and evaluating performance, and reporting cloud computing environment. Cloud simulators save cost and time of building real experiments on real environment. The current simulators (e.g. CloudSim, NetworkCloudSim, GreenCloud, etc.) deal with data as a workflow. According to our previous work, LocalitySim simulator has been proposed with considering data locality and its effect on the task execution time. This simulator deals with splitting and allocating data based on network topology. According to the work in this paper, LocalitySim simulator has been modified and extended to support extra feature (e.g. geographical distributed data centre(s), geographical file allocation, MapReduce task execution model, etc.) with friendly graphical user interface (GUI). This modified simulator is called GeoLocalitySim. The main issue of the proposed GeoLocalitySim simulator is that it could be extended easily to support more features to meet any future module(s). To validate the accuracy of the proposed GeoLocalitySim simulator, a comparative study has been done between our proposed GeoLocalitySim simulator and Purlieus simulator.

Keywords Cloud simulator · Cloud computing · Data locality · LocalitySim simulator · GeoLocalitySim simulator · Geographical distributed data centre · Geographical file allocation · MapReduce

---

A. H. Abase (✉) · F. A. Omara
Computer Science Department, Cairo University, Giza, Egypt
e-mail: Ahmed.expert.g@gmail.com

F. A. Omara
e-mail: mhk00@Fayoum.edu.eg

M. H. Khafagy
Computer Science Department, Fayoum University, Faiyum, Egypt
e-mail: f.omara@fci-cu.edu.eg

A. Z. Ghalwash et al. (eds.), *Internet of Things—Applications and Future*,
Lecture Notes in Networks and Systems 114,
https://doi.org/10.1007/978-981-15-3075-3_1

1 Introduction

A cloud is a pool of resources that was virtualized and dynamically provisioned to act as one or more computing resource(s) based on pay-as-you-go principle as a business model [1]. The resource(s) provisioning is based on Service-Level Agreements (SLAs) between the service provider and consumers [2]. The cloud provides three types of service models: Infrastructure as a Service (IaaS), Platform as a Service (PaaS) and Software as a Service (SaaS). The cloud deployment models are private, public, community and hybrid [3]. The Cloud Provider (CP) is responsible for providing possible services to specific consumers usually as Virtual Machines (VMs), while the Cloud Broker (CB) is responsible for delivering cloud services with suitable performance from CPs to consumers [4–6].

A cloud simulator is a framework which consists of a set of libraries that have been developed by a suitable programming language to achieve specific goals. Cloud simulators create configurable cloud modules and utilities to ease building, analysing and evaluating experiments (e.g. resources allocation, task scheduling, communication cost, file management, etc.) instead of using real data centre(s) to save cost and time. The cloud simulators differ from each other with respect to some features (e.g. underlying layer, programming language, availability, graphical user interface, communication model, etc.) [7].

On the other hand, Big Data concept appeared as a result of generating data from instrumented business processes, Internet of things, monitoring of user activity, archiving data and social network sites. This intensive data has some features as a multi-V model, variety, velocity, volume and veracity [3, 8]. Any large data growth changes rapidly from multi-source with validity could to Big Data. Figure 1 shows Big Data classification.

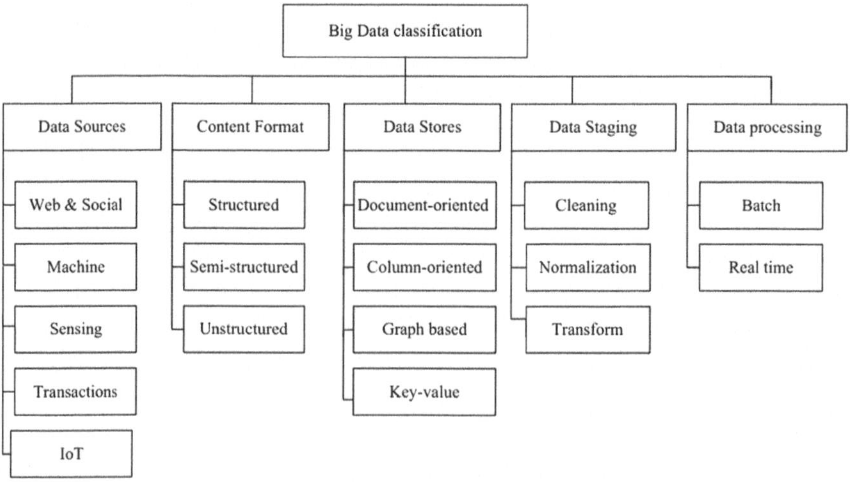

Fig. 1 Big Data classification [3]

MapReduce is a popular open-source programming model for Big Data processing and analysing across cluster(s) [9]. MapReduce deals with stored Big Data either in a file system (unstructured) or in a database system (structured). On the other hand, MapReduce deals with Big Data considering data locality to reduce communication traffic between processing units. Therefore, scheduling with MapReduce is considered a critical factor [10]. MapReduce has been implemented as a programming model paradigm by Hadoop which provides a Hadoop Distributed File System (HDFS) to split and manage Big Data across cluster(s) [11]. The most important feature of MapReduce is to hide the tangle of fault tolerance from the user. Basically, MapReduce has two main functions: (1) map function which gets its input from HDFS to produce a sequence of key–value pairs and (2) reduce function which combines the values of each specific key. Both Map and Reduce functions are controlled by master controller and executed in parallel on distributed environment (see Fig. 2) [12–17].

Geographical Distributed Data Centre (GDDC) concept is an expected paradigm to the cloud computing environment due to the growth of business and academia Big Data rapidly, and it is not acceptable to store these Big Data at single Data Centre (DC). Therefore, many companies have their public and/or private clouds with their branches located in many countries/continents. Therefore, the GDDC concept should be investigated in different issues (e.g. distributed processing frameworks, scheduler techniques, provisioning techniques, simulators, etc.).

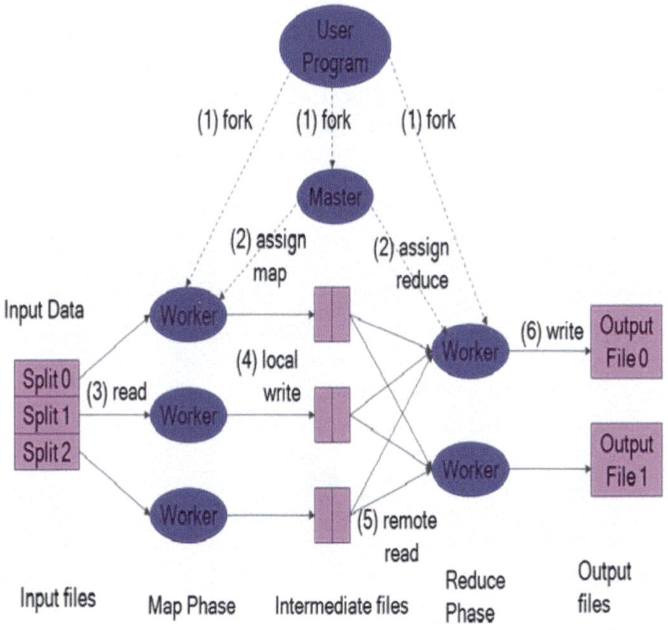

Fig. 2 MapReduce job paradigm [10]

Samples of GDDC Big Data are: (1) climate science, (2) Internet of things, (3) social network applications, (4) geographical information system, etc. [18]. A scenario for geographically distributed data processing is illustrated in Fig. 3.

CloudSim is an open-source simulator and considered the most common used to simulate cloud environment [19]. Many open-source cloud simulators like GreenCloud [20], NetworkCloudSim [21], DCSim [22], MDCSim [23], MR-CloudSim [24], CloudSimSDN [25] and LocalitySim [26] have been introduced based on CloudSim simulator to implement and evaluate different research problems such as task scheduling, resource provisioning and allocation, security, etc. Unfortunately, all existed cloud simulators except LocalitySim simulator do not support data locality and the effect of changing the file distribution across a data centre.

On the other hand, data locality is focused on data size and location in the storage devices. The data locality is classified into two types [27]:

(1) ***Temporal Locality***: it is the last data position accessed by a task and
(2) ***Spatial Locality***: it is the permanent location of the data.

Distributing Big Data across data centre(s) is achieved by placement techniques which is based on availability, reliability and Quality of Service (QoS). Any scheduling technique tries to allocate processing units near to the requested data (i.e. considering data locality) to reduce communication overheads [28, 29].

Unfortunately, the existed cloud simulators, as well as our previous LocalitySim simulator, do not support MapReduce programming model with data locality on geographical distributed data centres.

According to the work in this paper, LociltySim simulator has been extended to support geographical distribution data centres by adding new features to homogeneous/heterogeneous geographical distributed data centre, MapReduce module, geographical name node and replication module. In addition, a friendly

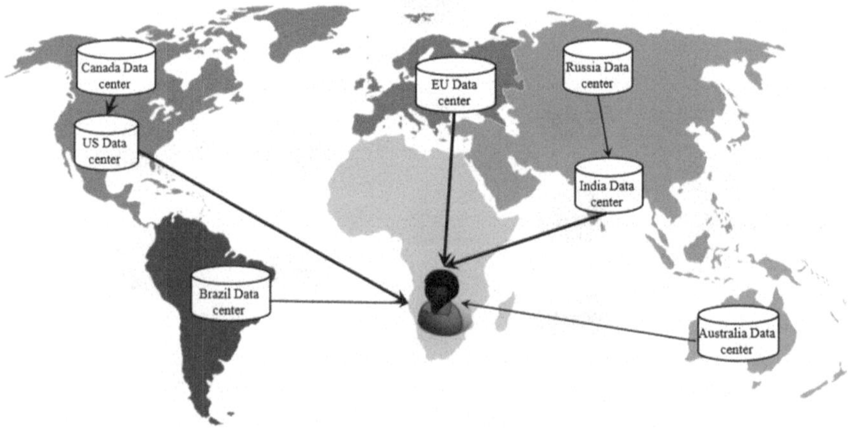

Fig. 3 A scenario for geographically distributed data processing [18]

Graphical User Interface (GUI) has been proposed to support: (1) configure MapReduce functions, (2) distribute split file(s) across geographical distributed data centre(s) and (3) choose workspace to run with appropriate technique(s). The workspace is represented as a folder which includes some XML files and text file to configure the geographical distributed data centre environments and the attached configurable scenario. This extended simulator is called GeoLocalitySim.

The remainder of this paper is organized as follows: In Sect. 2, a survey of related work is introduced. In Sect. 3, the proposed GeoLocalitySim simulator architecture is demonstrated. In Sect. 4, the validation and performance evaluation of the proposed GeoLocalitySim and its results are discussed. Finally, the conclusions and future work are illustrated in Sect. 5.

2 Related Work

CloudSim simulator is considered as an extension of GridSim simulator to simulate the cloud computing environment with many features such as (1) object-oriented programming, (2) simplicity and (3) common free programming language (Java) [19, 30]. It is an event driven simulator which modulates the cloud data centre at three layers (i.e. IaaS, PaaS and SaaS), in addition, the user level layer (see Fig. 4). CloudSim simulator supports data centre configuration, Virtual Machine (VM) provisioning and scheduling, and analysis method by tracking the history of elements of a data centre in the simulation process. Unfortunately, the CloudSim simulator does not support some features such as graphical user interface, data locality, MapReduce operation, etc. Therefore, other existed simulators have extended CloudSim simulator to inherit its features and overcome one or more drawback(s).

2.1 NetworkCloudSim Simulator

NetworkCloudSim simulator is considered an extension of CloudSim simulator by adding and modifying some features to overcome the limitation of the CloudSim simulator such as communication and application modules. The application module consists of multi-task and each task has multi-stage [21]. This module expresses actual workload as a group of tasks and each task has different states (i.e. send, receive, execute and end). Therefore, many real applications could be expressed such as multi-tier web application(s) (e.g. e-commerce), and any application consists of multi-task with multi-stage. Also, the communication module is modified by defining different types of switch modules (e.g. root, aggregate and rack).

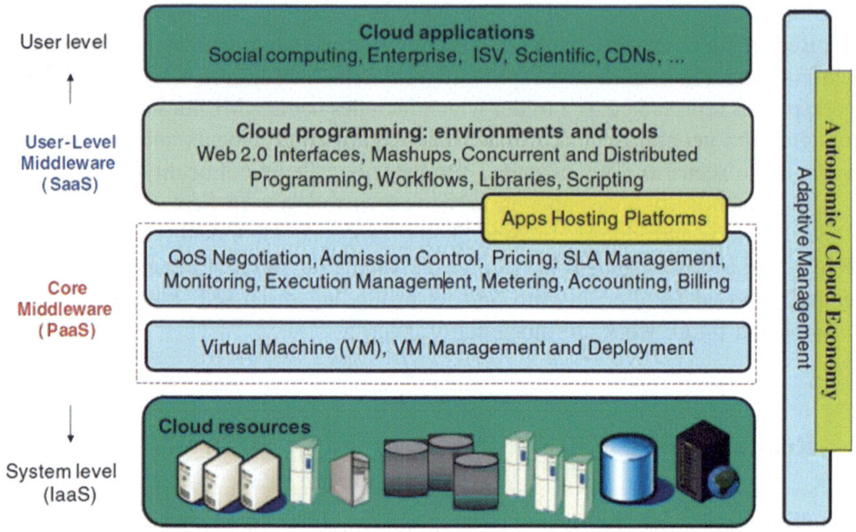

Fig. 4 CloudSim architecture [19]

2.2 CloudSimSDN Simulator

CloudSimSDN simulator is another extension of the CloudSim simulator which concentrates on VM provisioning to measure the data centre performance according to the user software-defined. CloudSimSDN simulator provides a GUI to configure the network data centre. According to CloudSimSDN simulator, different items such as number of cores to each task, virtual network to be supported, data transfer and needed computational unit are identified. The performance of CloudSimSDN simulator has been evaluated using different configurations of hosts in the data centre which guides the scenario of execution [25]. The network topology is considered the main limitation of CloudSimSDN simulator because its GUI is poor when dealing with large number of hosts.

2.3 DCSim Simulator

DCSim simulator is another extension of the CloudSim simulator which concerns with Infrastructure as a Service (IaaS). The main feature of this simulator is to manage VM provisioning, migration and sharing workload between them. The evaluated metrics which have been used in DCSim simulator are [22] as follows:

(1) ***Service-Level Agreement (SLA)***: This parameter is used to measure the SLA violation.

(2) ***Active Host***: It is used to keep tracking the host activity at any given time.
(3) ***Host Hours***: It is used to calculate the busy and idle time for each host.
(4) ***Activity Host Utilization***: Because resource utilization is considered one of the main issue of resource management, this metric is used to indicate it.
(5) ***Number of Migrations***: This parameter is used to measure the performance of the dynamic Virtual Machine allocation techniques.
(6) ***Power Consumptions***: Reducing power consumption is the main target of the majority of the dynamic Virtual Machine provisioning and migration techniques, so this metric is used to manage power consumption and
(7) ***Simulation and Algorithm Running Time***: It is used to report minimum, maximum and average execution time of the running algorithms.

2.4 MapReduce (MR)-CloudSim Simulator

MR-CloudSim simulator is also considered as an extension of CloudSim with MapReduce programming model consideration. According to MR-CloudSim, each Map function is associated with only one Reduce function. Therefore, MapReduce module might not depend on the size of the input file. MR-CloudSim simulator lacks communication features, distributed file systems (i.e. HDFS, GDFS) and has no GUI [24].

2.5 YARN Locality (YLocSim) Simulator

Apache Hadoop YARN (Hadoop 2.0) architecture is the new version of Hadoop. YARN supports MapReduce and non-MapReduce jobs. It increases efficiency and utilization of resources by separating cluster resource management component from the application management component.

YARN's components are (see Fig. 5) as follows:

1. **Resource Manager (RM)**: It monitors the cluster's resources and tests its liveness. RM is installed globally across the cluster to manage the resources between application and users. RM has two sub-components:
    - **Applications Manager (AsM)**: It is responsible for accepting applications and starting to provision the application resources across cluster by collaborating with other YARN components.
    - **Scheduler**: It is responsible for allocating the required resources to specific application only.
2. **Application Master (AM)** with resource manager is used to determine the application's required resources at each step of application.

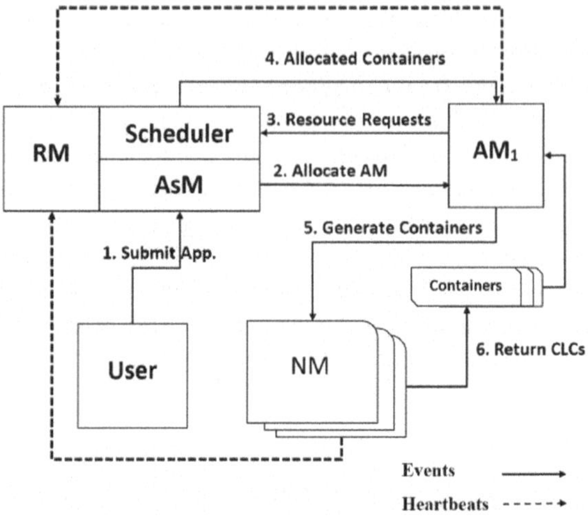

Fig. 5 YARN components interaction [31]

3. **Node Manager**: It is installed locally across each host to launch allocated resources to each application by collaborating with RM according to AM.

YLocSim simulator is introduced to YARN architecture to support locality analysis of real YARN architecture by extending Apache YARN Scheduler Load Simulator (SLS) features to support data locality analysis reporting with considering Map task only.

Figure 6 illustrates the interaction between YLocSim simulator, YARN architecture and Apache Rumen. First, Hadoop job history files are given as input to Apache Rumen to generate JSON files which contain the workload and network topology. Second, the JSON files are combined with the configurable information about YARN scheduler algorithm and are delivered to YLocSim simulator. Finally, YLocSim simulator reports the current data locality percentages to the user in real time [31].

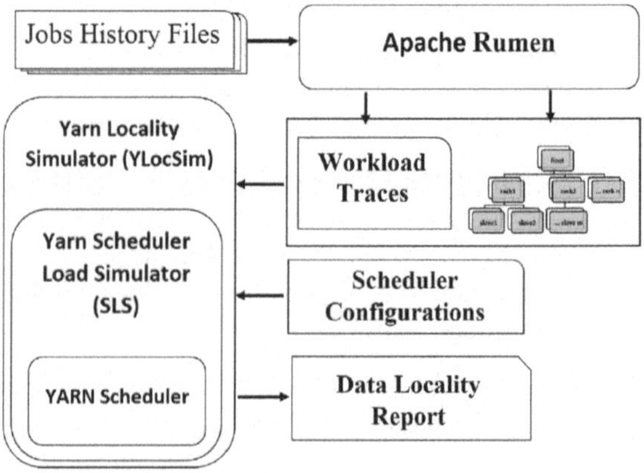

Fig. 6 YLocSim architecture [31]

2.6 *LocalitySim Simulator*

The main drawback of the existing simulators is that some of them deal with data workflow of the input file size and others consider network topology besides size of data workflow without any consideration about network topology in the data centres. According to our previous work, LocalitySim simulator has been proposed which, in turn, is considered as an extension of CloudSim and NetworkCloudSim simulators [19, 21]. The main issue of LocalitySim simulator is to support data locality. This has been satisfied by adding file allocation module to simulate the Name node at the data centre, modifying application workflow to achieve and consider the data locality effect, modifying network topology module to consider the data locality and extending data type to help simulating real data centre (see Fig. 7) [26]. Therefore, concerning data locality would help researchers to justify their results and introduce new techniques depending on data locality and would help cloud broker to make proper decisions about VM migration. LocalitySim simulator has been validated by constructing a mathematical model. To demonstrate the effect of changing data locality and network topology on the communication cost, two case studies have been implemented [26].

Generally, all existing simulators, besides our previous LocalitySim simulator, support only one data centre without any consideration about geographically distributed data centres. Table 1 presents features and limitations of the existing simulators.

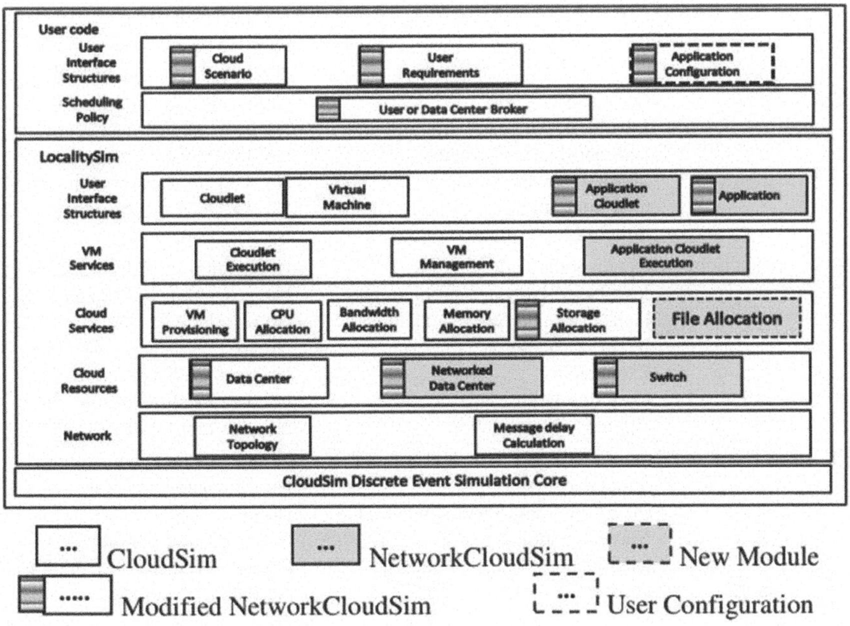

Fig. 7 LocalitySim architecture [26]

Table 1 Features and limitations of existed simulators

Simulators	Advantages	Limitations
CloudSim	Open-source code Java-based language programming IaaS and PaaS simulation Simple network topology Extendable Cost simulation	Application module limitation No geographical data centre No MapReduce programming model No distributed file system No GUI No workspace configuration No data locality consideration No data replication
DCSim	Open-source code Java-based language programming IaaS and PaaS simulation Simple network topology Extendable Cost simulation	Application module limitation No geographical data centre No MapReduce programming model No distributed file system No GUI No workspace configuration No data locality consideration No data replication
CloudSimSDN	Open-source code Java-based language programming IaaS and PaaS simulation Simple network topology Cost simulation GUI	Application module limitation No geographical data centre No MapReduce programming model No distributed file system No workspace configuration No data locality consideration No data replication Difficult to use at large number of hosts due to GUI
MRCloudSim	Open-source code Java-based language programming IaaS and PaaS simulation Simple network topology Cost simulation Simple MapReduce	Application module limitation No geographical data centre No distributed file system No workspace configuration No data locality consideration No data replication
NetworkCloudSim	Open-source code Java-based language programming IaaS, PaaS and SaaS simulation Network topology Extendable Cost simulation Application simulation	Application module limitation No geographical data centre No distributed file system No GUI No workspace configuration No data locality consideration No data replication
YLocSim	Apache YARN Scheduler Load Simulator (SLS), IaaS and PaaS simulation Network topology Data locality Data replication	No geographical data centre No distributed file system No GUI No workspace configuration No data locality consideration No data replication

(continued)

Table 1 (continued)

Simulators	Advantages	Limitations
LocalitySim	Open-source code Java-based language programming IaaS, PaaS and SaaS simulation Network topology Extendable Cost simulation Application simulation Data locality GUI	No Geographical data centre No MapReduce programming model No distributed file system No Workspace configuration No data replication

3 The Proposed GeoLocalitySim Simulator

GeoLocalitySim simulator is an extension of our previous proposed LocalitySim simulator with new features to support geographical distribution data centre(s), data locality and Big Data applications, where Geo indicates Geographic (see Fig. 8). According to Fig. 8, extra modules have been included to overcome limitations of other existing simulators such as Geographical Data Centres (GeoDCs), GeoNameNode, MapReduce, Workspace files and GUI. By including these modules, GeoLocalitySim simulator has earned an ability to express and simulate homogeneous/heterogeneous geographical distributed data centre(s), analyse Big Data using MapReduce, support GeoReplication and GeoNameNode modules.

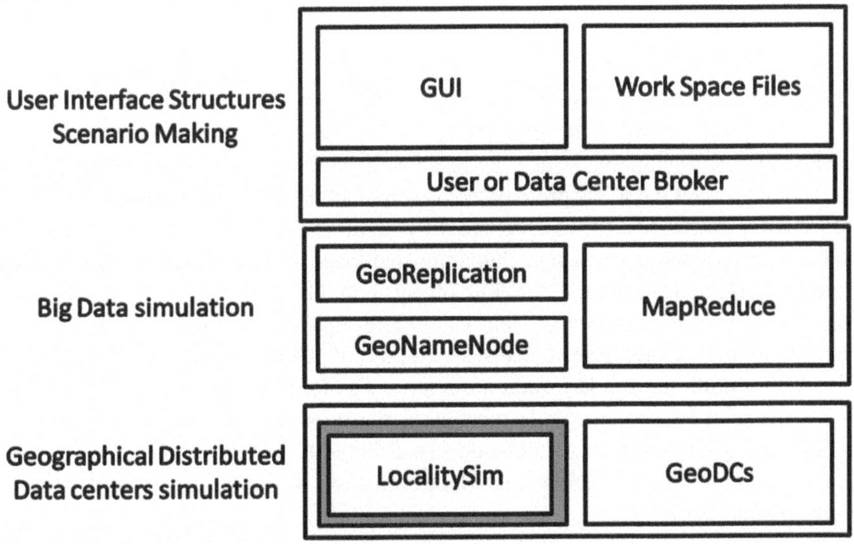

Fig. 8 GeoLocalitySim simulator modules

Also, GUI has been introduced to help configure multiple experiments using workspace files or XML files. The principles of these new modules will be discussed in the next sections. For more details, this proposed GeoLocalitySim simulator is documented and presented in [32].

3.1 Geographical Data Centres (GeoDCs) Module

GeoDCs module is considered the main contributed module to identify the data centre(s) and its network topology. It expresses the infrastructure of the GDDC, where heterogeneous data centres could be connected to each other with different bandwidths. GeoDCs module consists of three objects: NetworkDatacentre, DCRootLink and LS_XML_Geo. NetworkDatacentre object describes the internal network topology between hosts and switches and the configuration of each host (i.e. data storage, RAM and processing units). DCRootLink object is used to define the connectivity between all data centres in GDDC. LS_XML_Geo object is responsible of reading the user's configurations about the GDDC from XML file as an input method. The GeoDCs module contains list of NetworkDatacentre(s), list of DCRootLink(s) and LS_XML_Geo. To build the GeoDCs, some modifications have been done to some modules of our previous LocalitySim simulator such as Switches and NetworkDatacentre to support GDDC.

3.2 GeoNameNode Module

GeoNameNode module contains all information about Big Data files such as list of BigFile objects for different applications and/or list of ChunkFile objects for different tasks. GeoNameNode acts as a table of BigFile(s) and ChunkFile(s) with their addresses at GDDC. BigFile object is responsible for reading the Big Data files information from XML files to prepare it by dividing into ChunkFile(s). ChunkFile object manages the addressing, allocating, reallocating and removing chunk files from the system. Therefore, ChunkFile objects could be built by two ways: dividing Big Data files in BigFile objects into ChunkFile objects or the user customizes Big Data files into ChunkFile objects. Methods of GeoNameNode, BigFile and ChunkFile could manage Big Data file system. GeoReplication will be discussed in next section and GeoNameNode presents a simple simulation to Name and Data nodes with simple replication at Hadoop Distributed File System (HDFS) [11]. The extension of the GeoNameNode and GeoReplication would be considered as future work to simulate HDFS.

sub-screens provide the ability to generate, select, delete, read and write files from and to WSFs. Locality simulator Geo sub-screen is used to select appropriate job scheduler technique, VM scheduler, cost calculation and report some important information about the configuration or scenario. The communication cost could be determined using two bandwidth types: NoShare BW and Share BW. According to NoShare BW communication cost, data of different tasks will be transferred sequentially. By using Share BW communication cost, the bandwidth will be divided equally among the task's data to be transferred in parallel. This provides more flexibility to the proposed GeoLocalitySim simulator. Four files of type comma-separated values (CSV) contain the most important data and the results which are needed to analyse and evaluate the performance (e.g. time interval of each task, scheduler technique, number of job, etc.).

4 GeoLocalitySim Performance Evaluation

The proposed GeoLocalitySim simulator could be used to simulate multi-actions such as

- Building single and geographical distributed data centre(s),
- Implementing MapReduce and any other applications,
- Using replication/no replication and
- Distributing big files or chunk files across the GDDC.

These actions run under consideration of data locality and load aware. The performance of the proposed GeoLocalitySim has been evaluated using two experiments. In the first experiment, two case studies, which had been used to evaluate our previous proposed LocalitySim simulator, will be implemented after rebuilding the experiments using GeoLocalitySim [26]. Because the existed Purlieus resource allocation technique concerns data locality as the main factor for resource allocation, the performance of the proposed GeoLocalitySim simulator has been evaluated by implementing the existed Purlieus resource allocation technique as a second experiment [33].

4.1 Implementing Two Case Studies

Our previous LocalitySim simulator has been proofed experimentally by implementing two case studies which simulate the Map function of the MapReduce programming model, which reads data from storage across one data centre and determine the communication cost for each locality type (i.e. node locality, rack locality, aggregate locality and root locality) [28]. These case studies have been

Table 2 Implementation assumptions of two case studies on LocalitySim

First case study		Second case study	
Item	Value	Item	Value
All bandwidth of any node to another	Equal	All bandwidth of any node to another	equal
All bandwidth	100 MB	All bandwidth	1000 MB
Delay	0	Delay	0
Number of tasks	1000	Number of tasks	2000
Chunk file size	64 MB	Chunk file size	64 MB
Number of switch root	1	Number of switch root	1
Number of aggregate switches	4	Number of aggregate switches	6
Number of edge switch	16	Number of edge switch	24
Number of hosts	64	Number of hosts	96

implemented using different values of bandwidth, number of tasks, number of aggregate switches, number of edge switches and number of hosts (see Table 2).

To validate our proposed GeoLocalitySim simulator, the environment of these two case studies are rebuilt and implemented.

The implementation results prove that the GeoLocalitySim simulator is behaving exactly as our previous LocalitySim simulator (see Fig. 10).

4.2 Implementing Purlieus Resource Allocation Technique

Because the existed Purlieus resource allocation technique concerns with data locality as the main factor for resource allocation, the performance of the proposed GeoLocalitySim simulator has been evaluated by implementing this technique. The implementation results are compared with the results given in [33].

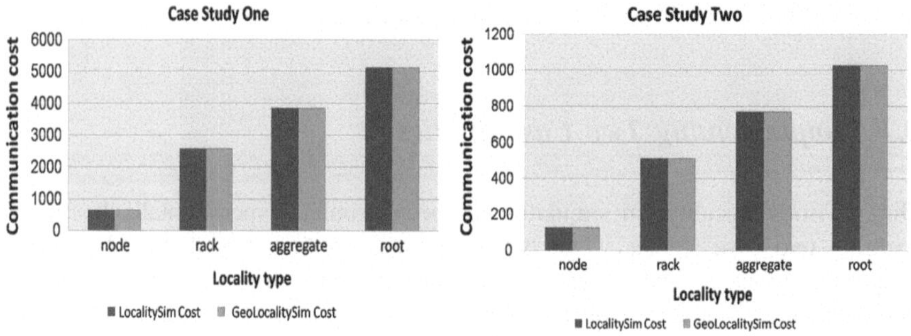

Fig. 10 Comparative communication cost of using LocalitySim and GeoLocalitySim

4.2.1 Purlieus Resource Allocation Technique

According to Purlieus resource allocation technique, data and Virtual Machine placement techniques have been integrated to reduce the network traffic with considering three types of workloads based on the input and output data size: Map-Input Heavy, Map-and-Reduce-Input Heavy and Reduce-Input Heavy. According to Map-Input Heavy workload, the input data size is larger than the output data size (e.g. word search job). By considering this workload, the Virtual Machine (VMs) should be placed close to the location of data to reduce traffic. In Map-and-Reduce-Input Heavy workload, the input and output data sizes are the same (e.g. sort job). So, the communication between Map and Reduce phases should be reduced by grouping hosts holding workloads. In Reduce-Input Heavy workload, the output data size is less than the input data size (e.g. permutation penetrator job), and then the communication reduction at the Reduce phase is considered more important than the Map phase. Therefore, the VMs should be close to each other to reduce traffic during the generation of data (i.e. shuffle and sort phases of MapReduce [34]). By considering the types of the workloads, three data placement techniques are introduced [33] as follows:

(1) **Placing Map-Input Heavy Data Technique**: It is coupled with Map-Input Heavy workload. So, data will be placed at the light load hosts.
(2) **Placing Map-and-Reduce-Input Heavy Data Technique**: It is coupled with Map-and-Reduce-Input Heavy workload. So, the data placement will occur at group of hosts which are closely connected to reduce the traffic between Mappers and Reducers.
(3) **Placing Reduce-Input Heavy Data Technique**: It is coupled with Reduce-Input Heavy workload by placing data at hosts which have maximum free storage.

In addition, three types of VM placements have been introduced based on the data locality placement:

(1) **VM Placement for Map-Input Heavy Jobs**: It places VMs on the hosts which contain needed data.
(2) **VM Placement for Map-and-Reduce-Input Heavy Jobs**: It places VMs on the hosts which are closely connected.
(3) **VM Placement for Reduce-Input Heavy Jobs**: It also places VMs on the closely connected hosts.

On the other hand, five VM placement techniques are suggested to study the effect of the data locality aware with load aware. These suggested VM placements are as follows:

(1) **Locality-Unaware VM Placement (LUAVP)**: It does not concern the data locality, but it chooses a set of least loaded hosts for placing VMs.

(2) **Map-Locality-Aware VM Placement (MLVP)**: It considers the locality of input data. So, VMs will be placed on the hosts containing the input data (i.e. Mapper host).
(3) **Reduce-Locality-Aware VM Placement (RLVP)**: It tries to place VMs on closely connected hosts with considering load aware only.
(4) **Map-and-Reduce-Locality-Aware VM Placement (MRLVP)**: In this type, VMs will be placed on hosts with considering both data locality and load aware.
(5) **Hybrid Locality-Aware VM Placement (HLVP)**: It concerns the workload type by placing VMs on the hosts based on type of the workload as MLVP, RLVP and MRLVP placements.

Therefore, the Purlieus technique tries to select the suitable VM placement corresponding to workload type. It selects MLVP placement to Map-Input Heavy jobs, RLVP placement to Reduce-Input Heavy jobs and MRLVP placement to Map-and-Reduce-Input Heavy job. In [33], Purlieus uses the suggested five VM placement techniques with Map-and-Reduce-Input Heavy workload job. Therefore, the second experiment will be implemented with concerning these issues to evaluate our proposed GeoLocalitySim simulator.

4.2.2 Implementing Purlieus Technique Using GeoLocalitySim Simulator

The Purlieus technique has been implemented using the proposed GeoLocalitySim simulator considering job execution time cost as communication cost and the obtained results are compared with the results in [33].

The configuration of the experiment is done using WSFs and GUI, while data and VM placement techniques are done at the SchedulerBag.java class and selected using GUI.

According to the results in [33], the reduction value of job execution time using MRLVP technique with LLADP technique considering 320 Mappers is reduced by 76% relative to MLVP technique with Random Data Placement (RDP) technique; in the same time, network utilization has been improved.

The same techniques have been implemented using our GeoLocalitySim simulator considering the same environment where 320 Mappers have been simulated using four applications with different numbers of Mappers for each application as follows:

1. 4 applications with 80 Mappers for each,
2. 8 applications with 40 Mappers for each,
3. 16 applications with 20 Mappers for each and
4. 32 applications with 10 Mappers for each.

Figure 11 presents the implementation results using our proposed GeoLocalitySim simulator and the result is given in [33].

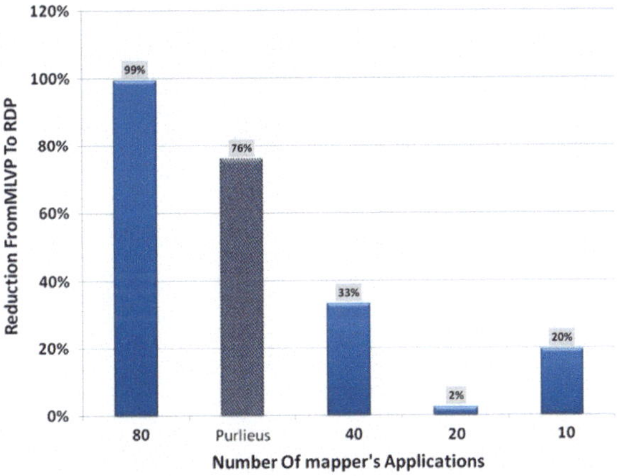

Fig. 11 Result comparative reduction values

According to the implementation results in Fig. 11, the job execution time using MRLVP technique with LLADP technique relative to MLVP technique with RDP technique has been reduced by 99%, 33%, 2% and 20% for experiments 1, 2, 3 and 4, respectively.

It should be noticed that the percentages of time reductions according to our proposed GeoLocalitySim simulator are different from that in [33], because the authors have not mentioned about number of applications, parallelism factor, VMs utilization and replication factor. Therefore, the accuracy of our proposed GeoLocalitySim has been proved.

5 Conclusions and Future Work

The data locality factor is considered a critical issue for implementing cloud framework because the time of data network traffic will be added to the whole time execution. The proposed GeoLocalitySim simulator could be used to simulate GDDC, MapReduce, replication and data locality supported with friendly GUI. Using the proposed GeoLocalitySim simulator, the researcher could adjust and add the effect of data locality as size, location and communication traffic cost at packet level to their work. In addition, the Big Data applications, file distribution, Map/Reduce and replication features could be simulated. In addition, WSFs module is used to configure multi-scenarios with no need to re-edit the code except adding new scheduler technique.

The comparison between the most popular simulators and the proposed GeoLocalitySim simulator is represented in Table 3.

Generally, the proposed GeoLocalitySim simulator is considered as a complete simulator which supports the features of geographical distributed data centres with considering and supporting data locality.

Table 3 Simulators comparison

Item	NetworkCloudSim	CloudSimSDN	MR-CloudSim	LocalitySim	GeoLocalitySim
Underlying platform	CloudSim	CloudSim	CloudSim	CloudSim and NetworkCloudSim	LocalitySim
Language	Java	Java	Java	Java	Java
Availability	Open source	Open source	Open source	Open source	Open source
GUI	No	Yes	No	Yes	Yes
Communication models	Full	Full	Limited	Full	Full
Data centres	Single	Geographical	Geographical	Single	Geographical
Data locality	No	No	No	Yes	Yes
Data replications	No	No	No	No	Yes

As a future work, some modules in GeoLocalitySim could be modified and extended considering data locality and load aware as follows:

GeoNameNode module to simulate the GFS or HDFS [11, 35].
Map/Reduce module to simulate join query [18, 36].

References

1. Buyya, Rajkumar, Chee Shin Yeoa, Srikumar Venugopal, James Broberg, and Ivona Brandic. 2009. Cloud computing and emerging IT platforms: Vision, hype, and reality for delivering computing as the 5th utility. *Future Generation Computer Systems* 25 (6): 599–616.
2. Sahal, Radhya, Mohamed H. Khafagy, and Fatma A. Omara. 2016. A survey on SLA management for cloud computing and cloud-hosted big data analytic applications. *International Journal of Database Theory and Application* 9 (4): 107–118.
3. Hashem, Ibrahim Abaker Targio, Ibrar Yaqoob, Nor Badrul Anuar, Salimah Mokhtar, Abdullah Gani, and Samee Ullah Khan. 2015. The rise of "big data" on cloud computing: Review and open research issues: 98–115.
4. Goiri, Íñigo, Jordi Guitart, and Jordi Torres. 2012. Economic model of a cloud provider operating in a federated cloud: 827–843.
5. Mezgár, István, and Ursula Rauschecker. 2014. The challenge of networked enterprises for cloud computing interoperability. *Computers in Industry* 65 (4): 657–674.
6. Grivas, Stella Gatziu, Tripathi Uttam Kumar, and Holger Wache. 2010. Cloud broker: Bringing intelligence into the cloud: 544–545.
7. Ahmed, Arif, and Abadhan Saumya Sabyasachi. 2014. Cloud computing simulators: A detailed survey and future direction. In *2014 IEEE international advance computing conference (IACC)*, 866–872. IEEE.
8. Assunção, Marcos D., et al. 2015. Big data computing and clouds: Trends and future directions. *Journal of Parallel and Distributed Computing* 79: 3–15.
9. Shi, Juwei, Yunjie Qiu, Umar Farooq Minhas, Limei Jiao, Chen Wang, Berthold Reinwald, and Fatma Özcan. 2015. Clash of the titans: MapReduce vs. spark for large scale data analytics. *Proceedings of the VLDB Endowment* 8: 2110–2121.
10. Thomas, L., and R. Syama. 2014. Survey on MapReduce scheduling algorithms. *International Journal of Computer Applications* 95 (23).
11. White, Tom. 2012. Hadoop: The definitive guide. O'Reilly Media, Inc.
12. Pakize, Seyed Reza. 2014. A comprehensive view of Hadoop MapReduce scheduling algorithms. *International Journal of Computer Networks & Communications Security* 2 (9): 308–317.
13. Dean, Jeffrey, and Sanjay Ghemawat. 2004. MapReduce: Simplified data processing on large clusters. To appear in OSDI (2004).
14. Dean, Jeffrey, and Sanjay Ghemawat. 2010. MapReduce: A flexible data processing tool. *Communications of the ACM* 53 (1): 72–77.
15. Chen, Quan, et al. 2010. Samr: A self-adaptive MapReduce scheduling algorithm in heterogeneous environment. In *2010 IEEE 10th international conference on computer and information technology (CIT)*, 2736–2743. IEEE.
16. Sun, Xiaoyu, Chen He, and Ying Lu. 2012. ESAMR: An enhanced self-adaptive MapReduce scheduling algorithm. In *2012 IEEE 18th international conference on parallel and distributed systems (ICPADS)*, 148–155.
17. Katal, Avita, Mohammad Wazid, and R.H. Goudar. 2013. Big data: Issues, challenges, tools and good practices: 404–409.

18. Dolev, Shlomi, Patricia Florissi, Ehud Gudes, Shantanu Sharma, and Ido Singer. 2017. A survey on geographically distributed big-data processing using MapReduce.
19. Calheiros, Rodrigo N., Rajiv Ranjan, Anton Beloglazov, César A.F. De Rose, and Rajkumar Buyya. 2011. CloudSim: A toolkit for modeling and simulation of cloud computing environments and evaluation of resource provisioning algorithms. *Software: Practice and Experience*: 23–50.
20. Kliazovich, Dzmitry, Pascal Bouvry, and Samee Ullah Khan. 2012. GreenCloud: A packet-level simulator of energy-aware cloud computing data centers. *The Journal of Supercomputing* 62 (3): 1263–1283.
21. Garg, Saurabh Kumar, and Rajkumar Buyya. 2011. NetworkCloudSim: Modelling parallel applications in cloud simulations. In *Fourth IEEE international conference on utility and cloud computing*.
22. Tighe, Michael, Gaston Keller, Michael Bauer, and Hanan Lutfiyya. 2012. DCSim: A data centre simulation tool for evaluating dynamic virtualized resource management: 385–392.
23. Lim, Seung-Hwan, Bikash Sharma, Gunwoo Nam, Eun Kyoung Kim, and Chita R. Das. 2009. MDCSim: A multi-tier data center simulation, platform: 1–9.
24. Jung, Jongtack, and Hwangnam Kim. 2012. MR-CloudSim: Designing and implementing MapReduce computing model on CloudSim: 504–509.
25. Son, J., A.V. Dastjerdi, R.N. Calheiros, X. Ji, Y. Yoon, and R. Buyya. 2015. CloudSimSDN: Modeling and simulation of software-defined cloud data centers. In *2015 15th IEEE/ACM international symposium cluster, cloud and grid computing (CCGrid)*, 475–484.
26. Abase, Ahmed H., Mohamed H. Khafagy, Fatma A. Omara. 2017. Locality sim: Cloud simulator with data locality.
27. Wang, Jianjun, Gangyong Jia, Aohan Li, Guangjie Han, and Lei Shu. 2015. Behavior aware data placement for improving cache line level locality in cloud computing. *Journal of Internet Technology* 16 (4): 705–716.
28. Wang, Guanying, et al. 2009. A simulation approach to evaluating design decisions in MapReduce setups. *MASCOTS* 9: 1–11.
29. Wang, Guanying. 2012. Evaluating MapReduce system performance: A simulation approach. Ph.D. diss. Virginia Polytechnic Institute and State University.
30. Buyya, Rajkumar, and Manzur Murshed. 2002. Gridsim: A toolkit for the modeling and simulation of distributed resource management and scheduling for grid computing: 1175–1220.
31. Elshater, Yehia, Patrick Martin, Dan Rope, Mike McRoberts, and Craig Statchuk. 2015. A study of data locality in YARN. In *2015 IEEE international congress big data (bigdata congress)*, 174–181.
32. Abase, Ahmed H., Mohamed H. Khafagy, and Fatma A. Omara. https://github.com/ahmedexpert/GeoLocalitySim.
33. Palanisamy, Balaji, Aameek Singh, Ling Liu, and Bhushan Jain. 2011. Purlieus: Locality-aware resource allocation for MapReduce in a cloud. In *Proceedings of 2011 international conference for high performance computing, networking, storage and analysis*, 58.
34. Verma, Abhishek, Ludmila Cherkasova, and Roy H. Campbell. 2011. ARIA: Automatic resource inference and allocation for MapReduce environments. In *Proceedings of the 8th ACM international conference on autonomic computing*, 235–244.
35. Xie, Jiong, Shu Yin, Xiaojun Ruan, Zhiyang Ding, Yun Tian, James Majors, Adam Manzanares, and Xiao Qin. 2010. Improving MapReduce performance through data placement in heterogeneous hadoop clusters. In *2010 IEEE international symposium on parallel & distributed processing, workshops and Phd forum (IPDPSW)*, 1–9.
36. Ji, Changqing, Yu Li, Wenming Qiu, Uchechukwu Awada, and Keqiu Li. 2012. Big data processing in cloud computing environments. In *2012 12th international symposium on pervasive systems, algorithms and networks (ISPAN)*, 17–23.

Trustworthy Self-protection for Data Auditing in Cloud Computing Environment

Doaa S. El-Morshedy, Noha E. El-Attar, Wael A. Awad and Ibrahim M. Hanafy

Abstract Cloud computing lineaments enable the users to deal with resources and get benefit from the services in a distinct economic manner. One of the vital cloud's services is storing the users' data on a remote location on a cloud's storage resource, which means that this storage has to be characterized by a high level of security and safety, for instance, to ensure the data integrity and preserve its confidentiality. In the existing cloud environments, these security issues are handled by a third party; neither user nor provider, known as the Cloud Third Party (CTP). Despite the evident important role of CTP, it may adversely affect the security and becomes untrustworthy insider employee (i.e., malicious insider). In this paper, we present an automated framework for enhancing the cloud security "Self-Protecting of Data Auditing (SPDA)" based on the concept of autonomic computing. This framework is divided into three phases in order to handle the two significant security issues: the integrity and the confidentiality. The first stage concerns with encrypting the user's data automatically to maintain its confidentiality. The second phase is for automated data auditing in order to keep the data integrity. And, the last one is used to automatically decrypt the data to its original correct form when the user requests to retrieve it.

Keywords Cloud computing · Security · Challenges · Cloud third party

D. S. El-Morshedy (✉) · W. A. Awad · I. M. Hanafy
Faculty of Science, Port Said University, Port Said, Egypt
e-mail: Doaa_morshady@yahoo.com

W. A. Awad
e-mail: Wael_abdelkader@sci.psu.edu.eg

I. M. Hanafy
e-mail: Ihanafy@hotmail.com

N. E. El-Attar
Faculty of Computers and Artificial Intelligence, Benha University, Benha, Egypt
e-mail: noha.ezzat@fci.bu.edu.eg

A. Z. Ghalwash et al. (eds.), *Internet of Things—Applications and Future*,
Lecture Notes in Networks and Systems 114,
https://doi.org/10.1007/978-981-15-3075-3_2

1 Introduction

Initially, the cloud computing is an Internet-computing model based on enabling appropriate access to shared pools of computing resources such as storage, applications, software, etc. [1]. The cloud computing has many essential merits to the individual users as well as the organizations beginning with accessing the cloud environment from anywhere and at any time via a secure channel ending with paying as you use [2].

In general, the cloud computing is characterized by many benefits such as **utility-based pricing** (i.e., cloud computing is based on "pay-as-you-go" pricing model), **shared resource pooling** (i.e., a pool of resources used by the infrastructure provider can be dynamically specified to multiple service requests), **multi-tenancy** (i.e., resources that are located in a single data center can be used by several service providers), **self-organizing** and **dynamic resource provisioning** (i.e., the resources can be allocated or de-allocated on-demand, service providers are delegate to control their resource consumption according to the service requests. The resources are usually released on the fly based on the current demand state) [3].

In spite of the evident benefits of cloud computing environment, there are some main challenges that the cloud computing systems still face which are presented in Fig. 1. In general, **security and privacy** are rated as one of the serious challenges in the cloud systems. Many issues and questions are still growing about whether cloud is a safe environment or not [4]. Each of them has to be compatible with the other to guarantee data confidentiality [5, 6]. Thus, data protection is always regarded as one of the most important challenges of the cloud computing environments.

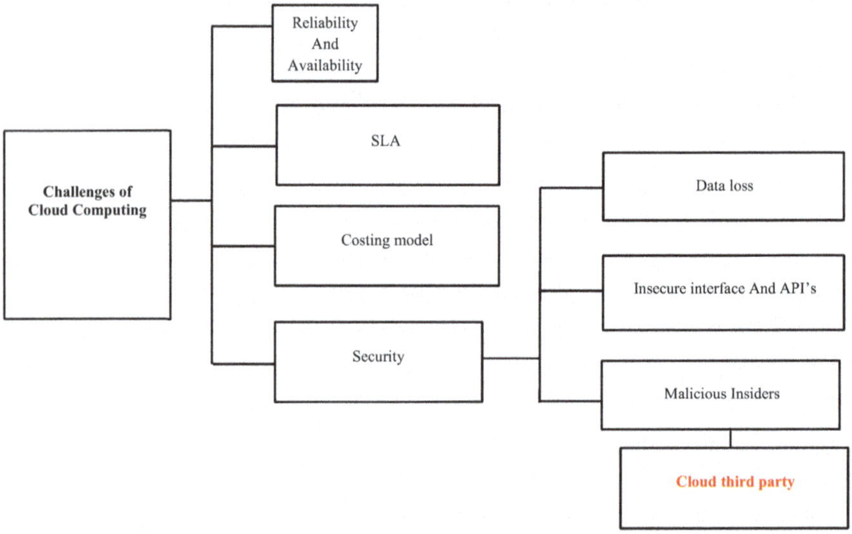

Fig. 1 Cloud computing challenges

The privacy mainly concerns with protecting all the "Personally Identifiable Information" and permitting the access for the authorized users only. On the other hand, securing data in cloud has essentially three main issues: availability, integrity, and confidentiality [7]. Hereby, the cloud computing system has to provide the policies and methodologies that ensure the security and confidentiality for the hosted or shared data and applications [8]. Cloud computing usually resorts to verify and manage user's requests through a login authentication, encryption methodologies, ensuring integration, permanent availability, and many other trusting techniques [7].

In most of the traditional cloud environments, these mentioned security issues are addressed by a cloud third party; neither the service's users nor the service's providers. This Cloud Third Party (CTP) usually does many important functions such as controlling and handling the identity of the user who requests services from the cloud [7]. It is also responsible for data auditing to verify the stored data in the cloud system, and many other permissions that make his role very significant in securing data [9].

In spite of the significant functions of CTP, it may turn to be a vulnerable point in the cloud environment in a form of malicious insider; in this case, it becomes untrustworthy entity. Undoubtedly, such kind of malicious insiders is considered a very dangerous threat on the cloud system because of their high permission level in dealing with the users' data which may lead to leakage or deletion in data without any authorization from the data owners [10]. Therefore, guaranteeing data confidentiality in the cloud during storing and processing is considered a great dilemma, but it may be handled by some techniques such as the data encryption [11].

In this paper, we handle the issue of data confidentiality maintenance in the cloud computing systems based on the concept of autonomic computing to overcome the CTP defects. The rest of this paper is organized as follows: Sect. 2 presents an introduction to cloud computing security issues, Sect. 3 displays and defines the security problem specification. Section 4 presents some of the significant related researches in the security problem, Sect. 5 proposes the Self-Protection for Data Auditing (SPDA) framework based on the autonomic computing concepts. Finally, the paper ends with the conclusion and the future work in Sect. 6 (Fig. 2).

2 Cloud Computing Security Issues

Security is the protection from the unauthorized access to, or dealing with, the system state [6]. In general, security relates to six main issues in the cloud computing environment: (1) Ensuring safety mechanism by monitoring or tracing the server. (2) Preserving data confidently especially for sensitive information. (3) Protecting the system against malicious insiders. (4) Avoiding the service attacker who previously dealt with vulnerabilities in the IT architectures. (5) Isolating all the instances from one another to manage and control multi-tenancy. (6) Placing appropriate laws and applying legal jurisdictions for the service user and provider when needed [6].

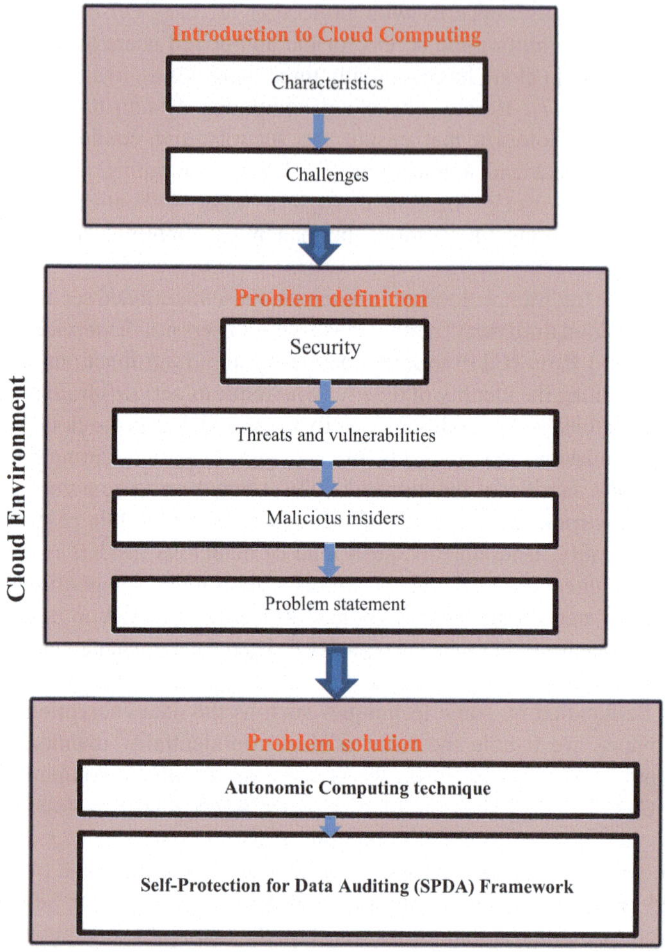

Fig. 2 The paper organization

All of the above risks make it important to establish more secure data repositories to keep the data which is stored, transmitted, or processed away from any unauthorized access.

Improving the security in the cloud computing environments depends on three main dimensions: the availability, the integrity, and the confidentiality [12].

A. Availability: It is one of the essential concerns for the cloud's users. The cloud providers should guarantee that the remote stored data will be available to the user all the times without any corruption when it is retrieved [13].
B. Integrity: It is another important type of data guarantees. Data integrity means that the remote stored data should be saved away from any intentional or unintentional modification from any unauthorized parties. Also, the cloud provider has to protect the long-time stored data from any deletion or corruption [14].

C. Confidentiality: Data confidentiality is considered the toughest challenge in the data protection process. Supposedly once the user stored his data in the cloud environment and on a remote server, his data must be in a secure place, and no one can access or deal with it without prior permission. But unfortunately, this cannot be guaranteed because of the existence of the cloud third party. Some cloud systems adopt the encryption techniques in attempt to save the desired confidentiality [14].

3 Security Aware—Problem Designation

The cornerstone of the cloud computing is to provide the required service with divesting the users from the infrastructure management. This feature usually isolates the users from their data, where the providers always store the users' data on the physical resources by the remote access in order to achieve the availability. Alongside the availability, the users constantly need to ensure the integrity and confidentiality of their data.

Cloud Third Party (CTP) is one of the common aids used by the cloud system to help the user in auditing and managing his data on the cloud environment. The existence of CTP makes dealing with data on cloud unpretentious, convenient and flexible especially if there is no data encryption [15]. In general, the Cloud Third Party (CTP) has multiple functions in the cloud environment which can be summarized as shown in Fig. 3 in [15, 16]:

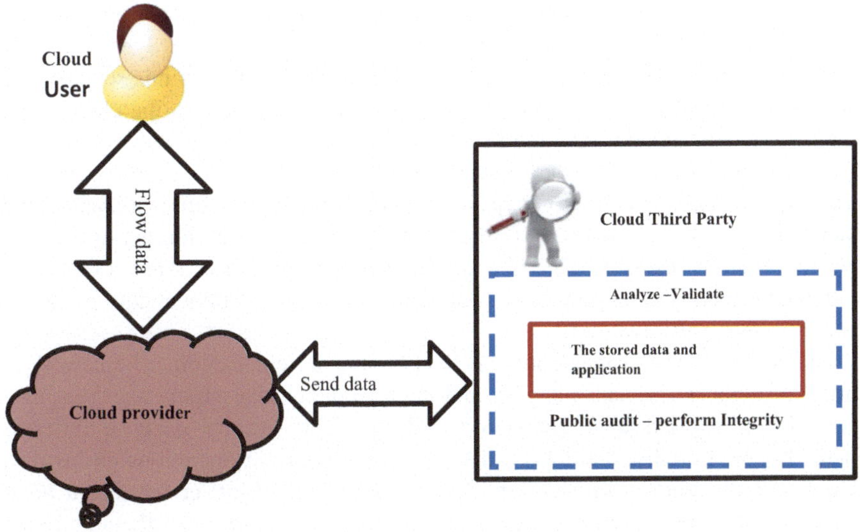

Fig. 3 Main roles of the cloud third party

1. Analyze the data storage;
2. Check the data correctness;
3. Perform and ensure the data integrity; and
4. Public audit for the stored data to ensure its protection from any outsider malicious attacks.

The vital permissions given to the CTP may transform it from a powerful aid means into a perilous hidden foe (e.g., it can exploit its ability of dealing with data in altering, deleting, leaking, or stealing the intellectual property of the user's data). Thus, without the awareness of the provider and the users the CTP itself may be an insider malicious that effects on the data confidentiality [17].

The **malicious insiders** are considered the most grave type of threats that face the cloud system from its inside environment; they may be employees, contractors, third parties, or other business persons who have the rights and permissions to access the data in the system [18]. However, many organizations overlook this threat because it is nearly impossible to detect and avoid their attacks during the actual service processing [19].

This paper aims at finding an alternative way based on the autonomic computing to perform the user's data public auditing away from the cloud third party in an attempt to curtail its functions on the data that may enhance the data security level. The next section presents some of the literature reviews and related works on solving and enhancing the problem of public auditing.

4 Literature Reviews

The researchers have exerted plenty of effort in order to handle the problem of CTP threats in case of turning to malicious insiders in the last few years. Syed et al. [20] have proposed a schema of data encryption based on the user. In this schema, the user must apply the encryption on the data before exporting it to the cloud. This schema did not abandon the third-party role completely, but it was based on a Trusted Third Party (TTP) to keep the secret keys that are needed in the decryption process as a main preliminary step to perform the required computations on the data. The significant drawback of this schema is increasing the burden of the data encryption on the users who are in many cases not IT specialists. Also, it still deals with the third party as a trustworthy way which makes the confidentiality problems from the third party still exist.

On another hand, Swapnali and Sangita [9] have presented another framework in order to preserve the data integrity and confidentiality. This framework has also used the user-based data encryption and the CTP but by using a different methodology. Initially, the data owner does some securing operations on his data which include ***splitting the file into blocks***, ***encrypting them***, ***generating a hash value for each one***, ***concatenating*** and ***generating a signature on them*** and then ***sending it to the Cloud***. The next stage is done by the CTP who is responsible for the public auditing to check the data integrity by generating the hash value for the encrypted blocks which are received from the provider and concatenating and generating signature on them in the cloud environment. Finally, the third party compares the two signatures to verify whether the stored data has been tampered or not. This technique has rather handled some of the fears related to CTP; however, it still increases the burden on user as in the previous schema.

Another solution for the mentioned security issues has been suggested by Anirudha and Syam [21]. They have optimized the third-party auditor role to resist replacing, replaying, and attacking on the cloud servers. They have established their solution relying on two verification ways on the cloud servers during the computation process. At first, during the dynamic data update phase, the client requests the data block from the Cloud Storage Server (CSS), the CSS checks the correctness of the data authentication path for this block and whether the previous update on this data has been successfully done or not. Second, the third party begins in public auditing and this phase is called the Third-Party Auditing (TPA). The TPA verifies the probabilistic proof of integrity that is generated by CSS. Rajat and Somnath [22] have designed the third-party auditing process for verifying the data integrity and batching the auditing tasks together in order to accomplish the load balancing task and perform the batch auditing by multiple third-party auditors.

From the above literature reviews, the drawbacks in the existing schemas and frameworks can be summarized as follows:

1. The CTP is usually defined as a person, so the data storing still suffers from some security concerns.
2. CTP deals usually with the original data during the public auditing process.
3. To decrease the CTP functions, the user may suffer from increasing the burden of securing his data by himself.

In consideration to overcome the threats of the Cloud Third Party (CTP) on the data security, especially on the data auditing, we propose a framework for "Self-Protection for Data Auditing (SPDA)." This proposed framework seeks to repeal the role of the CTP in data auditing operation by switching it to automated-based processes divided into three main phases: (1) automated data encryption, (2) automated data auditing, and (3) automated decryption.

5 Self-protection for Data Auditing (SPDA)

The main job of data auditing is how to verify the stored data to ensure its correctness, integration, and availability whenever the user requests to retrieve it. The proposed "Self-Protection for Data Auditing (SPDA)" framework is established based on the autonomic computing aspects to perform the data auditing process in an automated manner to keep the stored data protected away from any outside intervention and at the same time to repair any found faults. In general, the autonomic computing system can be defined as a self-managed computing system which is able to manage itself without any human interference [23]. The self-management has four essential characteristics: self-configuring, self-healing, self-protection, and self-optimization as illustrated in Table 1 [23–26].

The proposed SPDA framework stands essentially on the self-protection and self-healing from the autonomic computing environment. The self-protection can be a powerful aid in identifying and handling the malicious intentional actions in the cloud system by scanning all of the system activities to determine any illegal activities and take actions to deal with it accordingly. Although the existence of self-protecting helps in preventing it from the malicious threats, some other failure behaviors may occur in data such as loss of integration, errors of transferring, and unintentional deletion. In such cases, the self-healing is required, where it can return the system from any abnormal state to its correct and original form through three main phases: (a) detecting the existing problem, (b) diagnosing the problem by analyzing it and finding the suitable recovery plan, (c) and finally applying the recovery-adaptation plan [27].

Before proceeding in discussing the SPDA framework, the main architecture of the autonomic computing needs to be clarified as it is the backbone of the proposed framework. As shown in Fig. 4, the conceptual architecture of autonomic computing consists of two main parts: the ***autonomic manager*** and the ***managed***

Table 1 The characteristics of autonomic computing

Characteristics of autonomic computing	Description
1. Self-configuring	Self-configuring means that the system has the ability to adapt its state automatically according to the dynamic changes in the business/operation environments. For example, when a new component is presented, the system will reconfigure itself and incorporate this new component with the existing ones seamlessly and automatically
2. Self-healing	Self-healing is the system ability in detecting and diagnosing the errors and then repairing any disruptions in an automatic manner. The self-healing components allow the system to activate its valid state and take the proper corrective actions without disrupting the system functions
3. Self-optimization	The autonomic system always needs to optimize its resources status automatically to gain a high performance level for the user's requirements
4. Self-awareness	The self-awareness is an important preparing operation to the self-management phases. The system has to keep aware of its internal and around environment, and it should always be able to know its current state and behavior
5. Self-protection	Self-protection means that system ability in protecting itself from malicious attack threats that may happen from anywhere and do the appropriate actions to face them

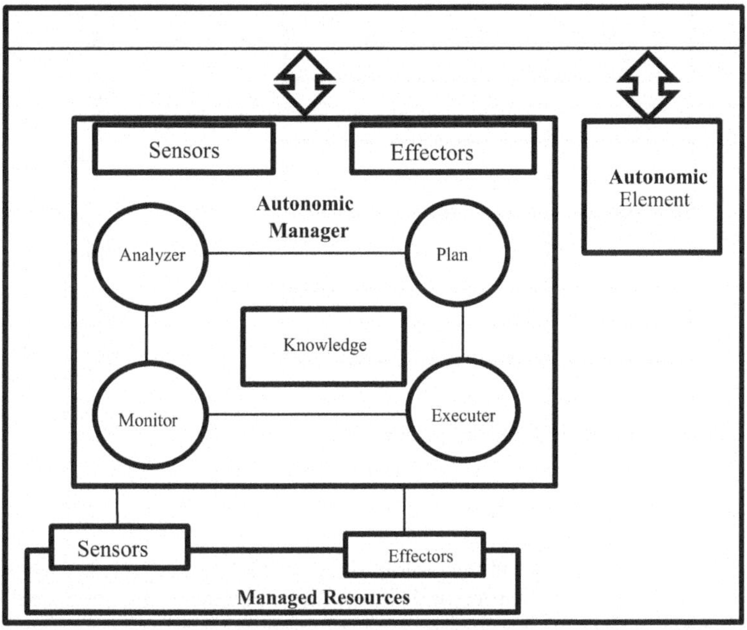

Fig. 4 The conceptual architecture of autonomic computing [27]

resources. The workflow between these parts is coordinated via six essential agents in the autonomic computing system as follows [28]:

(1) Sensor: collect resources.
(2) The monitor: search, report, filter, and manage the state of the resources.
(3) The analyzer: analyze the collected resources and data and determine the relationship between them.
(4) Plan: decide which action will be applied to accomplish the pre-defined goals.
(5) Executer: apply and manage the plan execution.
(6) Effector: adapt any changes occurring in the resources.

5.1 The Formulation of the Self-protection for Data Auditing "SPDA" Framework

As displayed in Fig. 5, the SPDA architecture contains three main phases in order to accomplish the protection and the public auditing processes without the mediation of the CTP: (1) automated data encryption, (2) public auditing and healing, and (3) automated data decryption.

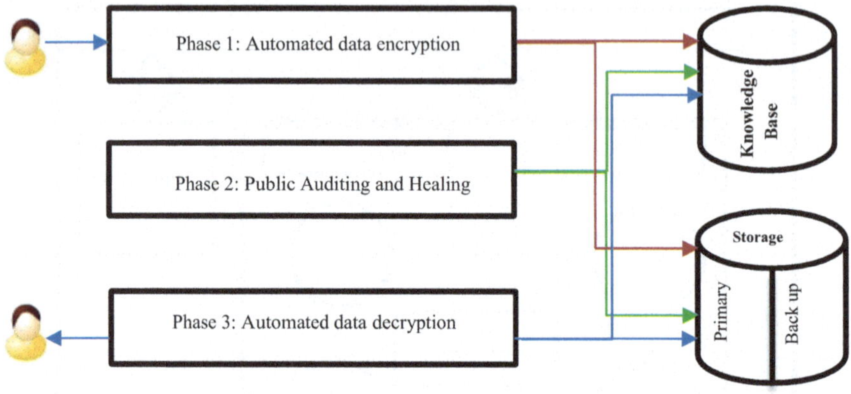

Fig. 5 The main phases of SPDA architecture

Automated Data Encryption

The SPDA framework aims at integrating the processes of auditing and security to deliver the user's data in a trusted and correct form. In general, the proposed SPDA system starts as soon as the user requests to store his data on one of the cloud computing platforms. The system starts with the automatic encryption, and whenever the user requests to retrieve the data, the automatic decryption and auditing processes begin to ensure the integrity and correctness of the data. The automated data encryption is also based on the autonomic self-protection through step-by-step procedure as shown in Fig. 6.

Step 1 (Log file): Initially, in order to maintain the confidentiality, the system needs to determine who are permitted to access this data through the log file that contains the user's information and service's details.
Step 2 (Services collector): Responsible for collecting the requests of data storing and the user's log file forms and transmitting them to the monitor by using a ***Digital Forensic Investigations*** technique for collecting and preserving data [29].
Step 3 (Monitor agent): It is the agent used for observing the service status and preparing the reports and then putting them with the log files in the knowledge base to be sent to the data analyzer.
Step 4 (Data analyzer): Starts analyzing the data received from the monitor agent and divides it into blocks by using ***Data Splitting*** algorithm (i.e., the file F will be split into m blocks (m_1, m_2, …, m_n) [30].
Step 5 (Task manager): Sends the blocks of data to the encryption operator and requests a new random key from the random key generator and then sends the hash function to the implementation.
Step 6 (Random key generator): It is responsible for generating two random keys: ***K1*** as a public key for encryption and ***K2*** as a private key for decryption based on ***Pseudorandom Number Generator (PRNG)*** algorithm [31].

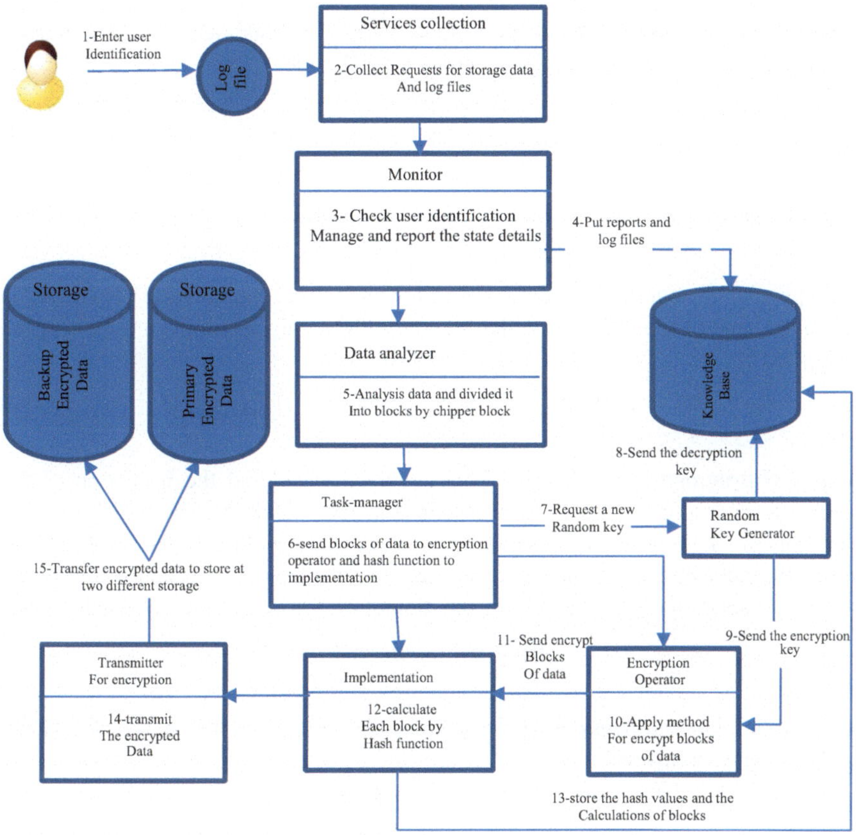

Fig. 6 Automated data encryption

Step 7 (Encryption operator): Responsible for applying ***RSA*** encryption algorithm [32] on the data blocks to find the cipher data according to (1) and then sends these encrypted blocks of data to the implementation agent

$$\boldsymbol{C} = \boldsymbol{M}^{\boldsymbol{K}1}(\mathrm{mod}\ \boldsymbol{L}) \tag{1}$$

where M is the original message and $(K1, L)$ is the public key.

For example, this text "Hello sir. Please send the required files" can be split and decrypted as follows:

1. Applying the ASCII code on the original data to be [33]:

(07210110810811111510511404608010810109711510111510111010011610 4101114101113117105114101100102105108101115).

2. Breaking the original block into small blocks; in this case, it splits into 35 blocks, each block has 3 digit to apply ASCII code.

$$\mathbf{M}_1 = 072, \mathbf{M}_2 = 101, \mathbf{M}_3 = 108, \ldots, \mathbf{M}_{35} = 115.$$

3. Calculating the encrypted text by Eq. (1) as follows:

By supposing the encryption public key $(K1, L)$ and decryption private key $(K2, L)$ is (17, 3599) and (1433, 3599) in order based on PRNG algorithm [31], the estimated ciphertext is

$$\mathbf{C}_1 = 2759,\ \mathbf{C}_2 = 581,\ \mathbf{C}_3 = 1172,\ \ldots, \mathbf{C}_{35} = 925$$

Step 8 (Implementation Agent): This agent is considered the backbone of the automatic encryption process. It is responsible for calculating the hash value for each data block based on the ***homomorphic hash value method*** based on (2) [34] to verify the data integrity. Then it sends the encrypted blocks of data to the transmitter and sends the calculated hash functions to be stored in the knowledge base.

$$\boldsymbol{h}(\boldsymbol{C}) = \boldsymbol{C} \bmod \varphi(\boldsymbol{L}) \tag{2}$$

4. Applying ***homomorphic hash value method*** on the proposed encrypted text can be done as follows:
   (a) Calculate hash value on each encrypted block

$$\boldsymbol{h}(\boldsymbol{C}_1) = 2759 \bmod 3480 = 2579.$$

   (b) Compute $\boldsymbol{m_p} = \boldsymbol{a}^{\boldsymbol{m}} \bmod \boldsymbol{L}$, where $\boldsymbol{a}$ is integer number chosen randomly $(1 < \boldsymbol{a} < \boldsymbol{L} - 1)$

$$\boldsymbol{m_p} = 15^{2759} \bmod 3599 = 2009.$$

Step 9 (transmitter for encryption): It is used for transmitting the encrypted blocks of data to be stored as two copies: one as a primary in the main storage device and the second as the backup in a safe backup encrypted data storage based on the secure hypertext transfer protocol (HTTPS) which is applied to transfer encrypted data [35].

Public Auditing and Healing

The second phase of SPDA framework is based on the self-healing component in the autonomic computing. Its essential job is to return the data to its correct state in case of any loss or illegal modifications in data. In general, the public auditing and the self-healing have to be interrelated to ensure the data integrity and confidentiality. The proposed framework depends mainly on performing the auditing process automatically without any intervention from the third party and without any additional burden on the user's shoulders. As shown in Fig. 7, the automated public auditing and healing processes are done through consecutive and interrelated steps as follows:

1. Checking the user's requests and his authority to access the data through his log file in the knowledge base and replying with a refused alarm in the case of unauthorized access. When the access is successfully done, the auditing request is then sent to the transceiver.

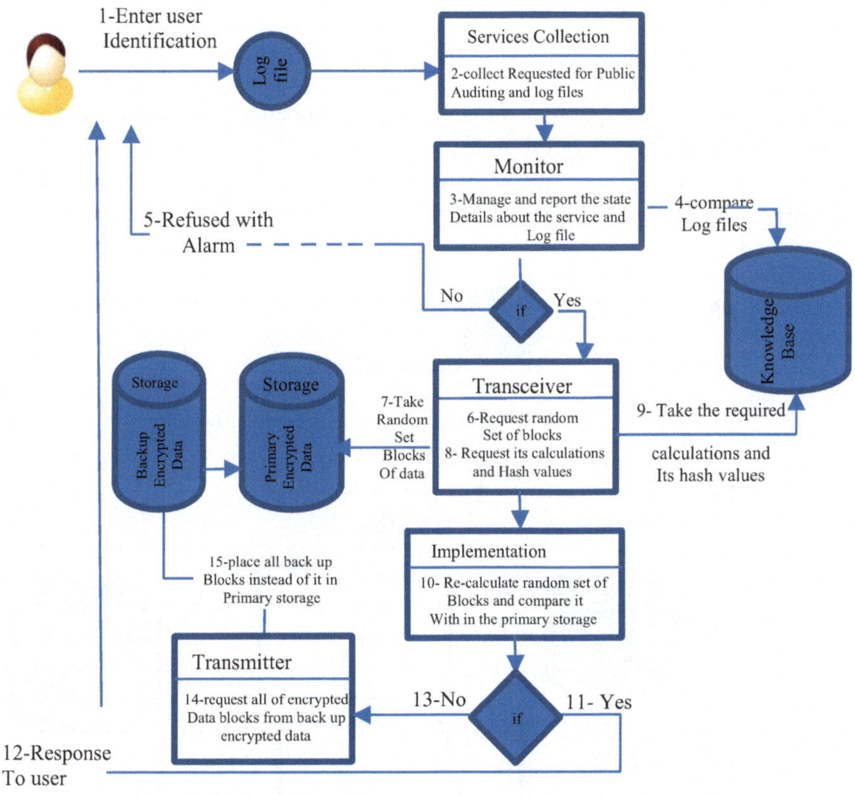

Fig. 7 The public auditing and healing

2. The transceiver chooses a random set of blocks to be audited from a primary storage device and recalculate them by their hash function which is stored in the knowledge base as in (3),

$$\boldsymbol{m}_p' = \boldsymbol{a}^{h(C)} \bmod \boldsymbol{L}. \tag{3}$$

3. In case of comparing the mismatch, the transceiver begins its work. It requests all of the encrypted data from the backup encrypted data storage and then place some backup blocks instead of them in the primary storage.

Automated Data Decryption

The proposed automated data decryption procedures are shown in Fig. 8. The decryption process is activated when the user requests to retrieve his original data. The significant component in this process is the ***decryption operator*** which is based on the RSA algorithm. As displayed in the encryption process phase, the RSA algorithm uses the private decryption key (***K2***, ***L***) which is randomly generated in the encryption stage to decrypt the ciphertext as follows in (4):

$$\boldsymbol{M} = \boldsymbol{C}^{K2} (\mathrm{mod}\ \boldsymbol{L}) \tag{4}$$

To clarify the decryption process, the same ciphertext from the encryption process will be decrypted as follows:

$$\mathbf{M}_1 = 2759^{1433} \bmod 3599 = 072,\ \mathbf{M}_2 = 101,\ \mathbf{M}_3 = 108, \ldots, \mathbf{M}_{35} = 115$$

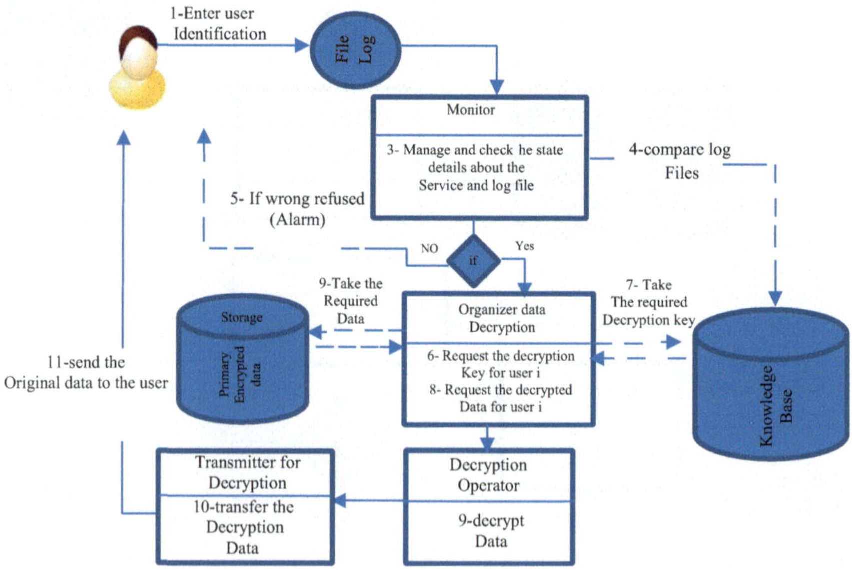

Fig. 8 The automated data decryption

6 Conclusion

A Self-Protecting of Data Auditing (SPDA) framework has been proposed in this paper as a security system based on the autonomic computing principle in attempting to protect the user's data which is hosted on the cloud storage resource from malicious insider and outsider. This framework works in an automated manner passing through three phases: (1) Automated Data Encryption Phase, which is used to encrypt the user's data and store it in the primary main storage and a safe backup storage. (2) Automated Data Auditing Phase, which is responsible for verifying the stored data to ensure its correctness, integration, and availability whenever the user requests to retrieve it. (3) Automated Decryption Phase which is beginning its work when the user requests to retrieve his original data.

Future Work: We would like to execute all of these phases without any interference from the traditional cloud third party in order to keep the data integrity and confidentiality with reducing the burden on the user's shoulders of securing his data by himself.

References

1. Tharam, D., W. Chen, and C. Elizabeth. 2010. Cloud computing: Issues and challenges. In *24th international conference advanced information networking and applications*, 27–33. IEEE.
2. Farrukh, S. 2014. State-of-the-art survey on cloud computing security challenges, approaches and solutions. In *The 6th international symposium on applications of ad hoc and sensor networks*, 357–362. Elsevier.
3. Noha, E.E., A.A. Wael, and A.O. Fatma. 2012. Resource provision for services workloads based on (RPOA). *International Journal of Computer Science Issues* 9 (3): 553–560.
4. Santosh, K., and R.H. Goudar. 2012. Cloud computing—Research issues, challenges, architecture, platforms and applications: A survey. *International Journal of Future Computer and Communication* 1: 356–360.
5. Amal, G., G. Mahmoud, and J. Mohamed. 2017. Privacy in cloud computing environments: A survey and research challenges. *Journal of Supercomputing* 73 (6): 2763–2800.
6. Dawei, S., C. Guiran, S. Lina, and W. Xingwei. 2011. Surveying and analyzing security, privacy and trust issues in cloud computing environments. Advanced in control engineering and information science. *Procedia Engineering* 15: 2852–2856. Elsevier.
7. Bharathi, C., V. Vijayakumarb, and K.V. Pradeep. 2015. An extended trust management scheme for location based real-time service composition in secure cloud computing. In *2nd international symposium on big data and cloud computing*, 103–108. Elsevier.
8. Passent, M.E., A.A. Azza, and F.S. Amr. 2015. Security issues over some cloud models. In *International conference on communication, management and information technology*, 853–858. Elsevier.
9. Swapnali, M., and C. Sangita. 2016. Third party public auditing scheme for cloud storage. In *7th international conference on communication, computing and virtualization*, 69–76. Elsevier.
10. Manish, M.P., C.A. Dhote, and H.S. Deepak. 2016. Homomorphic encryption for security of cloud data. In *7th international conference on communication, computing and virtualization*, vol. 79, 175–181. Elsevier.

11. Sabrina, C.V., F.E. Robert, F. Sara, J. Sushil, L. Giovanni, and S. Pierangela. Encryption and fragmentation for data confidentiality in the cloud. In *International school on foundations of security analysis and design*, 212–243. Springer.
12. Rao, R.V., and K. Selvamani. Data security challenges and its solutions in cloud computing. In: *International conference on intelligent computing, communication & convergence*, vol. 48, 204–209. Elsevier.
13. Rizwana, S., and S. Mukundan. Data classification for achieving security in cloud computing. In *International conference on advanced computing technologies and application*, vol. 45, 493–498. Elsevier.
14. Sulochana, M., and D. Ojaswani. Preserving data confidentiality using multi-cloud architecture. In *2nd international symposium on big data and cloud computing*, 357–362. Elsevier.
15. Abhishek, M., and K.A. Lalit. 2012. Cloud data security while using third party auditor. *International Journal of Scientific & Engineering Research* 3 (6): 1–4.
16. Zulkefli. M.Y., and H.A. Jemal. Analysis of insiders attack mitigation strategies. In *International conference on innovation, management and technology research 2013*, vol. 129, 611–618. Malaysia: Elsevier.
17. Songzhu, M., B. Haihe, T. Fang, R. Jiangchun, and W. Zhiying. 2014. In *Proceedings of international conference on soft computing systems*. China: Springer.
18. Kaaviyan, K., S. Deepak, and P. Prakash. In *Proceedings of international conference on soft computing systems*. Springer.
19. Naresh, V., and B. Thirumala Rao. 2016. A study on data storage security issues in cloud computing. In *2nd international conference on intelligent computing, communication & convergence 2016*, vol. 92, 128–135. India: Elsevier.
20. Syed, R., C. Katie, and G. Christopher. 2014. A trusted third party based encryption scheme for ensuring data confidentiality in cloud environment. In *Conference organized by Missouri university of science and technology 2014*, vol. 36, 381–386. Philadelphia: Elsevier.
21. Anirudha. P.S., and K.P. Syam. 2016. Optimized public auditing and data dynamics for data storage security in cloud computing. In *6th international conference on advances in computing & communications 2016*, vol. 93, 751–759. India: Elsevier.
22. Rajat, S., and D. Somnath. 2016. Cloud audit: A data integrity verification approach for cloud computing. In *Twelfth international multi-conference on information processing 2016*, vol. 89, 142–151. Elsevier.
23. Inderpreet, C., and S. Maninder. 2014. SHAPE—An approach for self-healing and self-protection in complex distributed networks. *Journal of Supercomputing* 67 (2): 585–613.
24. Naveen, I., and V. Sowjanya. 2015. Self-protection in autonomic computing. *International Journal of Emerging Research in Management &Technology* 4: 213–216.
25. Mona, A.Y., A.Y. Manal, and D. Ajantha. 2013. Autonomic computing: A framework to identify autonomy requirements. In *Conference organized by Missouri University of Science and Technology 2013*, vol. 20, 235–241, Baltimore: Elsevier.
26. Melvin, G., and R. Manuel. 2012. Autonomic computing drives innovation of energy smart grids. In *Conference on organized by Missouri University of Science and Technology 2012*, vol. 12, 314–319. Washington: Elsevier.
27. Tarek, G., and E.H. Aboul. 2014. *Bio-inspiring cyber security and cloud services: Trends and innovations*. Springer.
28. Noha, E.E., A.A. Wael, and A.O. Fatma. 2016. Empirical assessment for security risk and availability in public cloud frameworks. In *11th international conference on computer engineering & systems*, 17–25. IEEE, Egypt.
29. Stephen. O., and K. Anthony. 2013. *Advances in digital forensics IX*. Springer.
30. Danwei, C., and H. Yanjun. 2010. A study on secure data storage strategy in cloud computing. *Journal of Convergence Information Technology* 5: 175–179.
31. Michel, A., B. Sonia, P. David, R. Sylvain, and V. Damien. 2015. Robust pseudo-random number generators with input secure against side-channel attacks. In *13th international conference on applied cryptography and network security*, 1–23. Springer.

32. Nentawa, Y.G. 2013. Data encryption and decryption using algorithm in a network environment. *International Journal of Computer Science and Network Security* 7 (13): 9–13.
33. Larry, D.P. 2016. Modern assembly language programming with the ARM processor. Elsevier.
34. Kan, Y., and J. Xiaohua. 2012. Data storage auditing service in cloud computing: Challenges, methods and opportunities. *World Wide Web* 15 (4): 409–428.
35. Pankaj, P., and B. Arpit. 2011. Network security through SSL in cloud computing environment. *International Journal of Computer Science and Information Technologies* 2 (6): 2800–2803.

Review of Different Image Fusion Techniques: Comparative Study

Shrouk A. Elmasry, Wael A. Awad and Sami A. Abd El-hafeez

Abstract One of the most significant research fields in image processing is image fusion that means combining two or more images with different characteristics in order to integrate relevant information from the source images into a rich informative image. Generally, the fusion methods are categorized into two domains; the spatial domain methods that deal directly with the pixel value in the input images to obtain the fused image, and the frequency domain methods in which the source images are first transformed into the frequency domain by computing the Fourier Transform of the source images, where the fusion operations are all applied on the Fourier transform and then applying inverse Fourier transform to obtain the fused image. In this paper, a comparison among some of the recent image fusion techniques in both the spatial and transform domains using a medical image dataset of two different medical imaging modalities (CT and MRI). The techniques that have been used in the experiment include PCA and Simple Average in the spatial domain and DWT, SWT, LP, DCT, and LP+DCT in the transform domain. The performance parameters used in the evaluation step were: Standard Deviation (SD), and Peak Signal to Noise Ratio (PSNR). Experimental results illustrate some features of the spatial domain methods having a better peak signal to noise ratio and thus keep the similarity between the fused image and the source images, reduce image distortion, as well as they are simple and easy to implement. On the other hand, the multiresolution methods give better information and high contrast and enhance spectral information.

Keywords Image fusion · Spatial domain · Frequency domain

S. A. Elmasry (✉) · W. A. Awad · S. A. Abd El-hafeez
Faculty of Science, Port Said University, Port Said, Egypt
e-mail: shroukelmasry9@gmail.com

W. A. Awad
e-mail: Wael_abdelkader@sci.psu.edu.eg

S. A. Abd El-hafeez
e-mail: samyabdelhafeez.1952@gmail.com

A. Z. Ghalwash et al. (eds.), *Internet of Things—Applications and Future*,
Lecture Notes in Networks and Systems 114,
https://doi.org/10.1007/978-981-15-3075-3_3

1 Introduction

Image processing is the process of applying some operations on an image to get an enhanced image or to extract some useful information from it. Generally, image processing includes three stages: (1) obtain the digital image via image acquisition tools; (2) image manipulation and analysis; (3) produce the altered image or the image analysis report [1].

Image processing has a vast area of important applications like medical image processing, remote sensing, automatic character recognition, image fusion etc. [2]. The images obtained from a single sensor or modality does not enough to the accurate realizing of the real case of the object such as in the medical images or in the remote sensing applications. The process of integration two or more images with different characteristics in order to integrate relevant information from the source images into a rich informative single image is known as image fusion [3]. Image fusion has been applied to a wide area of applications such as remote sensing, machine vision, defect inspection, robotics field, medical imaging, and military areas [4].

According to the input data and the purpose, image fusion is classified into four types which are i. Multiview fusion, ii. Multimodal fusion, iii. Multi-focus fusion and iv. Multitemporal fusion. Multiview fusion: where the input images were taken by a single sensor at the same time but from different viewpoints. This type of fusion provides a high resolution fused image. Multimodal fusion: that combines the images from different modalities or sensors. This type of fusion is vastly used in applications like Medical diagnosis, Security, etc. Multitemporal fusion: that combines images taken at different interval times. Multi-focus image fusion: combines the images of different focal lengths from the imaging equipment into a single high-quality image [5]. According to the processing levels, image fusion can be divided into three categories, pixel level, feature level, and decision level; or low, middle, and high [4]. Pixel level fusion: that directly deals with the pixels at the source images. Feature level fusion: that fuses the similar extracted salient features from the source images. Decision level fusion uses the information extracted by processing the input images as inputs to the fusion algorithm, where the Extracted information is then combined by applying decision rules [3]. Because the pixel level is easy to implement and can reduce the probability of resulting of artifacts in the fused image comparing with the feature or decision level fusion scheme, it is the most commonly used [6].

There are many benefits in using image fusion where it reduces uncertainty, provides wider spatial and temporal coverage, improves reliability, and increases the robustness of system performance. The basic challenge of image fusion is determining one of the best procedures for combining the multiple source images. A perfect image fusion technique should have three fundamental factors, i.e. providing high computational efficiency, preserving high spatial resolution, and reducing color distortion. There are many image fusion techniques have been proposed in the past few years, such as PCA (principal component analysis), simple average, Laplacian pyramid, and DWT (discrete wavelet transform). Generally, fusion techniques are classified into two domains: spatial and transform domain techniques. Spatial domain techniques are based on manipulating directly the pixel value to get the desired results. In transform domain fusion, initially, the source images are transformed into another domain, i.e., the frequency domain by applying Fourier transform method on the input images, and then inverse Fourier transform method is applied to recover the fused image [7].

2 Related Work

Several image fusion methods and applications have been proposed earlier by many researchers in both the spatial and transform domains. In this paper [8] a novel approach of principal component analysis (PCA) image fusion method have been proposed to remove the image blurredness in two images and reconstruct a new de-blurred fused image. The proposed method is based on the calculation of Eigenfaces by using the principal component analysis (PCA). Vani M. et al. 2015, have proposed a multi-focus and multimodal image fusion based on the discrete wavelet transform (DWT) [9].

There are various hybrid image fusion methods and applications that have been proposed. Y. Yang et al. (2016), have introduced a new multimodal medical image fusion method based on multi-scale geometric analysis of the nonsubsampled contourlet transform (NSCT) with type-2 fuzzy logic techniques. First, the source images high-frequency and low-frequency sub-bands were obtained by performing the NSCT on pre-registered source images. Next, the high-frequency sub-bands have been fused using an effective type-2 fuzzy logic as a fusion rule. Finally, the fused image was obtained by the inverse NSCT with all composite sub-bands. They have tested their proposed technique on MR-PET, MR-CT, and MR-SPECT multimodal medical images. Comparing the state-of-the-art methods like GP, DWT, DTCWT, CT, and NSCT with the proposed technique, it was found that the proposed technique provided better contrast, accuracy, and versatility in the evaluation [10].

Another medical image fusion technique has been proposed in [11], by using a combination of stationary wavelet transform (SWT) and non-sub-sampled contourlet transform (NSCT). The proposed technique performed on two distinct medical imaging modalities (i.e. PET and MRI). This proposed technique increases the spatial information content and reduces the color distortion.

A new hybrid multi-focus image fusion method has been proposed in [12]. B. Panda et al. have proposed a combination of wavelet transform and Curvelet transform techniques by using the simple average and weighted average rules. The input images have been first filtered by Gaussian filter, and then DWT has applied to the filtered images. Then after applying the IDWT, the Curvelet transform was applied and the Curvelet coefficients obtained were fused by the simple average and weighted average rules. Finally, the inverse Curvelet transform was applied to obtain the fused image. From the results, they have proved that a weighted average fusion using Curvelet in comparison to simple Curvelet based image fusion will increase the PSNR and reduce the MSE. In 2017, M. Haddadpour et al. have studied the combination of the two dimensional Hilbert transform (2-D HT) and Intensity Hue Saturation (IHS) image fusion techniques for fusing Magnetic Resonance Image (MRI) and Positron Emission Tomography (PET) brain images. In this proposed method, three fusion rules have been tested: minimum, maximum, and average rules. The results showed that the proposed combination between the 2-D HT and HIS can preserve the spatial and spectral features [13]. The combination of the discrete wavelet transforms (DWT) and intuitionistic fuzzy sets (IFSs) have been proposed in [6], in order to provide a higher quality of the information in terms of contrast and physical properties as compared to the existing methods [14]. J. Reena Benjamin et al. (2018), have proposed an image fusion technique based on cascaded PCA and shift invariant wavelet transforms, where they used two different modalities (MRI and CT) collected from the whole brain atlas data distributed by Harvard University as input images. And the maximum fusion rule has been used in the dual-tree complex wavelet transform domain to enhance the average information and morphological details [15].

3 Image Fusion Methods

3.1 Principal Component Analysis (PCA)

The principal component analysis [8] comprises a mathematical procedure used to obtain a compact and optimal description of the data set where it transforms a number of correlated variables into a number of uncorrelated variables called principal components. The first principal component computes as much of the variance as possible in the data and each succeeding component accounts for as much of the remaining variance as possible. Where the variance is the measure of the variability of data in a data set. The second principal component points to the direction of maximum variance. The third principal component is taken in the maximum variance direction in the subspace vertical to the first two and so on. The principal component analysis fusion method can be described by the following steps:

1. Determine the column vectors form the input image $I_1(x, y)$ and $I_2(x, y)$, and then calculate the covariance matrix of the two column vectors that were computed before, then, calculate the Eigenvalues and Eigenvectors of this covariance matrix.
2. Normalize the column vector corresponding to the larger Eigenvalue by dividing each element with the mean of Eigenvector by using the following equations:

$$P_1 = V(1)/V \quad (1)$$

$$P_2 = V(2)/V \quad (2)$$

3. Multiply the normalized Eigenvector, respectively, with each pixel of the input images.
4. Obtain the fused image by computing the sum of the two scaled matrices through the following equation:

$$I_f = P_1 I_1(x,y) + P_2 I_2(x,y) \quad (3)$$

3.2 Simple Average Method

A simple average [16] is one of the most simple fusion techniques and it is easy to implement. In this method, the fused image is generated by computing the average intensity of corresponding pixels from both the source image. The fusion equation can be described as the following

$$F(i,j) = \frac{I_1(i,j) + I_2(i,j)}{2} \quad (4)$$

where $I_1(i,j)$ and $I_2(i,j)$ are two input images and $F(i,j)$ is the resultant fused image.

3.3 Discrete Wavelet Transform (DWT) Method

Wavelet transform [9] is a multiresolution analysis (MRA) tool that represents the image highlight by various frequency sub-bands at multi-scale. DWT technique is widely used in analyzing signals. Through the wavelet decomposition, the image is split into four frequency sub-bands such as LL (the approximation image), LH (the horizontal details), HL (the vertical details), and HH (the diagonal details). DWT is a multi-scale (multiresolution) approach valuable in various image processing applications including image fusion. The discrete wavelet transform (DWT) fusion method can be described by the following steps:

1. Apply the wavelet transform to the source images to obtain the wavelet decomposition (the approximation coefficients and detailed coefficients) at the desired level.

2. Combine each decomposition level using the fusion rules.
3. The new wavelet coefficients (the fused coefficients) transformed by applying the inverse wavelet transform (IDWT) to obtain the fused image.

3.4 Stationary Wavelet Transform (SWT) Method

The stationary wavelet transforms [17] is the same like discrete wavelet transform (DWT) but SWT except for the down-sampling process so it is translation-invariant. The process of up-sampling the filters by including zeros between the filter coefficients is called "àtrous" which means "with holes". As stationary wavelet transform (SWT) algorithm, the high and low pass filters are first applied to the rows and then to the columns. The four images obtained (one approximation and three detail images) are at half the resolution of the original image and are of the same size as the original image. At each level of decomposition, the spatial resolution becomes coarser and the size remains the same original image size.. The stationary wavelet transform (SWT) fusion method can be described by the following steps:

1. Apply the stationary transform to the source images at one level resulting in three detail sub-bands and one approximation sub-band (HL, LH, HH, and LL bands).
2. Combine the approximation parts using the average fusion rule.
3. For horizontal coefficients of the source images: the absolute values of them are computed and subtract the second part of the image from first: D = (abs (H1L2) − abs (H2L2)) $\geq$ 0, then the horizontal part is fused as follow: (HfL1 = D. *H1L1 + (D).*H2L1).
4. The same for vertical and diagonal parts to obtain the fused vertical and details of the image.
5. The fused coefficients transformed by applying the inverse stationary wavelet transform (ISWT) to obtain the fused image.

3.5 Laplacian Pyramid

The image pyramid [18] is a data structure designed to provide a scaled convolution by reducing image representation. The image Pyramid presents a series of copies of an original image in which both sample intensity and resolution are reduced in a sequence of levels. A Laplacian pyramid is a technique in image processing and it is very similar to the Gaussian pyramid with the difference that it uses a Laplacian transform rather than the Gaussian pyramid.

The Laplacian pyramid of an image is built first by a Gaussian image pyramid structure that is created. Based on the bottom of the Laplacian image Pyramid, the Gauss Pyramid image of each layer can be computed by layers. Finally, the original image can be reconstructed. Image fusion based on Laplacian Pyramid can be described by the following steps:

1. The images to be fused are decomposed into the Laplacian Pyramid.
2. The first layer image corresponding to the Laplacian Pyramid is used to generate the fused image using the pixel fusion rule.
3. The new fused image is taken as the new first layer image.
4. The new Laplacian image is used to construct the Pyramid layer by layer, and the final fused image is reconstructed.

3.6 Discrete Cosine Transform (DCT) Method

The discrete cosine transform (DCT) [19] is an appropriate technique in image processing to split the image into parts for different importance. The discrete cosine transform (DCT) method is identical to the discrete Fourier transform that decomposes the image into a series of waveforms, each with a specific frequency (low and high frequencies). Large coefficients mainly concentrate on the low-frequency region and high-frequency coefficient included the edges of images. The image fusion technique based on DCT can be described by the following steps:

1. Perform the decomposition on each source image by applying the discrete cosine transform (DCT).
2. The coefficients are separately fused by using certain fusion rules.
3. The fused image is obtained by performing the inverse discrete cosine transform (IDCT) for the combined high and low-frequency coefficients.

4 Performance and Evaluation

In order to evaluate the quality of the fused image obtained by using the previous fusion methods, we used performance parameters like standard deviation (SD), and peak signal to noise ratio (PSNR) [20], which illustrated in Table 1.

Table 1 Represents the performance parameters used in the experiment: standard deviation (SD), and peak signal to noise ratio (PSNR)

Performance metrics	Description	Equation
Standard deviation (SD)	Measures the image contrast. A high value of (SD) indicates the fused image as high contrast	$SD = \sqrt{\frac{1}{mn}\sum_{m=0}^{L-1}\sum_{n=0}^{L-1} f(m,n) - \mu}$
Peak signal to noise ratio (PSNR)	Represents the relationship between the fused and the reference image. PSNR is computed by dividing the number of gray levels in the image on the difference between the source image and the fused image. A high PSNR value means that the fused and reference images are similar	$PSNR = 20\log_{10}\frac{(255)^2}{\frac{1}{mn}\sum_{i=1}^{m}\sum_{j=1}^{n}\left(I_r(i,j)-I_f(i,j)\right)^2}$

5 Experiment Analysis

The comparative study experiment is built on a set of medical images consisting of eight pairs of images of two different medical imaging modalities of size 256 × 256 available in [21]. MATLAB 2015R is used as a platform to execute the experiment (Fig. 1).

From Table 2, it has been observed that the spatial domain methods have a better peak signal to noise ratio and thus keep the similarity between the fused image and the source images and reduce image distortion this is evident from the PSNR metric results where the (PSNR of PCA = −30.00) and (PSNR of simple average = −30.08) and these are the highest values for the PSNR in the table. The spatial domain methods are simple and easy to implement and understand. On the other hand, the multiresolution methods give better information and high contrast, enhance spectral information, that is illustrated through the standard deviation (SD) values, where (SD of DCT = 62.81) and (PSNR of LP = 65.26) and these are the highest values for the standard deviation in the table. More levels of decomposition in multiresolution methods (like SWT) give a better result. The two levels of the decomposition in the SWT method provided better contrast (SD of SWT: 1L = 53.55 < SD of SWT: 2L = 53.66), but reduce the PSNR value where (PSNR of SWT: 1L = −30.51 and PSNR of SWT: 2L = −30.54). The combination between LP and DCT methods achieved higher SD and SNR values than that were achieved by the DCT and LP methods independently: SD values of LP = 65.26, DCT = 62.81, while DCT+LP = 65.77, and PSNR values of LP = −34.74, DCT = −31.50, while DCT+LP = −31.09.

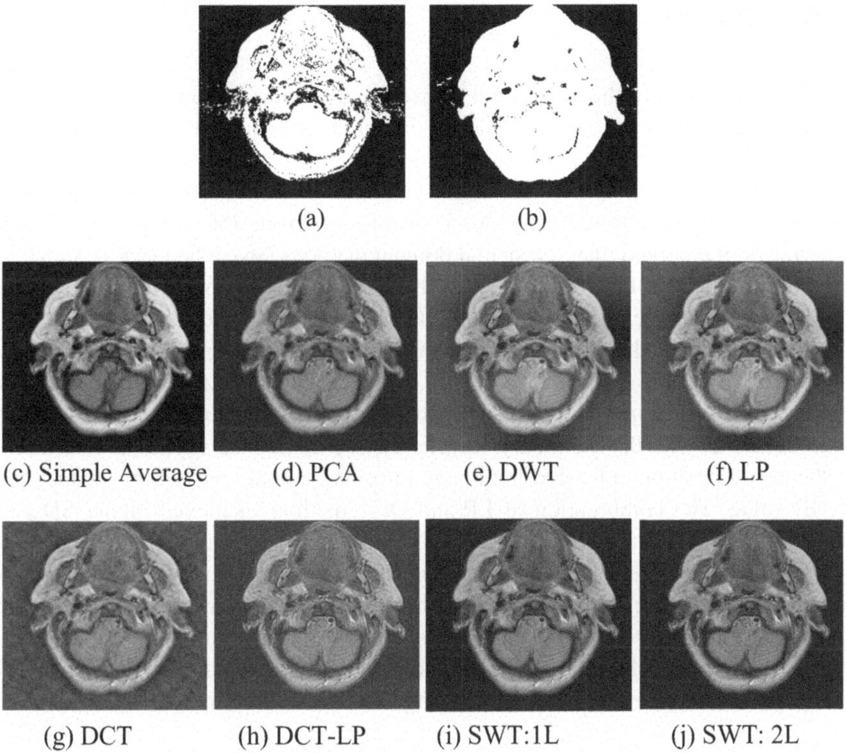

Fig. 1 **a** and **b** are one sample of the medical source images for normal axial brain images. **c**, **d**, **e**, **f**, **g**, **h**, **i**, **j** denote the fused images obtained by using the methods: simple average, PCA, DWT, LP, DCT, (DCT+LP), SWT (one level of decomposition), and SWT (two level of decomposition) respectively

Table 2 Comparison of performance metrics standard deviation (SD), and peak signal to noise ratio (PSNR), using: simple average, PCA, DWT, LP, DCT, DCT+LP, SWT (one level of decomposition), and SWT (two levels of decomposition) methods. The results represented in the table are the average values of quality metrics for 8 fused images by using the medical dataset

Methods	Metrics	
	SD	SNR
Simple average	53.14	−30.08
PCA	61.27	−30.00
DWT	53.25	−30.64
SWT: 1L	53.55	−30.51
SWT: 2L	53.77	−30.54
DCT	62.81	−31.50
LP	65.26	−34.74
DCT and LP	65.77	−31.09

6 Conclusion

Image fusion refers to the integration of two or more images from multiple sensors with different characteristics in order to integrate relevant information from the source images into a rich informative single image. A perfect image fusion technique should work on providing high computational efficiency, preserving high spatial resolution, and reducing color distortion. From the comparative study experiment, it emerged that: the spatial domain methods have a better peak signal to noise ratio and thus keep the similarity between the fused image and the source images, reduce image distortion, as well as they are simple and easy to implement. On the other hand, the multiresolution methods give better information and high contrast and enhance spectral information, which is illustrated through the standard deviation values. More levels of decomposition in multiresolution methods give a better result, as with the Stationary wavelet transform method, where the two levels of the decomposition enhanced the image information and contrast, but reduce the PSNR value. The combination of LP and DCT methods achieved higher SD and SNR values than that was achieved by the DCT and LP methods independently. A combination of spatial domain technique and transform domain technique would enhance the image information and quality and keep the similarity between the fused image and the source images. A combination of spatial domain technique and transform domain technique would enhance the image information and quality and keep the similarity between the fused image and the source images.

References

1. McAndrew, A. 2004. An introduction to digital image processing with Matlab notes for SCM 2511 image processing. School of Computer Science and Mathematics, Victoria University of Technology 264 (1): 1–264.
2. Deshmukh, M.D., and A.V. Malviya. 2015. Image fusion an application of digital image processing using wavelet transform. *International Journal of Scientific & Engineering Research* 6 (11): 1247–1255.
3. Dogra, A., B. Goyal, and S. Agrawal. 2017. From multi-scale decomposition to non-multi-scale decomposition methods: a comprehensive survey of image fusion techniques and its applications. *IEEE Access* 5: 16040–16067.
4. Sharma, M. 2016. A review: image fusion techniques and applications. *International Journal of Computer Science and Information Technologies (IJCSIT)* 7 (3): 1082–1085.
5. Kalaivani, K., and Y. Asnath Victy Phamila. Analysis of image fusion techniques based on quality assessment metrics. *Indian Journal of Science and Technology* 9 (31): 1–8.
6. Dileepkumar Ramlal, S., J. Sachdeva, C. Kamal Ahuja, and N. Khandelwal. 2018. Multimodal medical image fusion using non-subsampled shearlet transform and pulse coupled neural network incorporated with morphological gradient. *Signal, Image and Video Processing* 12 (6): 1479–1487.
7. Kumar Sahu, D., and M.P. Parsai. 2012. Different image fusion techniques—a critical review. *International Journal of Modern Engineering Research (IJMER)* 2 (5): 4298–4301.
8. Dep, S., S. Chakraborty, and T. Bhattacharjee. 2012. Application of image fusion for enhancing the quality of an image. (CS & IT) 6: 215–221.

9. Vani, M., and S. Saravanakumar. 2015. Multi focus and multi modal image fusion using wavelet transform. In *2015 3rd international conference on signal processing, communication and networking (ICSCN)*, pp. 1–6. IEEE, Chennai, India.
10. Yang, Y., Y. Que, S. Huang, and P. Lin. 2016. Multimodal sensor medical image fusion based on type-2 fuzzy logic in NSCT domain. *IEEE Sensors Journal* 16 (10): 3735–3745.
11. Shabanzade, F., and H. Ghassemian. Combination of wavelet and contourlet transforms for PET and MRI image fusion. In *2017 artificial intelligence and signal processing conference (AISP)*, pp. 178–183. IEEE, Iran.
12. Panda, B. 2016. Image fusion using combination of wavelet and curvelet fusion. *International Journal of Advanced Research in Computer Engineering & Technology (IJARCET)* 5 (8): 2316–2324.
13. Haddadpour, M., S. Daneshvar, and H. Seyedarabi. 2017. PET and MRI image fusion based on combination of 2-D Hilbert transform and IHS method. *Biomedical Journal* 40 (4): 219–225.
14. Soundrapandiyan, R., M. Karuppiah, and S. Kumari. 2017. An efficient DWT and intuitionistic fuzzy based multimodality medical image fusion. *International Journal of Imaging Systems and Technology* 27 (8): 118–132.
15. Reena Benjamin, J., and T. Jayasree. 2018. Improved medical image fusion based on cascaded PCA and shift invariant wavelet transforms. *International Journal of Computer Assisted Radiology and Surgery* 13 (2): 229–240.
16. Rani, K., and R. Sharma. 2013. Study of different image fusion algorithm. *International Journal of Emerging Technology and Advanced Engineering (IJETAE)* 3 (5): 288–291.
17. Pradnya, M., and S.D. Ruikar. 2013. Image fusion based on stationary wavelet transform. *International Journal of Advanced Engineering Research and Studies* 2 (4): 99–101.
18. Mukane, S.M., Y.S. Ghodake, and P.S. Khandagle. 2013. Image enhancement using fusion by wavelet transform and Laplacian pyramid. *IJCSI Journal* 10 (4).
19. Naidu, V.P.S. 2012. Discrete cosine transform based image fusion techniques. *Journal of Communication, Navigation and Signal Processing* 1 (1): 35–45.
20. Jagalingam, P., and A. Vittal Hegde. A review of quality metrics for fused image. *Aquatic Procedia* 4 (2015): 133–142.
21. The whole brain Atlas. http://www.med.harvard.edu/AANLIB/home.html. Accessed 23 Apr 2019.

Cloud Technology: Conquest of Commercial Space Business

Khaled Elbehiery and Hussam Elbehiery

Abstract Cloud-computing technology has far exceeded what business ever thought of, from significant cost saving to reducing deployment time. It has invaded into almost all types of businesses, and in the last decade, it simplifies the way business is running, growing, adapting to economic changes, and maximizing out the return of investment to immeasurable levels. Space has become the new era's business, and it is not anymore watching rocket launch. The real space money comes from direct home television, GPS products, and services, the things affecting our day-to-day lives or things that have global markets, which is where space is making money out of. Space used to be dominated by governments' agencies, but in today's world there is a democratization of space (Tim Fernholz in Quartz: SpaceX is about to take the lead in the satellite internet race. https://qz.com/1618386/spacex-launches-first-starlink-internet-satellites/). Commercial space industry is spending multimillion dollar on satellites and rockets, which are increasingly playing a part in our everyday lives. For satellites to function, they need to be able to communicate with ground station, which is a combination of signal processing digitization called "Headend" and computation of data at centers. Billionaires such as Jeff Bezos and Elon Musk and others have brilliantly and genuinely brought the cloud-computing and commercial space industry together and made them inseparable, from earthly profitable businesses, such as internet and video, communication infrastructure, autonomous vehicles like UBER, online mapping such as Google or Apple, energy industries, imagery, defense, and intelligence to the twenty-first century optimistic futuristic businesses of trillions of dollars such as Space Cloud, Moon Cloud, Space tourism, asteroid mining, and Mars exploration voyages (Wendover Productions, https://www.wendoverproductions.com/).

K. Elbehiery
DeVry University, Denver, CO, USA
e-mail: khaled.elbehiery@gmail.com

H. Elbehiery (✉)
Ahram Canadian University (ACU), Cairo, Egypt
e-mail: hussam.elbehiery@gmail.com

A. Z. Ghalwash et al. (eds.), *Internet of Things—Applications and Future*,
Lecture Notes in Networks and Systems 114,
https://doi.org/10.1007/978-981-15-3075-3_4

Keywords Moon Cloud · Space Cloud · CSPO · SCC · SSA · LSP · Serverless computing

1 Introduction

"We chose to go to the Moon in this decade and do the other things, not because they are easy but because they are hard"—US President John F. Kennedy. Sixty years ago, space-race was not only an exploration but "doing the other things" in the long term. It has been proven that it was an investment and has paid off because space now makes money. Cloud providers with their ground station, data centers, and resources, along with satellite rocket launchers and their unlimited application from communication to imagery, make combined effort in forming cross-businesses environment to concur with the commercial space industry (see Fig. 1) [1].

It is a tremendously fast-growing industry but by itself really isn't worth much. Customers are demanding more actionable information faster than ever, which means customers want data and data become very complex and the new oil. The combined effort is also stretching their goals through making access to orbital space and getting beyond the earth's orbit to outside of our cosmic neighborhood, into the deep space [2]. The publication is covering how public cloud providers and space industry are coming together and introducing new business opportunities. Beyond earth's orbit how to build a similar public cloud in the space. The business does not stop there by the earth's orbit but is going farther to the Moon. How to establish a

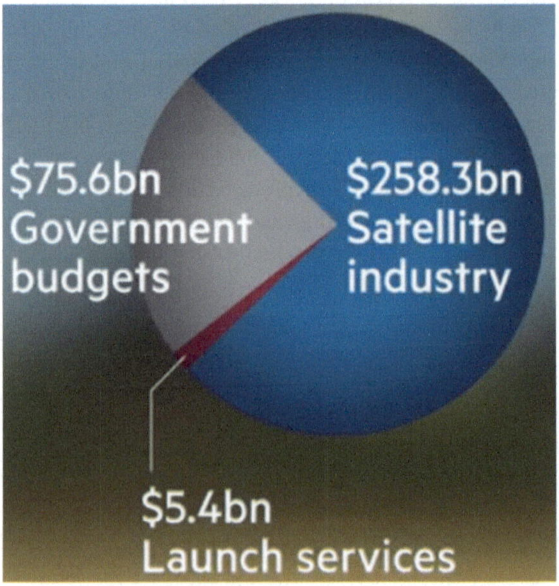

Fig. 1 Space industry business

cloud base and how would it obtain its power. Finally, every road you walk in has its own complications, and those goals will have all different types of challenges, from financial to gravitational effects, and interstellar impacts as well [3, 4]. Many aspects need to be covered, so let's begin.

2 Public Cloud Conquer Commercial Space Business

2.1 Amazon AWS Satellite Network Station

Today, Amazon is a titan of e-commerce. It is the go-to site for online shoppers and merchants alike. Adapting to cross-business environment of twenty-first century, they decided to dabble in plenty more industries. Amazon is expanding its retail sales and recently getting into satellites. The company has unveiled its plans to build a dozen satellite transmission facilities throughout the world. Announcement for that was made at the Amazon Web Services Annual Reinvent Conference in Las Vegas, Nevada, December 2018 for building satellite ground stations essentially equipped with antennae facilities that can send and receive data from the thousands of satellites orbiting the earth.

Iridium Communications announced a partnership with Amazon Web Services, the most widespread cloud-computing service in the world, to develop a satellite-based network called CloudConnect for internet of things (IoT) applications. Recently, Amazon is seeking talents to work on "interconnecting space system networks" which consequently and positively introduced new jobs to the society [5]. As much AWS has gained by bringing the rest of the world within reach of AWS and making tremendous profit, the stock shares of Iridium went up 7.1% in trading, hitting an all-time high. Amazon will let customers rent access to the satellite ground stations in the same manner that they lease access to their data computation centers. Amazon said that companies that traditionally not had the financial resources to build and operate their own satellite transmission infrastructure will be able to get access to satellite services on demand. This is basically building "Headend" everywhere available to consumers utilizing their own network to transport it anywhere around the globe (see Fig. 2).

For Amazon, it truly looks like that the space is the new frontier in their journey for global domination. AWS ground station has the capacity to reduce the processing time from one hour to less than a minute. During the test phase, it was found that in exact 55 s, data could be sent from a satellite to Amazon ground station, which is nothing short of a breakthrough given that normally downlinking a satellite image and getting it into the cloud takes around 60 min. Amazon ground station can be called Amazon Cloud customized for satellites, and the analogy holds true. Just like with the advent of Amazon Cloud, companies no longer needed their own mega cloud servers and data centers but what they can do is renting space on

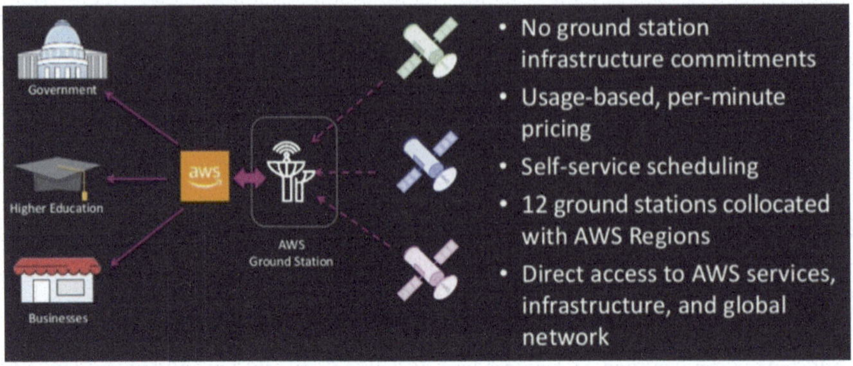

Fig. 2 AWS ground stations

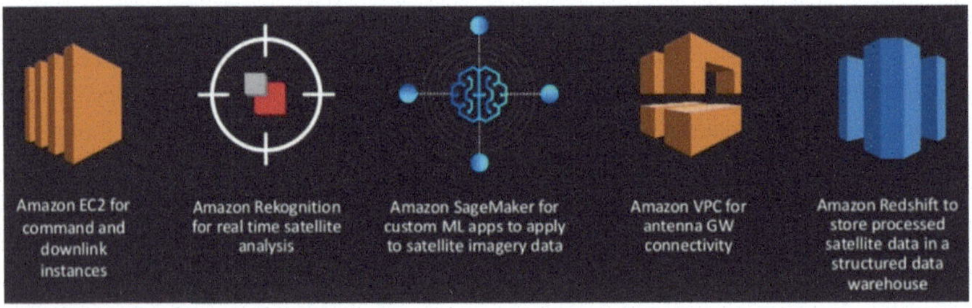

Fig. 3 AWS services innovation model

the Amazon Cloud. Similarly, AWS ground station satellite companies can rent a slot on demand and pay only for the service they want (see Fig. 3) [6].

The ground station would be fully operated, managed, and supervised by Amazon and its technical teams. Companies will not even have to build their own antennas or worry about technical glitches, as it would be handled by Amazon. This would not only save time, money, and efforts but would also provide greater liberty to the company to focus only on its core projects (see Fig. 4) [7].

2.2 *Social Media Providers Cross-Businesses*

Facebook may soon join SpaceX and OneWeb in the rush to deliver internet from orbit. Facebook plans to launch a satellite next year to test a system that would provide ultra-high-speed internet to the 3 billion and plus people not yet on the web. Elon Musk's SpaceX is working to launch more than 10,000 satellites into low earth orbit, intended to provide high-speed internet access to remote corners of the earth. A Softbank-funded company called OneWeb is planning a similar

Fig. 4 Global AWS ground stations

undertaking [8]. A filing with the Federal Communications Commission (FCC) revealed details of a multimillion-dollar experimental satellite from a stealthy company called PointView Tech LLC [9]. The satellite, named Athena, will deliver data 10 times faster than SpaceX's Starlink internet satellites. In fact, the tiny company seems to be a new subsidiary of Facebook, formed last year to keep secret the social media giant's plans to storm the space [10].

For Facebook to be the internet provider for all-new markets around the globe raises the possibility that those new internet users will become members of its social network, which in turn expands its global reach and further solidifies its online advertising empire (see Fig. 5) [11–13].

Fig. 5 Social media storms the space

2.3 *Space Agencies Partnership with Public Cloud Providers*

The project team that built and operates the Mars rovers, Spirit and Opportunity, has become the first NASA space mission to use cloud computing for daily mission operations. Cloud computing is a way to gain fast flexibility in computing ability by ordering capacity on demand—as if from the clouds—and paying only for what is used. NASA's Mars exploration rover project moved to this strategy for the software and data that the rovers' flight team uses to develop daily plans for rover activities. NASA's Jet Propulsion Laboratory is a center for the robotic exploration of space, having sent a robot toward every planet in the solar system. NASA's Jet Propulsion Laboratory manages the project and gained confidence in cloud computing from experience with other uses of the technology, including public participation sites about Mars exploration. The Mars exploration rover "Opportunity" is still roving on Mars after landing eight years ago, and Mars rover "Curiosity" landed on August 5, 2012 (see Fig. 6).

With Curiosity safely on Mars, the mission continues to use AWS to automate the analysis of images from Mars, maximizing the time scientists have to identify potential hazards or areas of particular scientific interest. As a result, scientists are able to send a longer sequence of commands to Curiosity that increases the amount of exploration the Mars science laboratory can perform on any given sol (Martian day).

2.4 *Long-Range Satellite Communication Relay Networks*

Direct communication between the Earth and the Mars can be strongly disturbed and even blocked by the Sun for weeks at a time, cutting off any future human missions to the red planet. The European Space Agency (ESA) engineers working

Fig. 6 Mars exploration rover

with counterparts in the UK may have found a solution using a new type of orbit combined with continuous-thrust ion propulsion [14].

The European researchers studied a possible solution to a crucial problem affecting future human missions to Mars: how to ensure reliable radio communication even when Mars and Earth line up at opposite sides of the Sun, which then blocks any signal between mission controllers on Earth and astronauts on the red planet's surface. The natural alignment, known as a conjunction, happens approximately every 780 days, and would seriously degrade and even block transmissions of voice, data, and video signals [15].

The research findings were released at the 60th International Astronautical Congress (IAC), the world's biggest space event, being held in Daejeon, South Korea. To counter the effects of gravity and remain in place, they would have to be equipped with cutting-edge electric ion propulsion. The ion thrusters, powered by solar electricity and using tiny amounts of xenon gas as propellant would hold the satellites in a B-orbit in full view of both Mars and the Earth. The satellites could then relay radio signals throughout the Mars–Earth conjunction season, ensuring that astronauts at Mars were never out of touch with Earth [16].

NASA is seeking proposals for commercial Mars data relay satellites and also opening the gates for industry input for a high-tech, state-of-the-art Mars orbiter, which is scheduled to be launched in early 2020s. This satellite business in the space is attracting SpaceX and many others, as mentioned earlier that space belongs to everyone in the twenty-first century and not governments or special agencies, and it is an open business field to grab (see Fig. 7) [17].

Fig. 7 Commercial Mars data relay satellites

3 Space Cloud Computing (SCC)

3.1 Infrastructure

Performing complex analysis tasks are usually performed by ground-based computing centers, and avoided to be done in the space due to the limited computing power of the space-qualified processors, long verification and validation process of flight software, complexity of integrating with space hardware, and space communications bandwidth and connectivity challenges. However, with more spacecraft fleets and constellations entering the market and the evolution of small satellite technologies using faster CPUs, performing more analysis in space is becoming more feasible to even further enable complex analysis tasks in the space to meet the needs for artificial intelligence, or machine learning, and as a result of that, the Space Cloud Computing (SCC) flight software (FSW) framework is being developed [18].

Similar to public and private cloud-computing platforms like Amazon EC2 and OpenStack, SCC provides a suite of services tailored for the space domain to expose computation, storage, sensors, and networking resources as services. While the concept of cloud computing matches, the implementation is quite different to address the resource-constrained space environments and cyber security concerns. Figure 8 shows the Space Cloud Computing FSW framework that enables Warfighters to task space assets to meet operational goals without the hindrance of long planning lifecycles or detailed knowledge of assets (see Fig. 8) [19].

The SCC project was designed to meet many mission objectives where imagination is the only limitation. Early adopters will have small isolated networks that

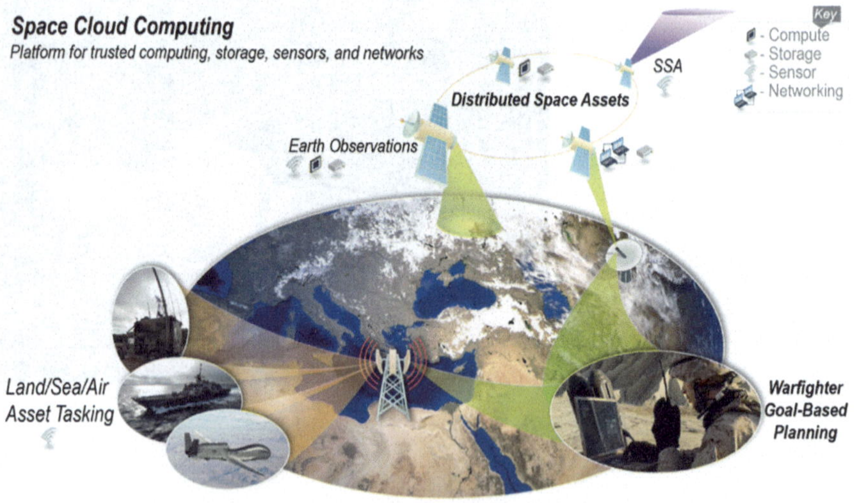

Fig. 8 Space Cloud Computing (SCC) flight software (FSW)

will meet a handful of specific tasks [20]. An example could include a fleet of small satellites operating in near proximity with cross-link capabilities. Future missions would include collaboration of assets from different providers that enter and exit communications ranges and support discovery and registration of assets that augment the cloud. The SCC FSW framework is initially being developed to meet Air Force and DARPA mission objectives. It can also be leveraged to meet future NASA and commercial space mission objectives, such as robotics exploration on the Moon and Mars or collaborative Earth and space science missions [21].

One of the early adopters' companies is Sat-Space that supply the customers a global, secured, controlled, and optimized environment to run their corporate communication and business applications (see Fig. 9) [22].

3.2 *Powering Up the Space Station*

If you think of the Sun as the energy source, well, the Sun is pretty much as good as it gets, says Lewis-Weber; and Elon Musk has stated, we have a fusion reactor in our sky. The problem with regular solar power is that the Sun isn't always up; we have nights, and cloudy days. The panels also take up a lot of land. What if, instead of sending thousands of solar panels into orbit, we could just send up one that's programmed to make copies of itself, and then each machine would make copies of itself, and so on. Like multiplying rabbits, the population of solar panel satellites would grow exponentially, covering the size of state of Nevada in a few months or years.

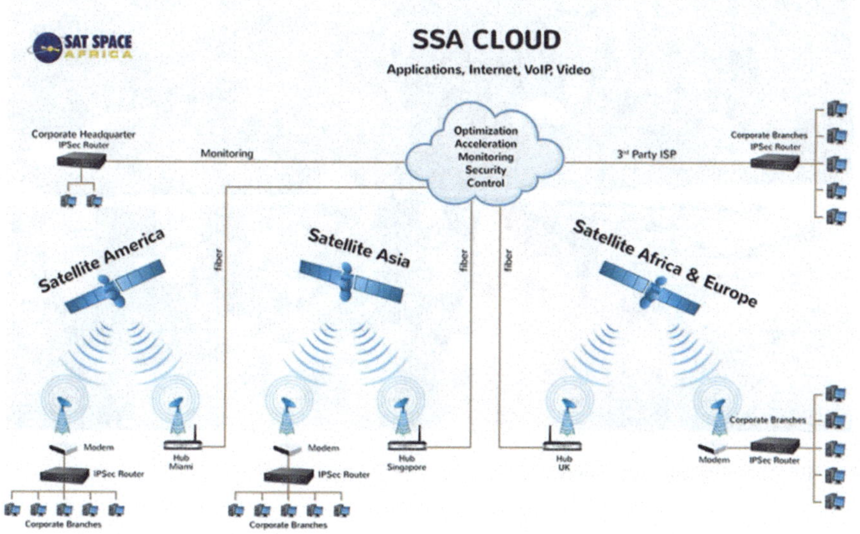

Fig. 9 Sat-space cloud computing

A far-out plan to create swarms of self-replicating array solar panel satellites not only to self-power the SCC but also could beam power down across vast portions of the globe wherever the receivers are set up. That opens up the possibility of sending electricity to villages in developing countries, or to disaster-stricken areas. The receiving equipment could fit into a couple of shipping containers, Jaffe says (see Fig. 10).

An international team would test the technology on the ground before bringing it to the International Space Station (ISS). After that, they would launch a "pathfinder mission"—a small-scale version of the array—into low earth orbit. This mission would be able to beam power to anywhere in the world. These steps can be accomplished by 2021 if we start now, and for about $350 million dollars, which is about the same amount of money that Americans spend annually on Halloween costumes for their pets.

Solar farms are becoming quite a familiar sign all over the world. However, China is taking it to a whole new level with its announcement of placing a solar power station in orbit by 2050. This would allow China to become the first country that will be harnessing the Sun's energy in space and sending it to Earth. What prompted this is the fact that the Sun is always shining in space. The project will be difficult and quite costly owing to the development of the hardware that is required for capturing and then transmitting the solar power. Furthermore, the launching of the system into space is going to be another challenge. China is currently constructing a test facility in the southwestern city of Chongqing for finding out about the most feasible way of transmitting solar power from the orbit to the ground (see Fig. 11).

Fig. 10 Space-based solar power

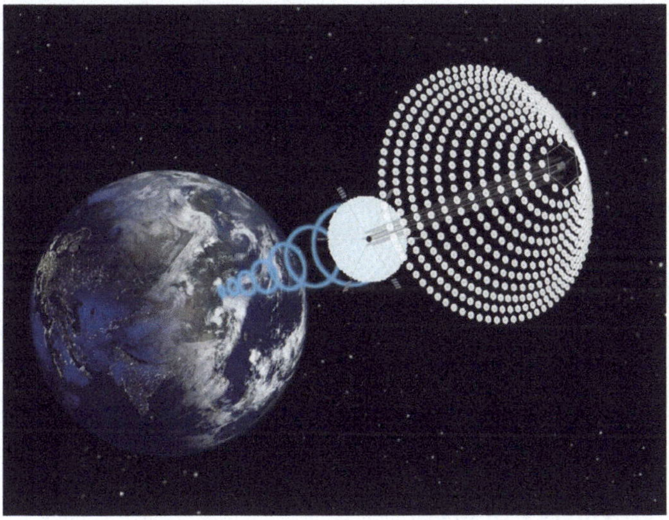

Fig. 11 China's space solar power station proposal

4 Moon Cloud

4.1 Architecture Establishment

Humans dream about leaving Earth and traveling through the galaxy or building a Moon base and in turn its computational center aka Moon Cloud. The reality is, we actually do have the technology, and current estimates from NASA and the private sector say it could be done for $20–40 billion, spread out over about a decade. The price is comparable to the International Space Station (ISS), not that big an investment really but the payoff would be immeasurable. The Moon is a sandbox to develop new technologies and exploit unlimited resources. It would start a new space-race and lay the foundation for us to spread out into the solar system and beyond. It would create a vast array of new technologies to benefit us on Earth. Only thing on the way of accomplishing it is the fact that it is hard to get governments interested in long-term investments in the future of humanity.

Moon is not a welcoming place for living things, since in Moon a day lasts 29 Earth days with a difference of maybe 300 °C between sunlight and shade and there is no atmosphere to shield us from meteorites, big and small, or cosmic radiation. Apollo missions have discovered tons of information about the Moon 60 years ago and for the last few decades, satellites like the American Lunar Reconnaissance Orbiter have mapped the Moon. Rovers like the Chinese Yutu have studied the composition of the lunar surface looking for water, ice, and metals. Astronauts will build the first light-weight Moon base which will be completed in a decade and deployed with a natural shelter such as caves or underground lava tube tunnels or craters near the poles where the days are six months long using the available lunar material. Remember there will be many trips during the first phase going to the Moon and taking off where it's much easier and cheaper to get things off the Moon into orbit.

The significant importance of the Moon base and its cloud computation center is supplying an orbital depot and necessary compute resources where scientific missions to Mars and the outer solar system can refuel. It is much easier and cheaper to get things off the Moon into orbit, colonizing Mars may mean starting from the Moon.

Fortunately, lunar soil has all the necessary ingredients to make concrete. Robotic mining rigs can sift the lunar dust for organic molecules and could be used to build huge structures while advances in 3D printing will make it possible to produce almost everything else the crews need [23].

4.2 Moon Energy Sources

The pursuit of clean energy to preserve the environment has become a necessity now. Non-renewable energy sources on the Earth are depleting at an alarming rate, causing scientists to look for alternative viable options. The Moon's energy has been a power source for several decades, wherein its gravitational pull has been harnessed to spin generators. Similarly, tidal power plants have been arranged like hydroelectric dams to trap water during high tide and then during low tide, and release it through turbines [24, 25].

Resources on the Moon are abundant; platinum, silicon, iron, titanium, ammonia, mercury, and even water have been proven to exist on the Moon. One possible energy source that is promising is the mining of helium-3, an isotope that could one day be used in nuclear fusion reactors, something the Chinese lunar exploration program is currently looking into. In addition, in the future helium-3 can be exported back to Earth. Scientists estimate that the Moon is likely to contain roughly 1 million tons of helium-3, which translates to a hypothetical 10,000 years' worth of energy. Since there is no atmosphere on the Moon, one thing could not be afforded is to create a polluted environment. Helium-3 can fuel non-radioactive nuclear fusion reactions to produce safe, clean, and large quantities of energy, and worth to remember that fusion reactions are much more efficient than fission reactions that are currently used in nuclear plants [26, 27].

Helium-3 allows energy to be generated with very limited amount of waste as well and when it is burnt with deuterium, it generates unbelievable amount of energy and to make it simple for our minds to understand, it is about 100 tons of helium-3 has the potential to power the entire population of Earth for a year. When all set and done, it will illuminate the life on the Moon and probably on Earth as well. Technology won't stop there, and when enough energy is available, asteroids could be pulled into the Moon's orbit and mined [28, 29].

Certainly, all these Moon ventures and mining concepts will be very expensive and require billions of dollars of investment. The extraction and refining of helium-3 also requires new technologies, and thus more costs (see Fig. 12) [30, 31].

Another approach for generating clean energy and more recently researched by the Japanese corporation "Shimizu" have been gearing up to develop solar power on the Moon. Shimizu took off with the idea in 2013 in the aftermath of Japan's 2011 Fukishima accident, which produced a political climate demanding alternatives to nuclear power plants. Shimizu's plans call for beginning construction of a lunar solar power base as early as 2035. The solar array would be 250 miles wide and span the lunar circumference of 6,800 miles. They're calling it the Luna Ring [31, 32].

Lunar solar power (LSP) arrays would receive higher energy density from sunlight than we get through Earth's atmosphere, avoid weather, and could even beam energy to any part of Earth facing the Moon. LSP could, theoretically, even satisfy 100% of our energy needs. That would be approximately 18 TW today and possibly 24 TW by mid-century. Figure 13 shows how to harvest Terawatts of solar power on the Moon using lunar–solar energy technology (see Fig. 13) [33].

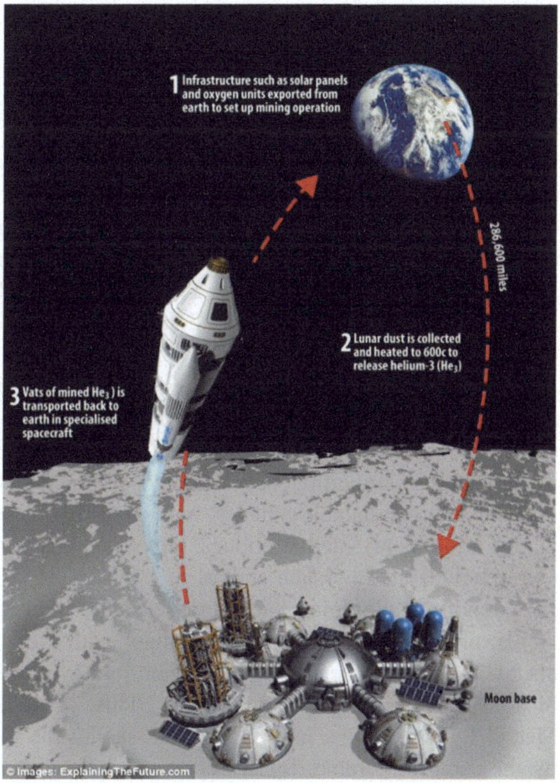

Fig. 12 Mining the moon for renewable energy resources

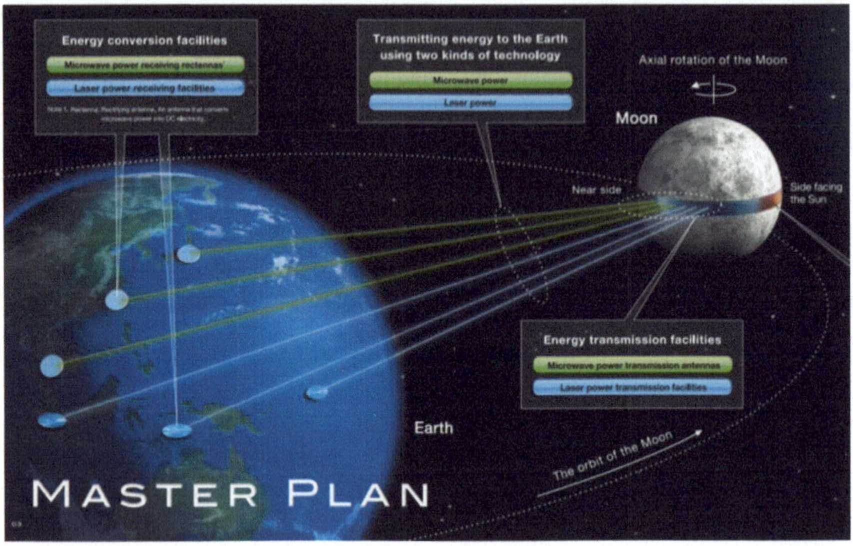

Fig. 13 Lunar ring solar power

5 Progress Challenges

5.1 Finance

It has traditionally been difficult for companies to make money from space due to the high CAPEX requirements and frequent delays. However, for investors with a truly long-term time horizon, we see it as one of the final frontiers of investing. The market was expected to grow from US $339 billion in 2016 to US $2.7 trillion by 2045. The space market is already well developed, with several opportunities including: defense and contractors, satellites (77% of market), launchers (US $5 billion + market), and insurance (US $30 billion in exposure) (see Fig. 14) [34].

The space economy is the full range of activities and use of resources that create and provide value and benefits to human beings in the course of exploring, understanding, managing, and utilizing space. It involves public and private sectors in developing, providing space-related products using the space infrastructure (ground stations, launch vehicles, and satellites), space-enabled applications (navigation equipment, satellite phones, meteorological services, etc.), and the scientific research generated by such activities.

A report by Goldman Sachs predicted the industry would reach $1 trillion by 2040s, noted Jeff Matthews; a consultant with Deloitte who moderated the panel discussion. A separate study by Morgan Stanley projected a “most likely outcome” of a $1.1 trillion space economy by 2040s. A third study by Bank of America Merrill Lynch was the most optimistic, seeing the market to grow to $2.7 trillion by the same timeframe. Space represents a new frontier for cloud infrastructure,

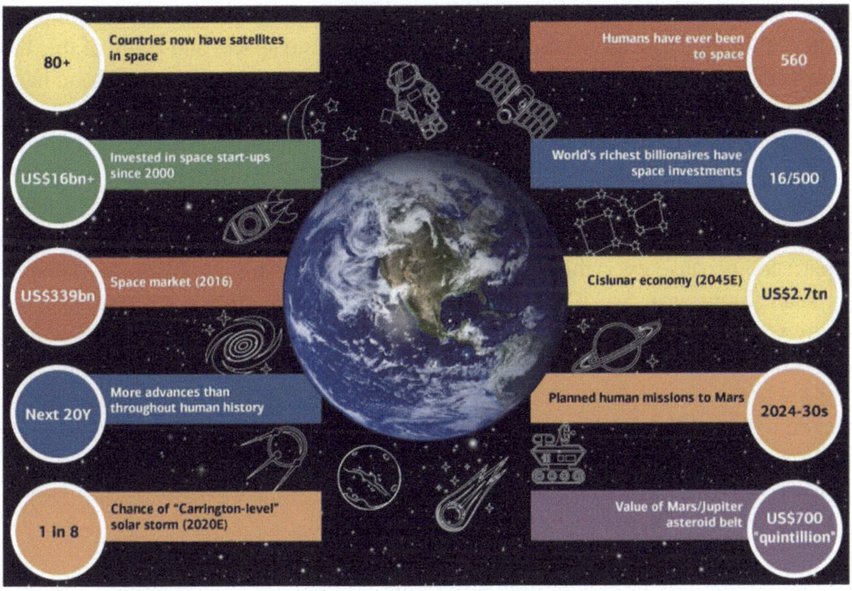

Fig. 14 Space age 2.0

such as Amazon, Microsoft, Google, IBM, and Oracle battle for business from companies that are offloading their data center computing and storage needs [35].

5.2 *Moon Low Gravity*

Natural global disaster could happen, and few known ways in which any celestial bodies can affect the Earth or its inhabitants are by direct contact as in a collision, by their gravitational forces, or by their electromagnetic radiation. The Sun affects the Earth by its gravity which keeps the Earth in its orbit, and despite it is in a direct physical contact with the Earth but highly unlikely to happen in our life time; however, it does send our way streams of charges particles in the form of solar wind or as high energy particles produced during solar flares. The solar wind affects the Earth's Van Allen radiation belts and the shape of the Earth's magnetic field, giving it a comet-like shape pointed directly away from the Sun. The charged particles affect the Earth by causing Aurora Borealis and upsetting the structure of the ionosphere which interferes with radio communication at low frequencies.

The Moon affects the Earth by its gravitational field, raising the very large water and earth tides, which are about twice as high as the solar tides. The Moon's small mass compared to the Sun is offset by the fact that the Moon is much closer to the Earth than the Sun, and from Newton's law of gravity, the gravitational force is proportional to mass, but inversely proportional to the square of the distance.

There are no lunar electromagnetic influences; however from time to time, impacts on the Moon by large asteroids have ejected matter into space, and some of this has shown up as lunar meteors and meteorites on the Earth so in a sense the Moon has a direct physical contact with the Earth [36].

The "low-gravity" end of things is not very interesting since it is indistinguishable from the "distant observer" case. In addition, we are a low-gravity world ourselves, so nothing to see there. This is, of course, why you are asking in the first place… you haven't seen any gravitational time dilation effects in your own experience because everything around you is at a very low gravitational potential (as such things go). So, it's down to how high your high-gravity case is. Even taking your low-gravity observer to be at basically zero potential, the effects will be very small for any reasonable range of gravity on the high-gravity world. Well, any range where you are likely to still be able to find living beings remotely like us. Maybe if some life form has evolved on the surface of a neutron star or the like you could get an extreme enough gravitational potential difference that you would notice it in regular communication. For example:

$$z = 1 - \frac{t_{\text{fast}}}{t_{\text{slow}}} = \sqrt{1 - \frac{2GM}{rc^2}} \tag{1}$$

which approximates for moderate changes to

$$z \approx \frac{GM}{rc^2} = \frac{rg}{c^2} \tag{2}$$

So, this depends on both g and r. But g is proportional to r for a given density ($r = \frac{3G}{4\rho} g$), and therefore

$$z \approx \frac{3Gg^2}{4\rho c^2} \tag{3}$$

For a density around Earth's (about 8000 kg/m^3) this means that to have a 10% difference in rate of speech your massive world would have a surface gravity of 1.2 × 1015 m/s^2…. I take it back, that can't be an even vaguely Earth-like planet. So, if your heavy gravity world is something like a neutron star then you'll have to speed up (for the low-gravity listener) or slow down (for the high-gravity listener). But otherwise it just isn't going to be enough to matter [37].

Additional fun fact: A quick "back of the envelope" calculation gives me about 25% time dilation for the surface of a neutron star.Additional very fun fact: There is a direct measurement of this; a paper was published in Nature, but available on ArXiv looked at spectral lines from a gamma ray emission from a neutron star to directly measure the red shift (though you have to allow for the magnetic field, which introduces a lot of uncertainty!) They found a "slow down" of about 35%.

5.3 *Interstellar Effect*

The universe is a cosmic cornucopia with endless possibilities, and it is an infinite universe that is full of extreme violence and potentially life phenomena. Black holes, warm holes, and white holes: are they science fantasy or science fact? The recent photographing of a black hole for the first time on May 2019 has at least proven it is not faraway phenomena. A black hole is a region space with extremely strong gravity that there is no way for objects to get near to it and breakaway from its gravitational forces, nothing can escape from close encounter with a black hole not even light. Anything that gets too close is doomed, planets, stars, even an entire solar system [38].

No surprise, black holes are on the loose right here on our own cosmic neighborhood; if a black hole found its way into our solar system it would rip us apart. Any kind of a black hole that could pass through the solar system would be pulling on all the system and all the planets harder than the Sun does and so it's just going to totally disrupt the gravitational balance of the solar system. The black hole would literally tear planets from their orbits and smash them into each other. It's just an epic disaster [39].

If it got close enough to say planet Jupiter, it could pull the Moons of Jupiter away from the planet itself. It would just be flinging planets left and right everywhere as it whipped through the solar system leaving disaster in its wake.

If a black hole approached earth all that gravity would rip asteroids from their orbits and hurl them toward our planet [40]. The earth's surface would become an inferno and it would be the beginning of the end. First, it would swallow up the atmosphere, then the planet itself, destroying an entire solar system eventually. It is more than just a big empty sucking piece of space; it is incredibly heavy to get an idea just how heavy and dense a black hole is, and imagine the earth now starts to crush it. I guess we all got the point [41].

6 Conclusion

"We stand at the birth of a new millennium, ready to unlock the mysteries of space and to harness the energies, industries and technologies of tomorrow"—US President Donald Trump. Government agencies like NASA and Russia Roscosmos have done a great job in quantifying the risks that could compromise any mission, identifying the cost, and from being used to be the only way to explore space; but now it is out for entrepreneurs and the risk takers in forming new markets. Also, with lower launch costs, meaning more satellites and in turn more data, more revenue could be generated. Cross-businesses' era has begun and commercial space business is the driving economic fuel that is opening unlimited opportunities on the earth, in the space, or at the Moon or even farther at Mars. Elon Musk's firm along with OneWeb, Telesat, and even Amazon are investing in plans to launch thousands of satellites that aim to deliver internet connections to customers below. All these plans will require billions of dollars in investment—for factories, satellites, and ground stations, which need to be built, and in addition, the cost of launches and operations. The brighter side of it is the business plan, which is attractive to investors; again, thanks to the growing demand for connectivity around the world, alongside with cheaper and more powerful satellites [32].

References

1. Fernholz, Tim. *Quartz: SpaceX is about to take the lead in the satellite internet race*. https://qz.com/1618386/spacex-launches-first-starlink-internet-satellites/. 14 May 2019.
2. Wendover Productions. https://www.wendoverproductions.com/. Last accessed 11 May 2019.
3. Transforming World. *Thematic investing, to infinity and beyond—global space primer*. Bank of America/Merrill Lynch, 30 Oct 2017.
4. Dialani, Priya. Analytics insight: AWS launches cloud computing in space. https://www.analyticsinsight.net/aws-launches-cloud-computing-in-space/. 8 Dec 2018.
5. Chaturvedi, Aditya. ©Geospatial World: Amazon launches AWS Ground Station and how it would turn earth observation industry on its head. https://www.geospatialworld.net/blogs/amazon-launches-aws-ground-station/. 29 Nov 2018.
6. Garrison, Justin. *Space computing and orbit native applications*. https://medium.com/@rothgar/space-computing-b9face985d04. 19 Oct 2016.
7. Gershgorn, Dave, Alison Griswold, Mike Murphy, Michael J. Coren, and Sarah Kessler. *Quartz: what is amazon, really?*. https://qz.com/1051814/what-is-amazon-really/. 20 Aug 2017.
8. Wochit, Y.T. *News: Amazon Announces Satellite Stations*. http://wochit.com, https://www.youtube.com/watch?v=sL1H48ARzr4. 27 Nov 2018.
9. Amazon Web Services, Inc. *AWS ground station: easily control satellites and ingest data with fully managed ground station as a service*. https://aws.amazon.com/ground-station/. Last accessed 17 May 2019.
10. Statt, Nick. The Verge: Facebook is developing an internet satellite after shutting down drone project; Facebook wants to bring the world online. https://www.theverge.com/2018/7/21/17598418/facebook-athena-internet-satellite-project-fcc. 21 July 2018.
11. Gregg, Aaron. Washington post Business; Amazon's plan to profit from space data. https://www.washingtonpost.com/business/2018/11/30/amazons-plan-profit-space-data/?noredirect=on&utm_term=.d422ffea101e. 30 Nov 2018.
12. Cooklev, Todor, Robert Normoyle, and David Clendenen. The VITA 49 Analog RF-Digital Interface. *IEEE Circuits and Systems Magazine* 12 (4): 21–32. https://ieeexplore.ieee.org/document/6362466.29 Nov 2012.
13. Amazon Web Services, Inc. and E.C. Amazon. 2017. Secure and resizable compute capacity in the cloud. *Launch applications when needed without upfront commitments*. https://aws.amazon.com/ec2/. Last accessed 29 April 2019.
14. Tracy, Phillip, and RCR wireless News. A cloud in space, NASA uses AWS for Mars image analysis, https://www.rcrwireless.com/20161111/big-data-analytics/nasa-aws-mars-tag31-tag99. 11 Nov 2016.
15. Amazon Web Services, Inc. Public Sector, AWS Government, Education, & Nonprofits Blog. *Earth is Just Our Starting Place: An Earth & Space on AWS Event Recap*. https://aws.amazon.com/blogs/publicsector/earth-is-just-our-starting-place-an-earth-space-on-aws-event-recap/, 2018/6/19.

16. Harris, Mark, and IEEE Spectrum. Facebook may have secret plans to build a satellite-based internet. In *Public filings suggest the social media giant is quietly developing orbital tech to rival efforts by SpaceX and OneWeb to deliver Internet by satellite*. https://spectrum.ieee.org/tech-talk/aerospace/satellites/facebook-may-have-secret-plans-to-launch-a-internet-satellite. 2 May 2018.
17. Kornfeld, Laurel and Space Flight Insider. *Industry Input Sought for Next Nasa Mars Orbiter*. https://www.spaceflightinsider.com/missions/solar-system/industry-input-sought-next-nasa-mars-orbiter/. 29 April 2016.
18. Brown, Mike, and INVERSE. *Solar energy-powered hydroponics could feed mars colonies, Elon Musk Says*. https://www.inverse.com/article/55541-best-laptop-accessories-5-products-you-need-for-your-laptop. 26 Feb 2019.
19. The Conversation Africa, Inc. Academic rigour, Journalistic flair. *Method of making oxygen from water in zero gravity raises hope for long-distance space travel*. https://theconversation.com/method-of-making-oxygen-from-water-in-zero-gravity-raises-hope-for-long-distance-space-travel-99554. 10 July 2018.
20. Luis Vazquez-Poletti, Jose, and Ignacio Martin Llorente. 2018. Serverless computing: from planet mars to the cloud. *Computing in Science & Engineering Magazine* 20: 73–79. doi: https://doi.org/10.1109/mcse.2018.2875315, https://www.computer.org/csdl/magazine/cs/2018/06/08625892/17D45VsBTVK. June 2018.
21. Moon Cloud, HOW IT WORKS. Università degli Studi di Milano, https://www.moon-cloud.eu/en/compliance-assurance-and-monitoring-everyone/how-it-works/. Last accessed 25 April 2019.
22. Emergent Space Technologies. Space cloud computing, https://www.emergentspace.com/capabilities/research-development/space-cloud-computing/. Last accessed 17 May 2019.
23. SAT-SPACE AFRICA LTD: Internet. All the time, anytime, sat-space cloud. https://www.sat-space.net/en/products/cloud/. Last accessed 30 April 2019.
24. Delgado, Rick. Cloud computing is moving to outer space?, © SmartData Collective ™. https://www.smartdatacollective.com/cloud-computing-moving-outer-space/. 21 June 2016.
25. Space Station Research Explorer on NASA.gov: NanoRacks-QB50, https://www.nasa.gov/mission_pages/station/research/experiments/explorer/Investigation.html?#id=7479. Last accessed 17 May 2019.
26. Brilliant. Math and science done right, https://brilliant.org/. Last accessed 17 May 2019.
27. Zolfagharifard, Ellie. Could the moon fuel Earth for 10,000 years? China says mining helium from our satellite may help solve the world's energy crisis. Daily Mail. Mail Online. https://www.dailymail.co.uk/sciencetech/article-2716417/Could-moon-fuel-Earth-10-000-years-China-says-mining-helium-satellite-help-solve-worlds-energy-crisis.html. 5 Aug 2014.
28. Clark, Josh. 2008. How can the moon generate electricity?, How stuff works. https://science.howstuffworks.com/moon-generate-electricity.htm.
29. AZoCleantech. Mining on the moon for renewable energy, https://www.azocleantech.com/article.aspx?ArticleID=347. 9 June 2014.
30. European Space Energy (ESA). Helium-3 mining on the lunar surface, https://www.esa.int/Our_Activities/Preparing_for_the_Future/Space_for_Earth/Energy/Helium-3_mining_on_the_lunar_surface. Last accessed 17 May 2019.
31. Nelson, Bryan, and MNN GALLERIES. 10 surprisingly easy sources of alternative energy, https://www.mnn.com/earth-matters/energy/photos/10-surprisingly-easy-sources-of-alternative-energy/mining-the-moon. 30 Nov 2009.
32. Plataforma SINC. Producing electricity on the moon at night, https://phys.org/news/2013-12-electricity-moon-night.html. 20 Dec 2013.
33. Fecht, Sarah. Popular science: solar panels grown on the moon could power the earth. https://www.popsci.com/for-nearly-infinite-power-build-self-replicating-solar-panels-on-moon. 17 Mar 2016.
34. Fecht, Sarah. Popular science: a robot army to build solar panels (on the moon). https://www.popsci.com/robot-army-to-build-solar-panels-on-moon. 17 Mar 2016.

35. Nelson, Patrick, and IDG Communications, Inc. Network World: Grow lunar-based solar panels to eliminate fossil-fuel reliance, says kid. https://www.networkworld.com/article/3049491/grow-lunar-based-solar-panels-to-eliminate-fossil-fuel-reliance-says-kid.html. 1 April 2016.
36. Warmflash, David. Discover science for the curious magazine: how to Harvest Terawatts of Solar Power on the Moon. http://blogs.discovermagazine.com/crux/2016/04/22/moon-lunar-solar-power-plants/#.XN8Kg9czbIU. 22 April 2016.
37. Wonderful Engineering. China will be setting up a solar power station in space. https://wonderfulengineering.com/china-will-be-setting-up-a-solar-power-station-in-space/?fbclid=IwAR2XSTqiaHepT6ibYWteYVSCSJKATjNAKhSXxPQfsp9XkosVzHY1xiMf4Tw. 17 April 2019.
38. Foust, Jeff. SPACENEWS: a trillion-dollar space industry will require new markets. https://spacenews.com/a-trillion-dollar-space-industry-will-require-new-markets/. 5 July 2018.
39. Novet, Jordan, and CNBC. Amazon's cloud-computing business is looking to space. https://www.cnbc.com/2018/09/11/amazons-cloud-business-is-looking-to-space.html. 11 Sept 2018.
40. Abadie, Laurie J., Charles W. Lloyd, and Mark J. Shelhamer, NASA Human Research Program. The human body in space. https://www.nasa.gov/hrp/bodyinspace. Last accessed 17 May 2019.
41. DuckDuckGo, Quora. Would gravitational time dilation affect communication in this situation?. https://www.quora.com/Would-gravitational-time-dilation-affect-communication-in-this-situation. Last accessed 17 May 2019.

Survey of Machine Learning Approaches of Anti-money Laundering Techniques to Counter Terrorism Finance

Nevine Makram Labib, Mohammed Abo Rizka and Amr Ehab Muhammed Shokry

Abstract Today's most immediate threat to be addressed is terrorism. Like any other business, terrorism requires financing, terror organizations use number of illegal methods to raise their funds and their sources are various like scamming banks and fraud, selling antiques of artifact, donation, taxation, ransom and oil. For example, ISIS gains up to $50 million a month from oil sales. This illicit money needs be laundered to be used within legal economy. Our study aims are to survey the technical aspects of anti-money laundering systems (AML), review the existing machine learning algorithms and techniques applied to detect money laundering patterns, detect unusual behavior and money laundering groups, and finally, pinpoint the study contribution in detecting money laundering groups.

Keywords Money laundering · Anti-money laundering · Machine learning

1 Introduction

Considering the rapid growing threat of terrorism, counter-terrorism financing is becoming the most pressing matter for the globe because terrorism financing, where even small flaws of money, can have a major impact by contribution to serious terrorist crimes. Terrorism financing became a major political issue after the September 11th terrorist attacks in the United States. After these attacks Financial Action Task Force (FATF) developed a series of recommendations [1] that became officially approved as the international standard for counter money laundering,

N. M. Labib
Sadat Academy for Management Sciences, Cairo, Egypt
e-mail: nevmakram@gmail.com

M. A. Rizka · A. E. M. Shokry (✉)
Arab Academy for Science Technology & Maritime Transport, Cairo, Egypt
e-mail: aeamehab@gmail.com

M. A. Rizka
e-mail: ccitdean@gmail.com

A. Z. Ghalwash et al. (eds.), *Internet of Things—Applications and Future*,
Lecture Notes in Networks and Systems 114,
https://doi.org/10.1007/978-981-15-3075-3_5

financing of terrorism and weapons proliferation. These recommendations were issued in 1990, and then were amended in 1996, 2001 and 2003, and lately in 2012 to guarantee that they are up-to-date and applicable. They are meant to be of universal application. Terrorism financing is often closely linked with money laundering, as money laundering acts as mechanism to aid terrorist financing.

Simply, money laundering is the process that disguises illegal profits without compromising the criminals who wish to benefit from the proceeds (United Nations Office on Drugs and Crime) [2]. The amount of money laundered globally in one year is approximately 2–5% of global gross domestic product (GDP) [3]. Criminals use financial institutions (FIs) to exploit weakness in the global financial system to blur the trail of illicit funds. FIs are delegated by law to actively monitor and report suspicious activities. These institutions combat terrorist financing using programs known as "counter-financing of terrorism" or CFT.

More recently, researchers have begun to venture on the feasibility of artificial intelligence and machine learning techniques such as support vector machines (SVM), clustering, sequence matching, outlier detection, neural networks, genetic algorithms, fuzzy logic, Bayesian network, to better detect suspicious transactions, activities and patterns.

Our survey objectives is to review and compare most recent researches that apply machine learning approaches and different techniques in money laundering detection area and sum up with future research directions. The organization of our survey is as follows: Sect. 2 deals with the principles of money laundering process, machine learning approaches and different money laundering detection techniques; Sect. 3 performs a comparative study of recent researches of money laundering detection. Finally, the survey will end with conclusion and future directions in Sect. 4.

2 Literature Review

2.1 Money Laundering

Money laundering at its simplest is converting black money which is earned illegally to white money to look like it was gained from legal sources by disguising the origins of money.

Criminal and terrorist organizations need to use illegally obtained money. They need to deposit the money in financial institutions so that they avoid drawing the attention of law-enforcement officials. They can only do that if it appears to be obtained through legal sources (laundered).

There are many techniques to launder money and that varies from simple to complex, like smurfing, overseas banks, underground banks, shell companies and investing in legal businesses. Launderers combine two or more methods while creating a schema or pattern.

Money laundering is a dynamic three-stage process that requires:

1. **Placement**: This is the riskiest stage as launderers move black money from its sources into financial institution, both local and abroad, and dealing with large amounts of money is pretty notable.
2. **Layering**: The purpose of this stage is to make it difficult to detect, uncover and trace original black money. It is the most sophisticated step in laundering scheme or pattern by making different financial transactions like wire transfers between different accounts, names and countries, various bank-to-bank transfers, deposits and withdrawals with different amount of money and changing the money's currency.
3. **Integration**: This is the easiest stage to launderer as it is difficult to detect a launderer with no documentation from the previous stages. Money renter legitimates the financial institution as it came from a legal transaction [4].

2.2 *Anti-money Laundering (AML)*

It is a software system designed to help financial institutions to combat money laundering and terrorism financing, by analyzing customer profiles data and detecting suspicious transactions and anomalies, which include any sudden increase in funds or a large withdrawal, and small transaction which is patterned also is flagged as suspicious. The term AML points out to all procedures, laws, policies, regulations and pieces of legislation that force financial institutions to monitor their clients and doing their best in order to prohibit money laundering and corruption. This requires financial institutions to report any financial offense they find and stop it.

3 Literature Survey

Detecting activities related to money laundering is necessary and inevitable for the economy, industries, banks and financial institutions. Because of the large volume of data and vast number of transactions involved, banking creates a convenient environment to hide the origin of money laundering. This makes money laundering techniques become more sophisticated and hard to trace. So the solution for detecting money laundering must be balanced between accuracy and the process time. Finding a suitable technique for money laundering detection in banks and financial institutions is the most important step for the overall solution of the problem. This originally depends on applying a suitable learning approach of techniques with perspective of dataset or provider of the data source. There are a variety of money laundering detection approaches and techniques within our survey that make comparisons difficult. However, analyzing, reviewing and comparing various methods of money laundering detection is essential to detect money laundering crime, patterns, unusual behavior and money laundering groups. In this part, different machine learning approaches, methods and techniques will be introduced which specifically do not belong to a particular approach of machine learning technique and they are a combination of several methods.

3.1 Machine Learning

Machine learning is a sub-sector of artificial intelligence (AI) that enables computers to automatically learn from experience without being explicitly programmed. Its main goal is to enable computers to self-learn, change and develop from the new data without any intervention or assistance (adapt independently to new data).

In recent years, machine learning has seen an increasing concern and popularity in the financial sector, as it can boost the efficiency of the existing anti-money laundering framework, already helping to drive a reduction in false positive rates and detect suspicious transactions in a timely manner. Machine learning approaches are categorized into three main groups: supervised, semi-supervised and unsupervised.

Supervised Machine Learning

An algorithm learns from a training dataset. A labeled training set contains both normal and anomalous transactions and patterns to build the predictive model, as the algorithm predicts after being trained on past data by supervisor. It is suitable for banks that have experience in detecting money laundering. The dataset must be well formed before applying machine learning algorithms and techniques.

Semi-supervised Machine Learning

Semi-supervised learning algorithms learn from both labeled datasets which contain outcome labels that help the algorithm to understand patterns and identifying relationships and unlabelled datasets that doesn't contain outcome labels but are used to train the algorithm to identify and detect new patterns and the trained ones. New transactions will be considered invalid if their behavior doesn't match with the training set. So this approach requires banks which already have a mechanism to distinguish normal and abnormal transactions.

Unsupervised Machine Learning

In unsupervised machine learning, algorithms act on dataset that is not labeled nor has any reference to know, depending on its own capability without any supervision. It does not require historic labels or training data. Unsupervised learning goal is discovering hidden patterns, similarities, hidden structure and grouping from unlabeled data without any prior training. This approach is suitable for banks without having any methods for reviewing data, which means that these banks don't have any training sets.

3.2 Money Laundering Detection Techniques

After we investigated machine learning approaches, below is a small description for some of machine learning techniques used for money laundering detection in pervious works:

Rule-Based Methods
One of the first anti-money laundering systems [5] was created in 1995 and was based on rules. Rules can be very complex and can be defined with the use of decision trees [6]. Rules are formulated by experts and can very accurately detect money launderer's schemes. However, this technique is human dependent, not flexible and not automatic. Furthermore, it cannot be used to recognize new typologies and schemas of money laundering transactions.

Decision Trees (DT)
This a very powerful supervised learning technique for classification and regression. Its objective is predicting the value of a target variable by learning decision rules derived from the data features. Here node stands for a test on an attribute, branch represents result of the test, and leaf node holds a class label. Prediction obtained from a tree is explained in a series rule in a decision tree; also it is not required for the data to be numerical in decision tree, unlike artificial neural networks where result is only expressed without being explained. Explanation is hidden in the network itself [7, 8].

Artificial Neural Networks (ANNs)
Artificial neural network is a technique that uses a set of connected nodes and imitates the human brain. It is used in both clustering and classification. Its objective is to detect and identify hidden patterns and features that too are complex for developers to extract and teach it to computer. ANN consists of three layers: The first layer consists of the input, the third layer consists of the output and the second is a hidden layer containing neurons which process the input and the results are sent as an output to the third layer [9, 10].

Support Vector Machines (SVM)
This is a supervised machine learning technique, which is used for classification and regression. The SVM objective is to find a separator super-vector of data points belonging to two classes, with a maximal margin. Margin is defined as the amount of space or separation between the two classes, which is defined by the super-vector. Maximizing the margin distance provides some reinforcement so that new data can be classified more accurate. From another perspective it is the distance between the super-vector and the nearest training samples. It's powerful and flexible class of supervised machine learning techniques [10–12].

Random Forest (RF)
Random forest builds forest with a collection of several decision trees that avoid instability and excessive risk education that may occur in a single tree. Another way that

causes less complexity in trees is pruning trees. Pruning a decision tree leads to removing parts of the tree that do not participate in the accuracy of test samples [10, 13].

Outlier Detection Methods (OD)
This method is used for anomaly detection and its objective is to detect and identify abnormal, unusual patterns, rare items, events or unusual observations in datasets by differing significantly from the majority of the data. Anomalies are classified into three categories: point anomalies, conditional outliers and collective outliers. Unsupervised outlier detection techniques do not need training data as it is useful when some abnormal behavior has not been demonstrated in the training sample. Finally, OD are capable of discovering new laundering patterns, however it has high rates of false positives [14, 15].

Social Network Analysis
This is the measurement and mapping of various aspects or relationships between people, organizations, and groups. It is also called link analysis or link mining. The nodes in the network represent the people and the links between the nodes represent the information flow. In the case of SNA, two nodes are said to be connected if they regularly interact or talk in some way. SNA objective is to discover and detect hidden pattern, schemas of interactions between social actors, detecting network structure and schema, as it is capable of detecting terrorist cells, patterns of interaction, leaders, followers and gatekeepers [16, 17] as it can clearly represent the participants in the network, their location, the node (person) that is at the core of the network and the nodes that are at the periphery. SNA provide three main techniques: degree centrality, closeness centrality and betweenness centrality [17, 18].

Naïve Bayes (NB)
This is a classification method that simply uses Bayes conditional probability rule, and it is a collection of classification algorithms based on Bayes theorem. It is not a single algorithm but a family of algorithms sharing same principles. Each class label is considered as a random variable, and assuming that the attributes are independent, the naïve Bayes finds a class to the new observation that maximizes its probability given the class label. It's beneficial to be used within large datasets as is easy to build, with no complicated iterative parameter estimation [19].

K-Nearest Neighbors (KNN)
One of the simplest and easy to implement supervised machine learning algorithms is KNN, which can be used for solving both classification and regression purposes. Its main objective is to search for closest match of the test data in feature space, thus it stores all available cases and classifies new cases which is similar to the pervious or trained ones. It works through three main steps as follows: calculate distance, find closest neighbors and vote for labels. K-NN proved to work very well in the financial sector [20].

Deep Learning
Deep learning is a subset of machine learning method that tries to teach computers to do what comes naturally to human brain, based on the layers used in artificial

neural networks because it has several deep layers that enable learning. Learning requires large amounts of data and labeled datasets; it may be supervised, semi-supervised or unsupervised, so deep learning models are really only practical for large financial institutions or those that generate a lot of data points. As models are so complex, it is hard to figure out why the model classified a transaction as suspicious or not. It loses to explain the ability that simpler models provide. It's so-called a black-box method [21].

Graph Mining
Graph mining objective is to extract patterns of interest from graphs by matching, and graph isomorphism can be an appropriate technique for searching abnormal sub-graphs in the transaction graph which is good for representing data in anti-money laundering systems. A comprehensive survey of graph mining-based anomaly detection techniques is presented, which describes the hidden patterns, interdependencies between transactions and their relations. Graph mining could also be used for classification or clustering [22].

K-Means Clustering
This is a simple and popular unsupervised machine learning algorithm. Its objective is to find groups of similar data, grouping it together and discovering hidden patterns. K-means achieve this objective, by learning from data properties, searching for a pre-set number (k) of clusters in unlabeled datasets. It identifies k number of centroids. That if a data point is closer to a specific cluster centroid than any other cluster centroid this point then is considered to belong to this cluster. K-means clustering, points out to averaging data points that are close to a specific centroid to find the right cluster for the data points [23].

One Class SVM (OCSVM)
The goal of classical support vector machine (SVM) is to distinguish test data between a number of classes, using training data. But if we have only data samples of one class (normal class) and the goal is to find out which samples are unlike the normal ones, in this case we use OCSVM where an outline is created around normal samples to be isolated from the anomalous ones. It is very beneficial for anomaly detection; however, if there are too many anomalies in the training dataset, the accuracy will be degraded [24].

4 A Comparative Study Between Approaches of Machine Learning Techniques with Dataset Perspective

In this section, we analyze the contribution of each machine learning approach of different techniques and its effectiveness with the perspective of dataset used, in order to find a promising contribution for future work.

4.1 Criteria

In Table 1, we review and summarize researches conducted in the field of detecting money laundering based on learning approaches, techniques and dataset in order to have the most complete comparison criteria according to learning approach which reveals whether the learning approach is supervised, semi-supervised or unsupervised, what we expect from data, dataset; size, type and algorithms applied.

4.2 Synthesis and Discussion

While financial institutions begin to adopt statistical and machine learning models in their automated AML/CFT programs, it's highly challenging, and anti-money laundering compliance became more difficult, due to the rising number of customer transactions and automated interaction with customers. Based on the findings of comprehensive and most recent surveys of published research results in AML solutions, it has not escaped our notice that current techniques in the literature have attention to the quality of datasets and suitable machine learning approach to be provided with data. At first, raw data obtained from financial institutions often result in extremely large volumes. Also, the datasets for AML are terribly imbalanced as measured in the number of transactions. Some recent studies suggest handling the challenge of large dataset size by applying supervised techniques refers to algorithms that learn from a set of labeled examples. The label can be either normal or suspicious, or represent a binary variable. Based on the training examples, a supervised learning model is built to classify new data into different label categories.

Furthermore, other studies suggest applying semi-supervised techniques after the clustering process, and that the entire input data will be categorized into smaller groups. Learning data can be reduced by randomly selecting representatives from the trivial normal group instead of using the whole data. As a consequence, the performance is enhanced significantly when the specific learning model is applied for a specific cluster.

Moreover, other recent studies suggest applying unsupervised techniques which consist of algorithms that try to separate data examples into different groups without having a labeled datasets. These groups, known as clusters, hold their unique characteristics or patterns and define by similar member instances. Although sometimes unsupervised learning technique also contains training set, the data label will be omitted during learning process. It will be used for the evaluation after forming the clusters. Since the financial operations may vary from time to time, the need of new anti-money laundering machine learning methods, techniques and models are more effective and can detect new money laundering patterns earlier, identifying all of the accounts associated with money laundering groups.

Table 1 Representing recently related work of learning approaches for money laundering detection technique with perspective of datasets

Learning approach	Purpose	Objective	Techniques	Dataset	Key findings
Supervised	Classification and regression	Validate the use of decision tree to create the detection rules of money laundering risk by customer's profiles of X bank	Decision tree	160,000 customer profiles Of a X bank in China	The study results specified the effectiveness of decision tree in generating anti-money laundering rules from company's customer profiles and determining money laundering risk of a bank customer [25]
		Using neural networks which are composed of three layers to detect money laundering	Neural networks Support victor machine Outlier detection	6,000 accounts, 1,000,000 records transactions	The RBF neural network model is compared against SVM and outlier detection, RBF neural network resulted the highest detection rate with lowest false positive rate [26]
		Optimizing selection of parameters of the support vector machine (SVM) model to improve the identification of suspicious financial transactions	Support vector machine Cross-validation	5,000 accounts, 1.2 million records transactions	The results show that training SVM using the selection parameters by cross-validation method is more effective than using the randomly selected parameters on enhancing performance and improving detection of suspicious transactions [27]
		Presenting a graph mining method to detect suspicious transactions and identifying sub graphs within a network that closely match known typologies	Graph mining Fuzzy matching	Synthetically generated by simulating a mini-economy	The study shows providing greater flexibility than a simple motif search by using fuzzy matching where the sub-graphs may deviate from the given typology [28]

(continued)

Table 1 (continued)

Learning approach	Purpose	Objective	Techniques	Dataset	Key findings
		Creating a model of the user's past activities (historical data) and it will be a criterion of future customer activities	Bayesian networks Rule-based methods	8.2 million transactions conducted by 100,000 accounts	New transaction of the user will be considered as suspicious money laundering activities. If the transaction or activity has significant variation to the pattern, but it may result in a high false positive rate [29]
		Analyzing group behavior, using a combination of network analysis and supervised learning for money laundering detection	Network analysis K-nearest neighbors Random forest Support vector machine	(AUSTRAC) Unpublished data	Evaluation of the system indicates that suspicious activity is successfully detected with a low rate of false positives and a suitable level of accuracy is achieved at high levels of precision [30]
Semi-supervised		A case study of applying knowledge-based solution that combines data mining and natural computing techniques in an investment bank for detecting suspicious cases of money laundering	K-means clustering Neural network Genetic algorithm	10,000,000 transaction records and around 10,000 Accounts from CE BANK	Experimental results show that study can detect money laundering cases within the investment activities, improving learning process and improve the performance from current CE bank's solution in terms of running [31]
		Applying combination of decision trees with BRICH and K-means clustering algorithms to detect money laundering	Decision tree BIRCH K-means	Unpublished data source	The results show identifying suspicious transactions more effectively but it is too dependent on the strong performance of IBM mainframe whereas it didn't gain popularity due to the high cost [32]

(continued)

Table 1 (continued)

Learning approach	Purpose	Objective	Techniques	Dataset	Key findings
		Exploring an approach applying the social network analysis for detecting money laundering	Social network analysis K-means Graph construction	Obtained from X banks disk files and web pages	The study showed that the SNA is a feasible and effective technique in money laundering detection domain also specified that the role finding algorithm is always faster than the interconnection analysis [33]
		A model for identifying association and relationships within transactions and customers using Social Networks Analysis (SNA)	Social network analysis Degree centrality Semantic network graph Social network clustering analysis	(AUSTRAC) Unpublished data	The study identified successfully specific relations among illegal transactions and suspicious customers, such as business, parent, spouse, friend, siblings [34]
Unsupervised	Clustering	Proposing a novel algorithm for money laundering detection based on improved minimum spanning tree clustering	Minimum spanning tree clustering	65,001 financial records from X bank	Experimental results show that the new algorithm has not been affected by noise data, simple, detection of suspicious transactions is complete and effective and decreased dependency on the parameters [35]
		Presenting a system for money laundering detection in Vietnam's banking industry	CLOPE	8,020 records of transferring transaction of X bank	Experimental results of the study proved that CLOPE is a suitable algorithm for money laundering detection but it's base on the ability of analysts to provide a criteria to validate clusters after clustering [36]

(continued)

Table 1 (continued)

Learning approach	Purpose	Objective	Techniques	Dataset	Key findings
		Using expectation maximization for detecting suspicious financial transaction	Expectation maximization K-means	30,000,000 transactions From local X bank in Malaysia	The results show that EM defeated traditional clustering method K-means for AML in detecting true suspicious transactions with low false positive rate. Giving the advantage to EM to be employed in this field [37]
		Presenting an unsupervised machine learning model for detecting fraud suspects in exports	CRISP-DM Deep learning Auto-encoder Linear principal component analysis (PCA)	819,990 records from Foreign trade of the Secretariat of Federal Revenue of Brazil	Tests and Experimental results reduced Dimensionalities about 20 times faster using deep learning Auto-encoder in contrast with PCA and it is also proved to be promising in selecting suspicious cases of fraudulent exports [38]
		Using unsupervised learning methods without the use of ground truth to analyze financial transactions from different sources	One class support vector machine Isolation forest Gaussian mixture models	Ripple transaction data X private company	The resulting user ranking leads to suspicious findings, detecting not only expected anomalies but also cases that were not known prior. And confirms that anomaly detection on user behavior is a must in both traditional and modern payment systems [39]

5 Conclusion and Future Direction

Normal user behavior is chaotic, but money launderers work in patterns, whether they recognize it or not. This necessitates financial institutions to apply new anti-money laundering, machine learning methods, techniques, models and practices to detect new money laundering activities and managing its risk for financial institutions. According to what was mentioned in the previous sections and based on findings and contributions it has been study inferred that:

Supervised learning approaches for money laundering detection must be fed historical labeled data to determine what the suspicious accounts, activities and patterns are look like, versus normal ones. Then it will be able to consider all features linked to this account to make a decision. Therefore, model can only find suspicious accounts, activities and patterns that is similar to previously detected or training data. Many sophisticated modern-day money launderers are still able to get around these supervised machine learning models, as they fabricate new money laundering techniques, literally every day.

Semi-supervised approaches are diverse from the supervised. It just requires to be fed with a valid training dataset that contains information that has significant labels to be understood by algorithms while. New transactions will be considered invalid if their behavior doesn't match with the training set. So this approach requires banks which already had a mechanism to distinguish normal and abnormal transactions.

Unsupervised approaches search for similarities, hidden patterns, structures or grouping across all transactions, accounts and customers without any prior training, supervision or historical labeled dataset, instead of comparing transactions to previous ones. It detects suspicious transactions without knowing what a money laundering pattern looks like and this is a huge differentiator from supervised, semi-supervised and rule-based models, which necessitate knowledge of previous patterns to detect similar ones.

To sum up, each approach has its own strengths and weaknesses, and we can benefit from each according to the circumstances of cases. Rule-based methods depend on rules and reputation lists that can be implemented without AI expertise. However, it has to be permanently updated and will only detect and block the most naive launderers. Supervised machine learning and semi-supervised are nowadays considered as an out of box approach. Even though they showed promising results by considering all the attributes into account, but still limited in that it can't find new money laundering patterns and have a high rate of false positives.

Therefore, unsupervised machine learning is the next evolution, as it can detect new money laundering patterns, prevent it at the earliest opportunity and identify all of the accounts and groups involved in money laundering while keeping the ratio of false positives at a minimum. Future directions are mainly concerned with proposing an unsupervised machine learning technique for money laundering detection to counter terrorism financing and comparing its results with other machine learning techniques.

References

1. FATF. 2001. FATF IX Special Recommendations. Finance. Action Task Force, vol. 2001, no. Oct 2001, pp. 1–27.
2. Official website of United Nations Office on Drugs and Crime [Online]. Available: https://www.unodc.org/unodc/en/moneylaundering/laundrycycle.html.
3. UNODC. 2011. Estimating illicit financial flows resulting from drug trafficking and other transnational organized crimes. Res. Rep. October, pp. 1–140.
4. Salehi, A., M. Ghazanfari, and M. Fathian. 2017. Data mining techniques for anti-money laundering. *International Journal of Applied Engineering Research* 12 (20): 10084–10094.
5. Senator, T.E., H.G. Goldberg, and J. Wooton et al. 1995. Financial crimes enforcement network AI system (FAIS) identifying potential money laundering from reports of large cash transactions. *AI Magazine* 16(4): 580–585.
6. Wang, S.-N., and J.G. Yang. 2007. A money laundering risk evaluation method based on decision tree. Machine Learning and Cybernetics, 2007 International Conference on 2007.
7. Zhang, D., and L. Zhou. 2004. Discovering golden nuggets: data mining in financial application. 34(4): 513–522.
8. Han, Jiawei, Micheline Kamber, and Jian Pei. 2011. Data mining: concepts and techniques: concepts and techniques. Elsevier.
9. Bhattacharyya, S., S. Jha, K. Tharakunnel, and J.C. Westland. 2011. Data mining for credit card fraud: A comparative study. *Decision Support Systems* 50 (3): 602–613.
10. West, J., and M. Bhattacharya. 2016. Intelligent financial fraud detection: A comprehensive review. *Computers Security* 57: 47–66.
11. Farhat, N.H. 2002. Photonic neural networks and learning machines. *IEEE Expert* 7 (5): 63–72.
12. Song, S., Z. Zhan, Z. Long, J. Zhang, and L. Yao. 2011. Comparative study of SVM methods combined with voxel selection for object category classification on fMRI data. *PLoS One* 6 (2).
13. Cutler, A., D.R. Cutler, and J.R. Stevens. 2012. Random forests. *Ensemble Machine Learning Methods Applications*:157–175.
14. Kingdon, J. 2004. Applications : Banking AI fights money laundering. *IEEE Intelligent Systems*.
15. Omar, S., A. Ngadi, and H.H. Jebur. 2013. Machine learning techniques for anomaly detection: an overview. *International Journal of Computer Appllication* 79 (2): 33–41.
16. Jamali, M., and H. Abolhassani. 2006. Different aspects of social network analysis in Web Intelligence. In *IEEE/WIC/ACM International Conference on 2006*. IEEE.
17. Shaikh, K.A., and A. Nazir. 2018. A model for identifying relationships of suspicious customers in money laundering using social network functions. *Proceedings of the World Congress on Engineering* 1: 4–7.
18. Drezewski, R., J. Sepielak, and W. Filipkowski. 2015. The application of social network analysis algorithms in a system supporting money laundering detection. *Information Sciences (NY)* 295: 18–32.
19. Raza, S., and S. Haider. 2011. Suspicious activity reporting using Dynamic Bayesian Networks. *Procedia Computer Science* 3: 987–991.
20. Lee, Y.H., C.P. Wei, T.H. Cheng, and C.T. Yang. 2012. Nearest-neighbor-based approach to time-series classification. *Decision Support Systems* 53 (1): 207–217.
21. Goodfellow, I., Y. Bengio, and A. Courville. 2016. *Deep learning*. MIT Press book. Available: http://www.deeplearningbook.org.
22. Akoglu, L., H. Tong, and D. Koutra. 2015. Graph based anomaly detection and description: A survey. *Data Mining Knowledge Discovery* 29 (3): 626–688.
23. Kharote, M. 2014. Data mining model for money laundering detection in financial domain. 85 (16): 61–64.
24. Zhang, R., S. Zhang, Y. Lan, and J. Jiang. 2008. Svm Scada 2008. I: 19–21.

25. Wang, S.N., and J.G. Yang. 2007. A money laundering risk evaluation method based on decision tree. In *Proceedings of the Sixth International Conference on Machine Learning and Cybernetics ICMLC 2007*, vol. 1, 283–286.
26. Lv, L.T., N. Ji, and J. L. Zhang, "A RBF neural network model for anti-money laundering," *Proc. 2008 Int. Conf. Wavelet Anal. Pattern Recognition, ICWAPR*, vol. 1, no. 1, pp. 209–215, 2008.
27. Keyan, L., and Y. Tingting. 2011. An improved support-vector network model for anti-money laundering. In *Proceedings of 2011 International Conference on Management on e-Commerce e-Government, ICMeCG 2011*, 193–196.
28. Michalak, K., and J. Korczak. 2011. Graph mining approach to suspicious transaction detection. *Computer Science Information System*:69–75.
29. Khan, N.S., A.S. Larik, Q. Rajput, and S. Haider. 2014. A Bayesian approach for suspicious financial activity reporting. *International Journal Computer Application* 35(4).
30. Savage, D., Q. Wang, P. Chou, X. Zhang, and X. Yu. 2016. Detection of money laundering groups using supervised learning in networks:43–49.
31. Le Khac, N.A., and M.T. Kechadi. 2010. *Application of data mining for anti-money laundering detection: A case study*, 577–584. ICDM: Proceedings of the IEEE International Conference on Data Mining.
32. Liu, R., X.L. Qian, S. Mao, and S.Z. Zhu. 2011. Research on anti-money laundering based on core decision tree algorithm. In *Proceedings of the 2011 Chinese Control Decision Conference CCDC 2011*, 4322–4325.
33. Drezewski, R., J. Sepielak, and W. Filipkowski. 2015. The application of social network analysis algorithms in a system supporting money laundering detection. *Information Science (NY)* 295: 18–32.
34. Shaikh, K.A., and A. Nazir. 2018. A model for identifying relationships of suspicious customers in money laundering using social network functions. *Proceedings World Congress Engineering* 1: 4–7.
35. Wang, X., and G. Dong. 2009. Research on money laundering detection based on improved minimum spanning tree clustering and its application. In *2009 2nd International Symposium on Knowledge Acquisition Modeling KAM 2009*, vol. 2, 62–64.
36. Cao, D.K., and P. Do. 2012. Applying data mining in money laundering detection. *Intelligent Information Database System*, 207–216.
37. Chen, Z., L.D. Van Khoa, A. Nazir, E.N. Teoh, and E.K. Karupiah. 2014. Exploration of the effectiveness of expectation maximization algorithm for suspicious transaction detection in anti-money laundering. *ICOS 2014-2014 IEEE Conference Open System*, 145–149.
38. Paula, E.L., M. Ladeira, R. N. Carvalho, and T. Marzagão. 2017. Deep learning anomaly detection as support fraud investigation in Brazilian exports and anti-money laundering. In *Proceedings of the 2016 15th IEEE International Conference on Machine Learning Application ICMLA 2016*, 954–960.
39. Camino, R.D., R. State, L. Montero, and P. Valtchev. 2017. Finding suspicious activities in financial transactions and distributed ledgers. *IEEE International Conference Data Mining Work. ICDMW*, vol. 2017, 787–796.

Enhancing IoT Botnets Attack Detection Using Machine Learning-IDS and Ensemble Data Preprocessing Technique

Noha A. Hikal and M. M. Elgayar

Abstract The need of IoT technologies and their applications are increasing every day. While getting benefits from this technology, these vast numbers of non-smart connected cyber-physical devices have several properties that led to critical security issues, such as; nodes mobility, wireless communications, lack of local security features, scale, and diversity. IoT botnets attack is considered a critical attack that affects IoT network infrastructure that launches a distributed denial of service (DDoS). Current solutions face some limitations in achieving the trade-off between light weighted performance and high detection accuracy. In this paper, a light weighted machine learning anomaly-based intrusion detection system (IDS) is introduced based on applying an ensemble data preprocessing stage in advance. The preprocessing stage aims to clean and rank the collected IoT sensor's reading into a subset of candidate features that highly improve the detection accuracy of learners. The proposed framework was tested using a standard dataset, evaluated and compared for different learners; it has achieved detection accuracy up to 99.7% and a detection time within 30–80 s.

Keywords IoT botnets attack · DDoS · Anomaly-based IDS · Machine learning

1 Introduction

The massive increase in usage of the internet of things (IoT) applications has been creating challenging issues in terms of security. Transferred data, as well as edge nodes, are attracting malicious people to cause harm, ranging from spying up to IoT network failure. Data privacy, integrity, and availability are the main goals of IoT security.

N. A. Hikal (✉) · M. M. Elgayar
IT Dept., Faculty of Computers and Information Sciences, Mansoura University, Mansoura, Egypt
e-mail: dr_nahikal@mans.edu.eg

M. M. Elgayar
e-mail: Mostafa_elgayar@mans.edu.eg

A. Z. Ghalwash et al. (eds.), *Internet of Things—Applications and Future*, Lecture Notes in Networks and Systems 114,
https://doi.org/10.1007/978-981-15-3075-3_6

Different approaches were presented and discussed in the IoT security field. However, challenges in the IoT domain are still an open area for researchers, since there are a lot of restrictions when dealing with IoT networks. Among several suggested security solutions, IDS gained more attraction. It can separately watch all the traffic to detect abnormal IoT network behavior. In some advanced cases, it can specify the type of attack and its initiator. Integrating machine learning techniques with IDS has proved great improvements in terms of detection accuracy and time compared with traditional IDS. The key idea is to build a complete security structure that helps Machine learning-IDS to get the right decision and works in a light weighted manner. This paper proposes an enhanced structure that mainly starts from a reliable data preprocessing step. Since sensor readings contain a lot of noise, redundancies, and inconsistent and superfluous data, data cleaning and preprocessing are highly required. The proposed framework is tested against two of the most common IoT botnet attacks. The first type is BASHLITE which is one of the famous malware attacks that infect Linux systems by sending virtual data through the open telnet. The second type is Mirai, which works on the scan server to detect and identify vulnerabilities on IoT devices that can be hacked through the IP or Mac address and then load malicious programs to these devices, which leads to isolating these devices from the network. This paper firstly reviews the major security challenges of IoT environment followed by a literature review of recent machine learning-IDS used in IoT security. Secondly, in Sect. 2, a proposed framework for machine learning IoT-IDS is introduced in detail. Finally, Sect. 3 presents the analysis of the experimental outcomes, discussion, testing, and comparing against the benchmark results.

2 Security Challenges in IoT Networks

Researchers identified the IoT devices architecture by layers similar to the TCP/IP model. Each one has its own threats that must be analyzed separately, equally, and then prioritized them over each other's to get better improvement in the security level. This architecture is classified into three operational layers, which are: Perception layer, Application layer, and Network layer [1]. Each operational layer is connected to the other layers and relies on them to function. Perception layer is responsible for data collections using different types of sensors. Most of IoT devices are autonomous and can be easily compromised by attackers; this results in falsification of transmitted data. At this layer, security investigation mainly targets the detection of abnormal sensor readings [2]. The application layer poses a major challenge for data security. Since there are several IoT devices software's which differ according to manufactures and device type, it is impossible to get a standard construction of the application layer for these diverse collections. Therefore, security threats at this layer are vulnerabilities, as with all software threats. Security investigation mainly concerns identity authentication, data access permissions, malfunctions, and device manipulation [3]. Network layer is responsible for data transmission and its functionality is the same as the network layer in TCP/IP model. Therefore, it has the same traditional security threats which are: DoS attack, Man-in-the-middle attack, eavesdropping, identity theft, and illegal access, etc. [4]. Generally, in the IoT network, the attack domain has several vulnerabilities at each layer that can easily be exploited, therefore, securing one layer requests that the other layers must be secured also. Taxonomy of IoT security attacks is categorized into three domains corresponding to each layer, which are: Physical attacks, Application software attacks, and Network attacks [5]. Table 1 summarizes the taxonomy of these attacks and their effects.

Trusting an IoT environment is a great researching challenge. Different approaches have been introduced and discussed in this field. Expensive IoT devices offer a more secure environment, but it is not available for all consumers. Encryption mechanisms were introduced to preserve security [6], however, encryption/decryption of transmitted data are not affordable at IoT edge devices, since most of these devices are non-smart and have small storage and processing capacities. Moreover, securing communication links using secure communication protocols such as; TLS/SSL or IPSec. TLS/SSL can provide integrity, confidentiality, and authenticity, but can't prevent several attacks like DoS, DDoS or botnets attack that causes severe availability attack to IoT networks [6]. Therefore, it must be a monitoring system to detect any abnormal behavior in the IoT environment during its whole lifetime. Due to the limited resources of edge devices, embedding monitoring systems into IoT edge devices is almost impossible. Therefore, deploying an intrusion detection system (IDS) for securing IoT is considered a great defense mechanism that has the ability to automatically scan network activities and detect traffic abnormality in a distributed manner away from edge devices.

Table 1 Taxonomy of IoT network attacks

Attack category	Types included	Effect
Physical attacks	Reverse engineering Jamming Radio interference Tampering	Device damage, device misuse for malicious intent, record, block, or transmit false messages to other IoT devices
Application software attacks	SQL Injection Cross site scripting Exploitation of a Misconfiguration	Data breaches, attacks of integrity and confidently, data falsifications, and it could be exploited to cause network attack
Network attacks	DoS and Distributed DoS (DDoS), Spoofing Man-in-the-middle Network injecting Flooding attacks	These attacks can cause extraordinary damage with a network of vast number of IoT devices

Wider IoT topologies and data types introduce challenges in deploying and designing effective IDSs. The trade-off between detection accuracy and light weighted IDS is the key idea for most researchers. Applying Machine Learning methods for the development of IDS is considered more reliable compared to the classical approaches [7, 8]. Unfortunately, dealing with data gathered from IoT edge devices is difficult, the collected data are likely huge, imperfect, and contains negative factors such as noise, redundancies, and missing values. Since machine learning-IDS works online and iteratively depending on high quality data, data preprocessing plays an essential role in preparing collected data to fit learning and testing stages hence ensure high detection and accuracy. Preprocessing includes two major consecutive stages which are data preparation followed by data reduction.

Data preparation stage includes different process such as; data integration, data cleaning, normalization, while data reduction stage mainly aim to reduce the dimensionality and complexity of data features [9]. The final obtained data can be reliable and suitable for feeding machine learning-IDS. In this paper, an IoT security framework based on machine learning-IDS is introduced, tested, and discussed. The proposed framework mainly depends on data preprocessing to eliminate negative factors and obtain useful data for further processing. The final obtained data has a great effect on the detection accuracy of machine learning-IDS, as well as its light performance. Moreover, integrating different machine learning techniques with anomaly-based IDS is investigated for detecting accuracy and detection time.

3 Proposed Framework

Considering the data collected from IoT devices [10], it's very huge data, differs in structure, and contains a lot of redundancy, as well as missing data, in addition to the need for real-time processing. Therefore, there are many challenges that will be encountered while dealing with such type of data. The proposed framework deploys different machine learning-IDS techniques to reveal any abnormal activities by analyzing the traffic and classifying this activity in relation to previously recorded normal activities of the various IoT devices. However, depending on such data in its received form inevitably leads to low detection accuracy if not a false one. Thus, a new high-performance data preprocessing stage is highly required. The proposed framework introduces comprehensive data preprocessing steps before feeding machine learning-IDS to better improve its response. Figure 1 presents the general architecture of the IoT communication system. The communications are mainly conducted all the time between IoT devices and cloud servers (M2S, machine-to-server communications), communications between edge computing devices and servers (E2S, edge computers-to-servers communications) [8]. Considering the nature of processing requirements, position, and functionality of each member in this architecture, it can be concluding that cloud servers play a very important role in securing the whole communications. These servers are mainly

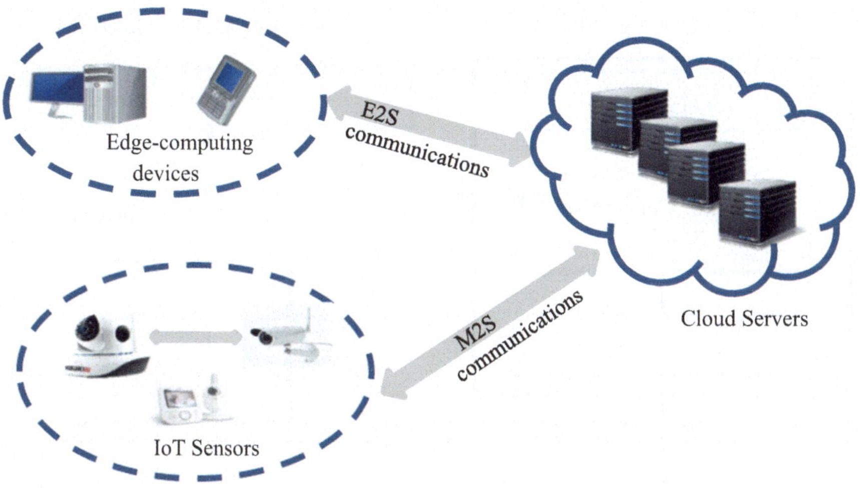

Fig. 1 Communications into IoT networks

responsible for analyzing data and abnormality detection. It has a dominant position, as well as highly equipped processors, and communications privileges with all terminals. Thus, cloud-based IDS is highly preferred.

Figure 2 shows a schematic diagram of the proposed framework of IDS cloud server operation. The whole operation passes through three phases, as shown in the Fig. 2: (i) Data Aggregation, (ii) Data Preprocessing, (iii) Machine Learning-IDS for final behavior evaluation. Different phases will be explained in detail through the next subsections.

3.1 Data Aggregation

This phase starts with IoT devices and ends at cloud servers. It aims to collect a normal activity traffic dataset, which is free from any attacks. This phase included sub-phases that participate in detecting some physical attacks by analyzing M2S packets and simple data collection. Three kinds of servers are involved in this phase; DHCP server, Scan and Load server, and Processing server. Each one has its own unique role. Firstly, the DHCP server is responsible for automatically assigning and providing IP addresses for the involved IoT devices. It is considered the default gateways to these IoT devices. Its operation relies on DHCP. Secondly, the scan and load server is responsible for IP-MAC address retrieving in a fast and robust way. It can scan the IP's and quickly obtain access to IoT devices for packet capturing and parsing. Also, this server works as a malware scanning and vulnerabilities tool to detect physical attacks. Processing server is responsible for: data collecting, data preprocessing, analyzing, and intrusion detecting.

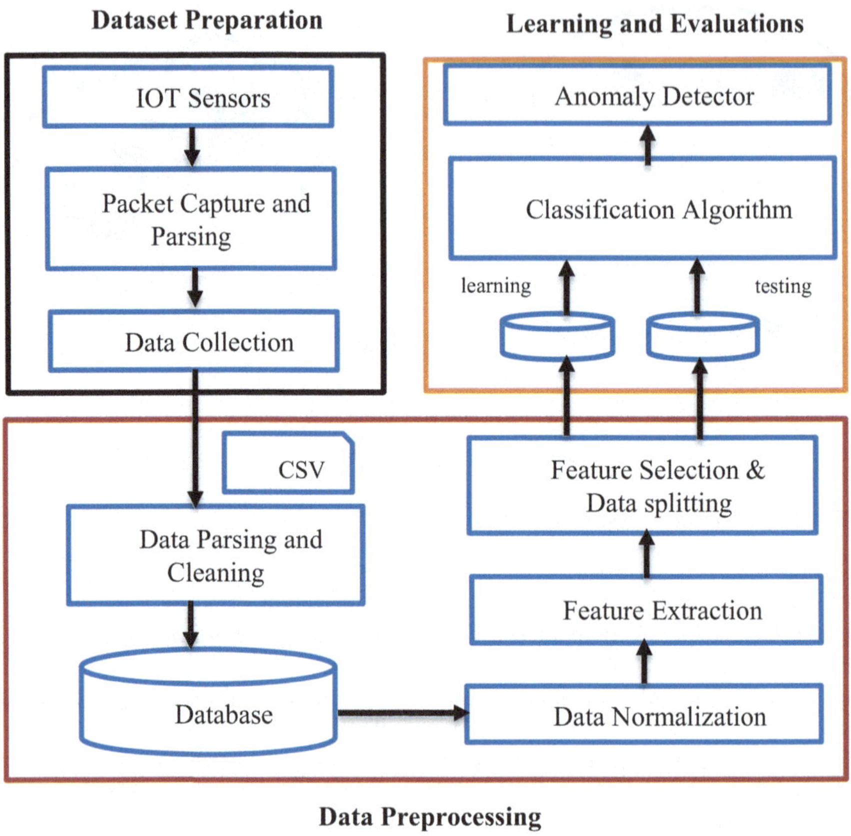

Fig. 2 Schematic diagram of the proposed framework of cloud-based IDS operations

3.2 *Data Preprocessing*

As mentioned earlier, the nature of the collected readings from IoT sensors has different structures in addition to a lot of noisy readings and redundancies. The preprocessing phase aims to prepare these large-scale complicated data for learning processes and extracting distinctive patterns. This phase includes many cycles such as data cleaning, data normalization, feature selection, feature extraction, respectively.

3.2.1 Data Cleaning

The data cleaning cycle is responsible for revealing duplicate records, incomplete data, and noising data. Therefore, it is necessary to unify the different structure of the collected data before starting the cleaning process. At this stage, the data is divided into a set of different features. Each feature contains many data, then data is stored in the database to find the minimum, maximum, mean, and standard deviation of each attribute values. These parametric values are used later to replace empty and loss using the average reading of each attribute. Moreover, these parametric values are also used to delete and exclude any noisy or redundant data. Table 2 illustrates the cleaning algorithm.

3.2.2 Data Normalization

Data may contain disparate numbers, different means, and different variances which in turn introduces difficulties in learning hence results in reducing the efficiency and accuracy of the detection techniques. Therefore, data normalization is considered an important preprocessing stage, especially when dealing with soft computing. It has the ability to scale-up or scale-down data to fit into further proceedings without affecting their attributes. Different data normalization methodologies were introduced by researchers. The min–max scaling technique has proved superiority in terms of reducing the negative impact of marginal values and independencies of data size [11]. This technique transforms the data linearly to fit into a predefined boundary that suits the next proceedings. The mathematical transformation equation used in normalizing input data into a predefined boundary is calculated from

$$\grave{x} = \frac{x - x_{\min}}{x_{\max} - x_{\min}} * (D - C) + C \tag{1}$$

where: x is the original data value, $x_{\min}$ and $x_{\max}$ are the minimum and the maximum values of data readings, respectively, D and C are the upper and lower boundaries of the data range, respectively. Table 3 illustrates the applied algorithm. All data values range from zero to one.

Table 2 Algorithm for data cleaning

```
Function:
  Input ← (DS) raw dataset
  Output  ← Return cleaned Datasets
Data Cleaning:
   array of data ← DS.csv
     for i: array of data. Length
            Compare with parametric values
            Remove noisy and Redundant data
            Insert missing data
     End for loop
End of Data Cleaning
```

Table 3 Min-max algorithm for data normalization

```
Function:
        Input ← (DS) cleaned dataset
        Output ← Return normalized Datasets
Data Normalization:
    Array of Features ← feature extraction
        for i: Array of Features. Length
              for j: array of data[i]. Length
                    min ← minimum value
                    max ← maximum value
                  new transformed value calculated from Equ. (1)
              End for loop
        End for loop
End of Data Normalization
```

3.2.3 Feature Selection

The purpose of this stage is to obtain a subset of the basic features that reduce data dimensionality and facilitate the process of learning and classification. Feature selection methodologies aim to identify a subset of features by removing irrelevant and redundant features as much as possible. This subset is used to train machine learning-IDS for normal activities. Applying feature selection as a pre-learning stage reduces the search space thus speeds up the learning process and provides less memory consuming. At this stage, it is necessary to delete all irrelevant features, or correlated features, which can be obtained from others or inappropriate and retain only the main basic features. For the proposed framework, feature selection and reduction are done, based on applying the ensemble feature selection technique [12]. This technique combines multiple feature selection techniques to get more stable and robust features subsets. Pearson correlation coefficient feature (PCC) has proved to be a robust tool used for feature weighting. For feature ranking, the Spearman rank correlation coefficient is used. Finally, for subset selection, Jaccard index is applied to measure the similarities. For measuring feature similarity and their mutual-dependencies, PCC is used. Feature weighting upon PCC can be mathematically computed using the formula

$$W(f_i, f_j) = \frac{\sum_l (f_i^1 - \mu_{f_i})(f_j^1 - \mu_{f_i})}{\sqrt{\sum_l (f_i^1 - \mu_{f_i})^2 \sum_l (f_j^1 - \mu_{f_i})^2}} \tag{2}$$

where: l is the total number of features, f is the feature vector, and μ_f is the feature mean value. Upon the obtained results, irrelevant features can be excluded easily. The next step is feature ranking using Spearman rank correlation. It can be computed mathematically

$$R(f_i,f_j) = 1 - 6\sum_{l}\frac{(f_i^l - f_j^l)^2}{N(N^2 - 1)} \tag{3}$$

The features are ranked from 1 to N, where 1 is assigned to the worst feature and N is assigned to the best one. Finally, subset selection is done by applying Jaccard index formula, in which it returns 1 if the feature is included in the subset, while returns zero otherwise. This formula is as follows:

$$S(f_i,f_j) = \frac{|f_i \cap f_j|}{|f_i \cup f_j|} \tag{4}$$

3.3 Machine Learning-IDS

Recalling the variety in IoT devices software's and its different data structures, as explained in earlier sections, It's difficult to get a fixed signature for IoT attacks. Therefore, anomaly-based IDS is highly suggested in comparison with signature-based IDS. Since anomaly-based IDS monitors network activities to classify normal and abnormal ones it mainly targets network layer attacks. Integrating machine learning algorithms with anomaly-based IDS improves the detection rate, as well as achieving high accuracy. In this paper, a number of reputable machine learning techniques are applied and tested, such as Support Vector Machine (SVM), Back propagation Neural Networks (BPNN), Random Forest (RF), and Decision Tree (J-48) [13, 14].

4 Experimental Results and Discussion

For testing the proposed framework, a simulation environment is implemented using PHP programming language version 5.3.13 on Intel Core I7 processor at 2.40 GHz, 2 GB of RAM running windows10. Apache Server version 2.2.22 is used to implement servers, and MYSQL version 5.5.24 is used to implement database engine. Machine learning algorithms are implemented using WEKA 3.8 platform. WEKA is an open-source software for machine learning and data mining [15]. A Benchmark raw data collected from different IoT devices for normal and abnormal activities are used in this experiment [16]. The proposed framework is tested against two types of IoT botnet attacks which are: Mirai and BASHLITE attacks [16]. Both attacks mainly target online IoT devices, such as IP camera. It compromises a set of IoT devices and turns them into bots that are used to initiate DDoS (distributed denial of service) attacks on the victim networks. The first type is Mirai, which scans servers to detect and identify vulnerabilities on IoT devices that can be hacked through the IP or Mac address and then load malicious programs to these devices, to isolate these devices from the network. The second type is BASHLITE which is one of the famous malware attacks that infect Linux systems by sending virtual data through the open telnet. We use three standard sets of benign (normal) data and (abnormal or malicious) data captured from three different types of IoT camera devices connected via Wi-Fi. These camera devices are Philips (Baby Monitor), Provision (Security Camera), and Simple Home (Security Camera). These datasets are downloaded from [16], for normal activity and after exposition to Mirai and BASHLITE attacks. The data collected from each device has 115 features. These features have been aggregated from different traffic such as *i.* using the same source host IP address, *ii.* using the same source host IP and MAC address, *iii.* using the source host IP and destination host IP (channel) *iv.*using different protocols such as TCP and UDP (sockets). Applying ensemble data pre-processing techniques, these features were reduced from 115 down to 40 basic features. For training and testing, 10-fold cross validation was used. 10-fold cross validation divides the data set into 10 parts. Nine parts are then used for training and the 10th part is used for testing. This procedure is repeated again using another part as the testing piece. Each data part is used once for testing and nine times for training. After ten repetitions, with a new part being the testing piece, an average result is generated from the ten runs. The investigation here is true in the presence of attack and false otherwise. Standard accuracy metrics are used to measure and compare the performances of the different learners, these metrics are computed mathematically as

$$Precision = \frac{TP}{TP + FP} \tag{5}$$

$$Recall = \frac{TP}{TP + FN} \tag{6}$$

$$F\text{−}measure = 2 \times \frac{Precision \times Recall}{Precision + Recall} \tag{7}$$

$$MAE = \frac{\sum_{i=1}^{n} |y_i - x_i|}{n} \tag{8}$$

where: *TP* equals the total number of correctly classified positive alarms, *TN* equals the total number of correctly classified negative alarms, *FP* equals the total number of misclassified negative alarms, *FN* equals the total number of misclassified positive alarms. High *TP* and *TN* (low *FP* and *FN*) prediction rates are acceptable and desirable. Results are shown in numerical details in Tables 4, 5, and 6 and illustrated Figs. 3, 4, and 5, respectively. It can be concluded that, for the same dataset, there is a trade-off between accuracy and detection time. J-48, BPNN, and RF achieve the highest *TP* compared with LSVM. Moreover, the J-48 and RF

Table 4 Classification result of Philips—Baby monitor device (DS1)

	LSVM	NN	J-48	RF
True positive rate (TPR %)	89.9	98.4	99.7	99.7
False positive rate (FPR %)	15	0.7	0.3	0.3
Precision	91.2	99.4	99.7	99.7
Recall	89.7	99.4	99.7	99.7
F-measure (%)	88.5	99.4	99.7	99.7
Mean absolute error (MAR)	6.8	0.83	0.31	0.37
Time (seconds)	280	393	96	113

Table 5 Classification result of provision—Security camera device (DS2)

	LSVM	NN	J-48	RF
True positive rate (TPR %)	89.2	99.3	99.5	99.7
False positive rate (FPR %)	11.6	0.6	0.4	0.3
Precision	90.2	98.3	99.5	99.7
Recall	89.2	98.3	99.5	99.7
F-measure (%)	87.9	98.3	99.5	99.7
Mean absolute error (MAR)	7.9	0.47	0.42	0.34
Time (seconds)	112	224	80	106

Table 6 Classification result of simple home—Security camera device (DS3)

	LSVM	NN	J-48	RF
True positive rate (TPR %)	99.4	99.6	99.5	99.7
False positive rate (FPR %)	0.4	0.3	0.3	0.3
Precision	99.5	99.6	99.6	99.7
Recall	99.4	99.6	99.5	99.7
F-measure (%)	99.4	99.6	99.5	99.7
Mean absolute error (MAR)	0.43	0.37	0.36	0.35
Time (seconds)	98	186	30	86

consumed minimum time compared with BPNN, using the same number of iterations (300) and the threshold value (0.005). Moreover, the ensemble data preprocessing technique has the ability to reveal the variations and redundancies in raw datasets and results in a high true positive ratio. In comparison with recent IoT botnets attacks detection techniques [17, 18]. The proposed framework is characterized by simple and light weighted performance while providing high detection accuracy.

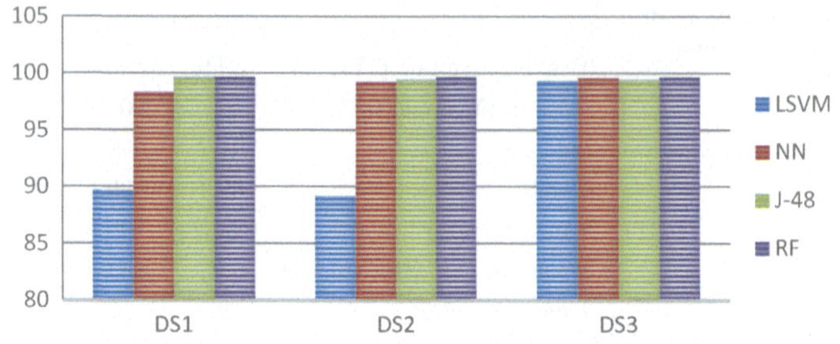

Fig. 3 TPR chart for different tested datasets using different learners

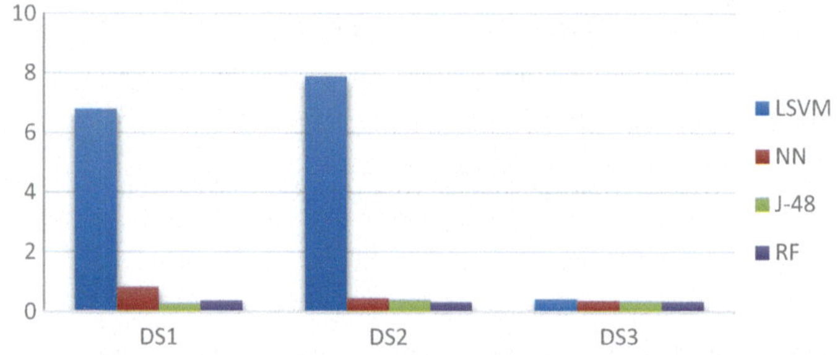

Fig. 4 Mean absolute error chart for different tested datasets using different learners

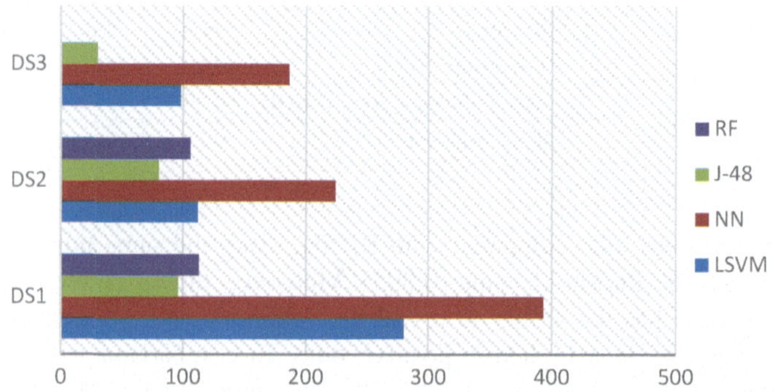

Fig. 5 Total detection time in seconds for different tested datasets using different learners

5 Conclusion and Future Work

In this paper, we present a framework for detecting botnets attacks in IoT networks mainly based on ensemble data preprocessing technique followed by machine learning anomaly-based IDS. Results have proved a significant enhancement in terms of detection time and accuracy. Since ensemble data preprocessing step has a great influence in data cleaning, ranking, and selecting candidate subsets of features that could be used in enhancing the performance of machine learning phases. To extend this work in the future, the signature of different attacks can be predicted, saved and applied to signature-based IDS. Furthermore, integrating ensemble data preprocessing and nonlinear classifiers could be tested and compared with the current solutions, also applying deep learning techniques.

References

1. Iqbal, A., R. Saleem, and M. Suryani. 2016. Internet of Things (IOT): On going security challenges and risks. *International Journal of Computer Science and Information Security* 14: 671.
2. Yang, Y., L. Wu, G. Yin, L. Lifie, and Z. Hongbin. 2017. A survey on security and privacy issues in Internet-of-Things. *IEEE Internet of Things Journal*: 1250–1258.
3. Swamy, S.N., D. Jadhv, and N. Kulkarni. 2017. Security threats in the application layer in IOT applications. In *I-SMAC (IoT in social, mobile, analytics and cloud) (I-SMAC)*. Palladam, India: IEEE.
4. Zhao, K., and L. Ge. A survey on the Internet of Things security. In *Proceedings of 9th international conference computer intelligence and security (CIS)*, 663–667.
5. Rizvi, Syed, Joseph Pfeffer, Andrew Kurtz, and Mohammad Rizvi. 2018. Securing the Internet of Things (IoT): A security taxonomy for IoT. In *17th IEEE international conference on trust, security and privacy in computing and communications/12th IEEE international conference on big data science and engineering*.
6. Suoa, Hui, Jiafu Wana, Caifeng Zoua, and Jianqi Liua. 2012. Security in the Internet of Things: A review. In *International conference on computer science and electronics engineering*. https://doi.org/10.1109/iccsee.2012.373.
7. Debar, M. Dacier, and A. Wespi. 1999. Towards a taxonomy of intrusion detection systems. *Computer Networks* 39 (9): 805–822. IBM Technical Paper.
8. Jan, Tony, and A.S.M. Sajeev. 2018. Ensemble of probabilistic learning networks for IoT edge intrusion detection. *International Journal of Computer Networks & Communications (IJCNC)* 10 (6).
9. Garcia, Salvador, et al. 2016. Big data preprocessing: Methods and prospects. https://doi.org/10.1186/s41011-01600014-0.
10. https://archive.ics.uci.edu/ml/datasets.
11. Panda, Sanjaya K., Subhrajit Nag, and Prasanta K. Jana. 2014. A smoothing based task scheduling algorithm for heterogeneous multi-cloud environment. In *3rd IEEE international conference on parallel, distributed and grid computing (PDGC)*, 11th–13th Dec. Waknaghat: IEEE.
12. Saeys, Yvan, Thomas Abeel, and Yves Van de Peer. 2008. Robust feature selection using ensemble feature selection techniques. In *ECML PKDD 2008: Machine learning and knowledge discovery in databases*, 313–325. Berlin Heidelberg, Springer.
13. Armah, G.K., G. Luo, and K. Qin. 2014. A deep analysis of the precision formula for imbalanced class distribution. *International Journal of Machine Learning and Computing* 4 (5): 417–422.
14. Vapnik, V.N. 1998. *Statistical learning theory*. New York: A Wiley-Interscience Publication John Wiley & Sons, Inc.
15. https://www.cs.waikato.ac.nz/ml/weka/.

16. https://archive.ics.uci.edu/ml/datasets/detection_of_IoT_botnet_attacks_N_BaIoT.
17. Ceron, João Marcelo, et al. 2019. Improving IoT botnet investigation using an adaptive network layer. *Sensors* 19 (3) (Basel, Switzerland).
18. Meidan, Yair, et al. 2018. N-BaIoT—network-based detection of IoT botnet attacks using deep autoencoders. *IEEE Pervasive Computing* 17 (3).

Mobile Application for Diagnoses of Cancer and Heart Diseases

Hoda Abdelhafez, Nourah Alharthi, Shahad Alzamil, Fatmah Alamri, Meaad Alamri and Mashael Al-Saud

Abstract Many medical applications for smartphones have been developed and widely used by health professionals and patients. In Saudi Arabia, the hospital appointment takes long time and also has to wait in a queue to see a doctor because of the huge number of patients. The goal of our application is to save time for doctors and patients as well as discomfort of the patients. The proposed mobile healthcare system provides new methods of engagement with patients and can help them in diagnosing their symptoms, especially in cancer and heart diseases. The system was developed using Android Studio, Flutter Framework, WAMP and Git client. The benefits of this proposed system are enabling patient to query his/her symptoms, getting the expert response from the system in the form of diagnoses of the disease and making the patient to have easy contact with the specialist doctors.

Keywords Mobile app · Android · Cancer · Symptoms · Diagnose · Heart

1 Introduction

Nowadays, smartphones have reached every hand and every home. Using mobile applications in healthcare helps people to get numerous benefits, like finding hospital information in the city and finding a doctor. These applications assist those people who find it difficult to select hospital and contact doctor to diagnose the symptoms they have [1].

Mobile healthcare is a significant element of electronic health. Mobile Health is important because it makes healthcare practices accessible to the public through applying mobile communication technologies [2]. Mobile healthcare has the pro-

H. Abdelhafez (✉)
Faculty of Computer & Informatics, Suez Canal University, Ismailia, Egypt
e-mail: hodaabdelhafez@gmail.com

H. Abdelhafez · N. Alharthi · S. Alzamil · F. Alamri · M. Alamri · M. Al-Saud
College of Computer & Information Sciences, Princess Nourah University, Riyadh, Kingdom of Saudi Arabia

A. Z. Ghalwash et al. (eds.), *Internet of Things—Applications and Future*,
Lecture Notes in Networks and Systems 114,
https://doi.org/10.1007/978-981-15-3075-3_7

spect to save both patients' and doctors' valuable time and money if used properly. Due to the rapid increase of the mobile devices attached to the internet, many researches had been developed to manage and maximize the benefit of such integration. The general concern about mobile devices is user looking for various services offered by mobile device. But now, providing applications that will look forward healthcare, provide the patient monitoring and remote diagnosis to patients have become more significant [3].

According to the World Health Organization, cancer is the second leading cause of death in the world in the year 2018 and has killed 9.6 million people [4]. Cancer is a serious illness; in 2017, an estimated 15,270 children and adolescents aged 0–19 were diagnosed with cancer and 1,790 died of the disease [5]. Moreover, heart diseases are at the top of the causes of death worldwide: the number of deaths from these diseases is higher than deaths due to any other causes of death. The 17.9 million people died of cardiovascular disease in 2016, representing 31% of all deaths in the world in the same year [6].

The aims of our proposed healthcare system are to: (1) help patients in Saudi Arabia save their lives through diagnosing their symptoms and provide contacts for specialist doctors and (2) save time and mobility of both doctors and patients. The developed solution focuses on cancer and heart diseases because they are the highest causes of death worldwide.

This paper has been organized as follows: Sect. 2 discusses the related works. Section 3 discusses the proposed system analysis and design. Section 4 describes implementation of our proposed system. Section 5 concludes the discussion and the limitation.

2 Related Works

Owing to the rapid increase of the mobile devices attached to the internet, various systems have been developed using smartphones. For healthcare, there are many mobile applications that help patients save their time, and some of these applications are mentioned below.

Authors in [7] present an Android application Mr. Doc. Their work is focused on taking appointment from doctors and resolving the problems that the patients face while making an appointment. The patient can schedule an appointment by selecting the preferred doctor, date and time. The appointments are managed by the admin through a website. All data of the registered doctors and patients and the data regarding the appointments are placed on the server. The data is shared by using APIs between the website and the Android application.

Authors in [1] present mobile application to help providing an effective healthcare system. The proposed system consists of two major modules: one for administration and one for the general users. The administrator module is for creating and updating information about hospitals and cabins. The other module provides several prominent features for the general users that will enable them to get quick and effective healthcare. The application is developed in Android OS. The startup page offers users a list of features. It has administrator section, hospital information section, cabin booking in hospital section, section for appointment making with doctor, section for emergency health care, section for aid and medication information, BMI index calculation section, medicine reminder section and hospital suggestion section.

Authors in [8] proposed an application using Android-based mobile phones connecting to the server managed by hospitals and uses GPS and GSM network. The client is part of the system that includes emergency alarm and healthcare management system. The server is another part deployed on a computer that could be in a hospital operated by a doctor. With the help of GPS and GSM network, the system can make sure the location of the users when they are in medical trouble. It triggers the emergency alarm and can also display all nearest hospitals to the user. When the doctor or family received the alarm message, they can immediately take medical measures to rescue the user.

3 Proposed System Analysis and Design

The system is focused on two types of diseases: cancer and heart. Common cancer diseases are classified into nine categories and their types as shown in Fig. 1. The general symptoms of cancer diseases are determined according to American Cancer

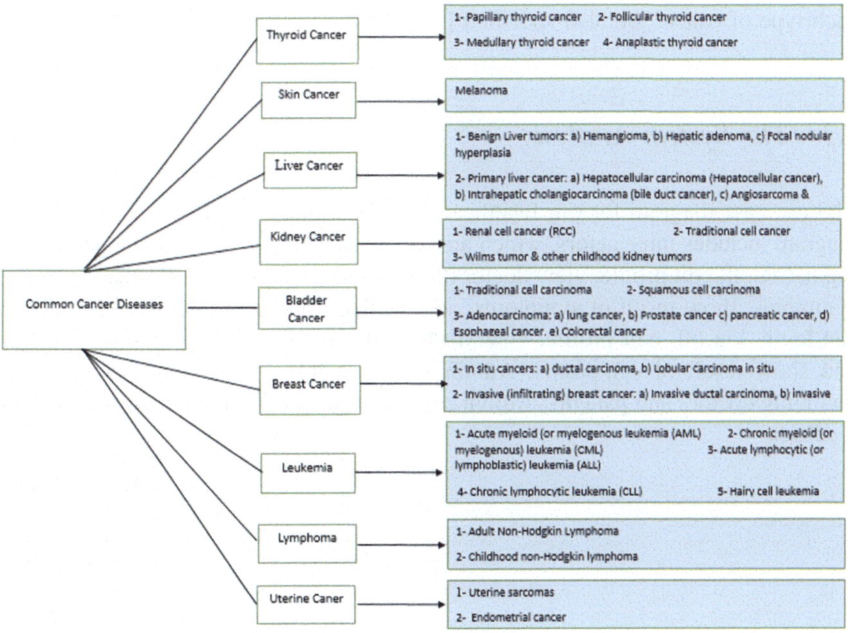

Fig. 1 Common cancer diseases and their types

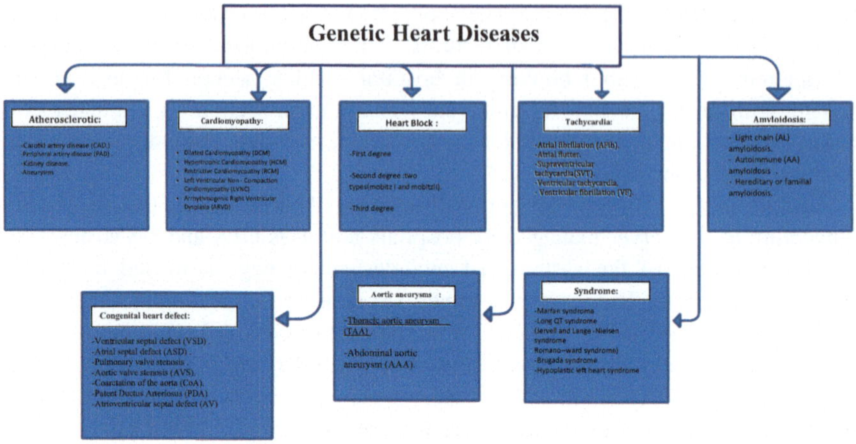

Fig. 2 Heart diseases and their types

Society such as fatigue, skin changes, unexplained weight loss and change in bowel habits or bladder function [9]. The specific symptoms of each type cancer are also specified [10, 11].

The heart diseases are classified into eight categories and the types in each category are shown in Fig. 2. The general symptoms of heart disease are determined according to American Heart Association such as chest discomfort, pain that spreads to the arm, irregular heartbeat and sweating [12]. The specific symptoms of each type of cancer are also specified [13–21].

3.1 *The System Analysis*

The use case diagram for our healthcare system is shown in Fig. 3. The use case diagram includes three actors, which are patient, doctor and admin. The patient can login, log off, edit profile, view doctor information, view patient information, select symptoms, show result of symptoms, add feedback and select doctors. The doctor can login, log off, edit profile, view his/her information, view patient information and show result of symptoms. Admin is responsible for updating all resources including doctors and patients. Admin can also add symptoms or doctors and show feedback.

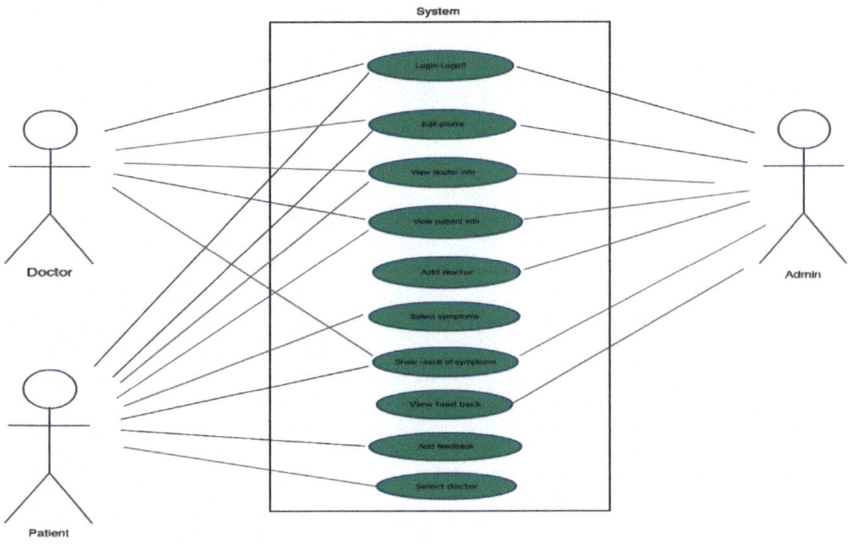

Fig. 3 Use case diagram

3.2 The System Architecture

The main idea of our application is to show the diagnoses after selecting the symptoms from the database. Rule-based system is used to diagnose the patient's diseases depending on the selected symptoms. The system consists of three major modules: one for admin, one for the patient and one for the doctor. The admin module is viewing the patient and doctor information and adding doctor and other symptoms. In the patient module, he/she selects the symptoms and finds the contact information for specialized doctors. In the doctor module, the doctor views the result of its patients and the diagnoses of specific patient.

Figure 4 shows the system flowchart, including the basic elements of the application: administrator, doctor and patient. The administrator's features can find information, add or update symptoms and diagnoses, and can add the doctors. The patient can login, enter his or her own information, select the symptoms, and then the system will show the diagnosis of the symptoms; then he can choose a doctor or logout. As for the doctor, he can sign in and the patient information is displayed.

4 Implementation of the Proposed System

The system consists of three major parts: admin, patient and doctor. The software tools that were used to develop the proposed system are: (1) Android Studio 3.4 to develop Android apps and to emulate them, (2) Flutter Framework (Google's

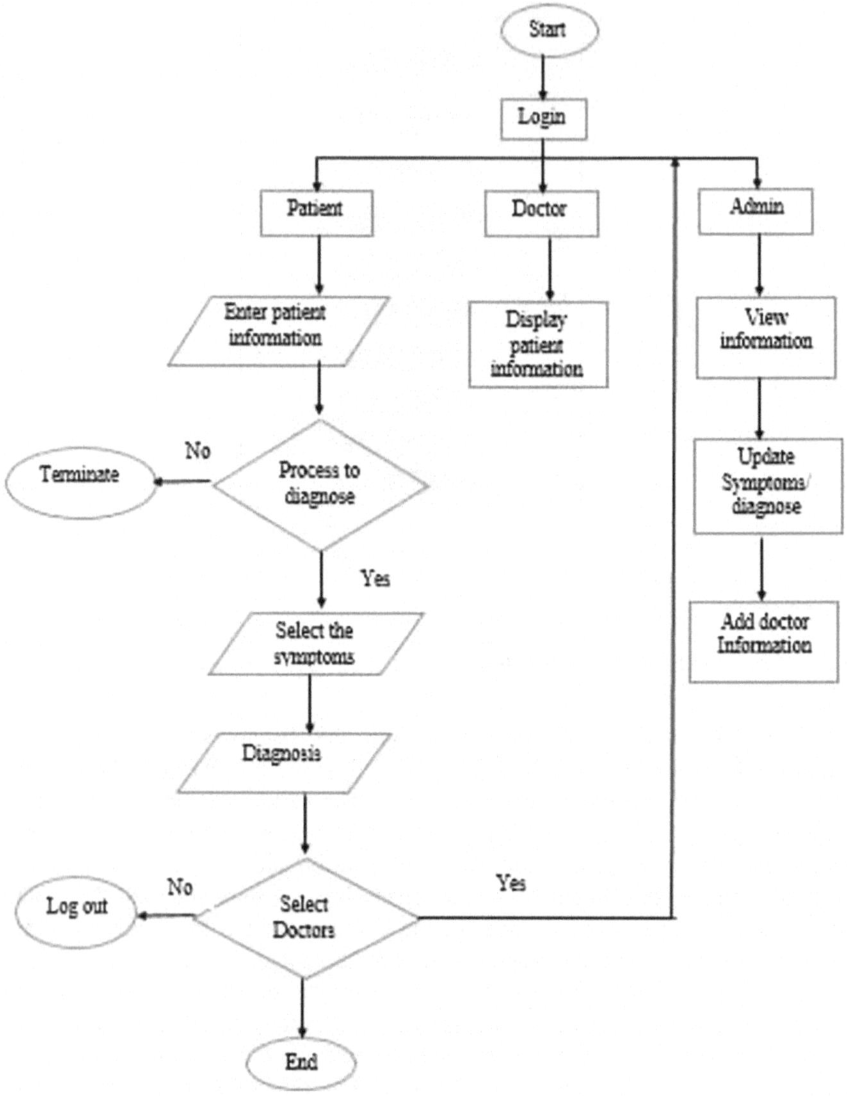

Fig. 4 Flowchart of the proposed system

mobile UI framework), (3) WAMP Server (stands for "Windows, Apache, MySQL and PHP"), (4) Git client and (5) Emulator.

Figure 5a shows the application's main screen. It presents three modules: patient, doctor and admin. Figure 5b is an example for patient login. The patient or doctor should register first to login to the application. To register, the user must fill the given fields, such as first name, last name, email, password, and confirm password and then press the register button. All the information provided by the

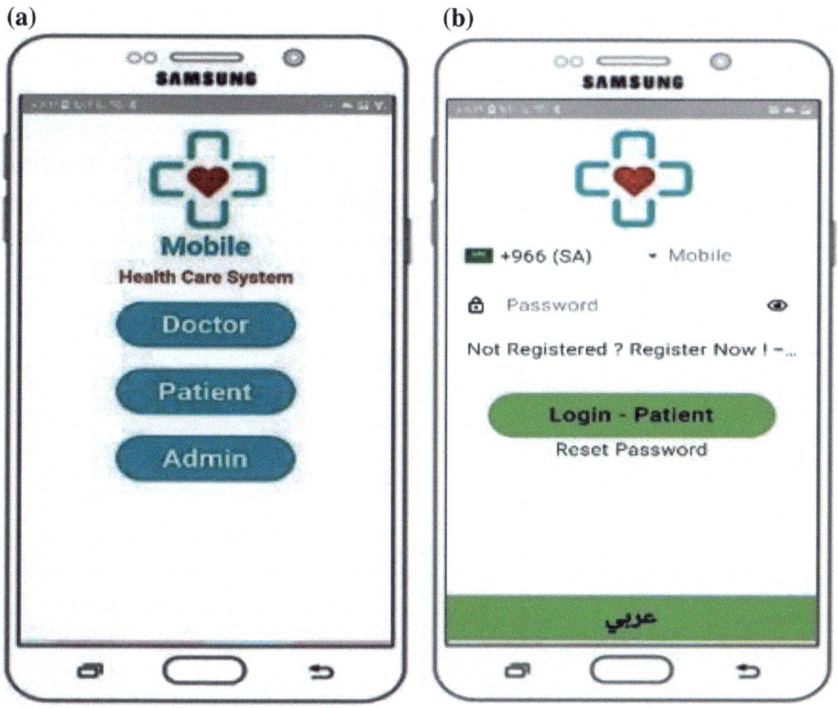

Fig. 5 **a** Main screen and **b** login for patient

user are saved in the database located on the cloud. After registration, the user could enter mobile and the password to run the application. Figure 6 shows the interface screen for patient.

Fig. 7 shows how the doctor can edit his or her profile to update the information. In the patient module, after signing up the patient interacts with the family history screen to select if there is a family history in cancer or heart disease. After that he/she interacts with the general symptoms screen in Fig. 8a, b to answer the symptoms questions. Depending on the patient choosing from the general symptoms, the patient specifies the symptoms related to cancer or heart, as shown in Fig. 8c. As a result of the patient's answers, the next interface is the diagnoses screen (output screen), as shown in Fig. 9a. The diagnoses screen shows the patient disease that is determined by the system.

The system determines the disease based on the symptoms specified by the patient using rule-based system. The system also provides the patient with contacts information of the doctors' specialists in his/her disease, as shown in Fig. 9b.

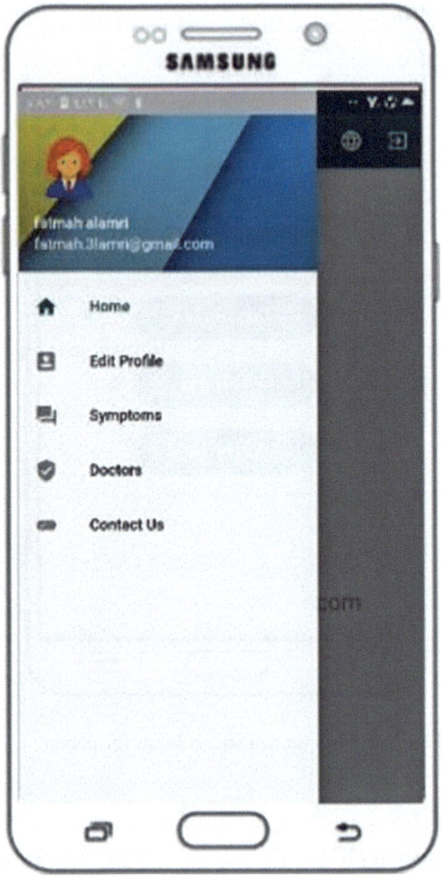

Fig. 6 Patient screen after login

5 Conclusion and Limitation

As the mobile communication technology is developing rapidly, patients can be equipped with powerful tools and support systems that can help them in their everyday health management. The purpose of this paper is to assist patients and easily contact with a doctor instead of going to the clinic or hospital. The proposed system using the Android app can provide advantages to patients, enabling them to query their symptoms and get the expert response from the application in the form of diagnoses of the disease. The nature of disease is changed over time, so that we continue to update the symptoms and contact information of the doctors to cope with the advanced discovering of the new symptoms of cancer and heart diseases. The limitation of this research is that the system could not cover all diseases because they have different type and many symptoms. Therefore, the system includes only two diseases: heart and cancer.

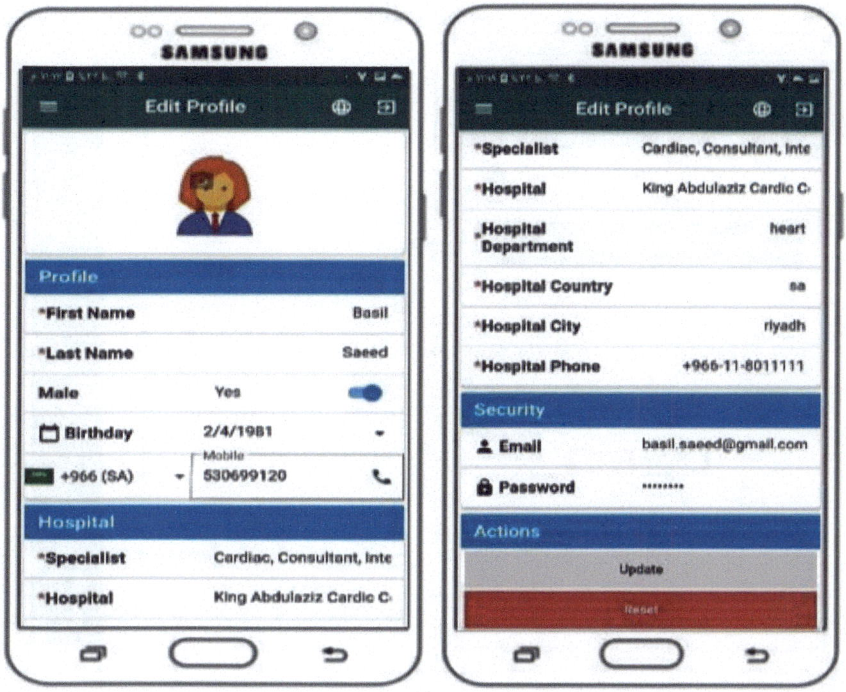

Fig. 7 Update doctor profile

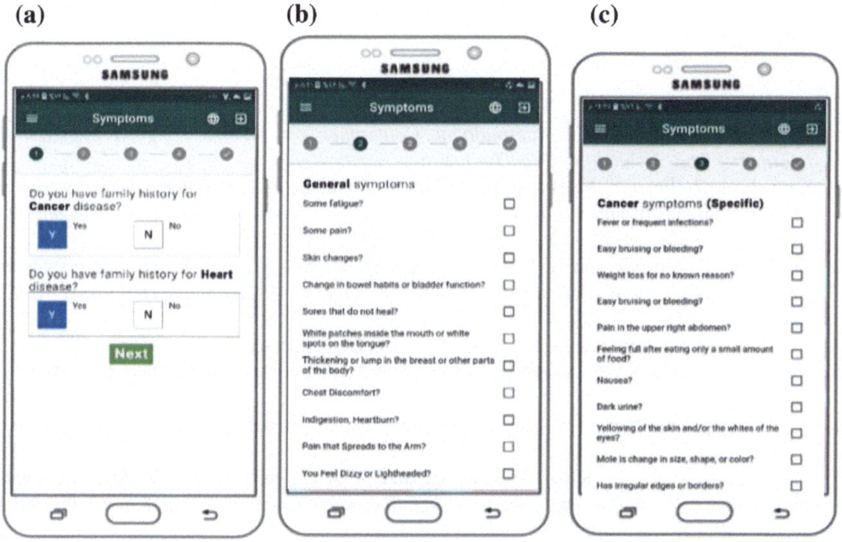

Fig. 8 **a** Patient history, **b** general symptoms, **c** specific symptoms

Fig. 9 **a** Diagnoses of symptoms, **b** doctors' contacts information

References

1. Imteaj, A., and M. Hossain. 2016. Smartphone based application to improve the health care system of Bangladesh. In *International Conference on Medical Engineering, Health Informatics and Technology (MediTec)*, 1–6.
2. Kayyali, R., A. Peletidi, M. Ismail, Z. Hashim, P. Bandeira, and J. Bonnah. 2017. Awareness and use of mHealth apps: A study from England. *The National Center for Biotechnology Information* 5 (4): 33.
3. Bodhe, K., R. Sawant, and A. Kazi. 2014. A proposed mobile based health care system for patient diagnosis using android OS. *International Journal of Computer Science and Mobile Computing* 3 (5): 422–427.
4. WHO homepage. https://www.who.int/cancer/en/. Last accessed 1 Nov 2018.
5. NCI homepage. https://www.cancer.gov/about-cancer/understanding/statistics. Last accessed 12 Nov 2018.
6. WHO homepage. https://www.who.int/cardiovascular_diseases/en/. Last accessed 1 Nov 2018.
7. Malik, S., N. Bibi, S. Khan, R. Sultana, and S. Abdulrauf. 2016. Mr. Doc: A doctor appointment application system. *International Journal of Computer Science and Information Security* 14(12): 452–460.

8. Chandran, D., S. Adarkar, A. Joshi, and P. Kajbaje. 2017. Digital medicine: An android based application for health care system. *International Research Journal of Engineering and Technology (IRJET)* 4(4): 2319–2322.
9. American Cancer Society homepge. https://www.cancer.org/cancer/cancer-basics/signs-and-symptoms-of-cancer.html. Last accessed 20 Feb 2019.
10. American Cancer Society homepage. https://www.cancer.org/cancer.html. Last accessed 16 Nov 2018.
11. NCI homepage. https://www.cancer.gov/. Last accessed 16 Nov 2018.
12. American Heart Association homepage. http://www.heart.org/en/health-topics/congenital-heart-defects/about-congenital-heart-defects/commontypes-of-heart-defects. Last accessed 17 Nov 2018.
13. Mayo Clinic, Marfan syndrome homepage. https://www.mayoclinic.org/diseases-conditions/marfan-syndrome/symptoms-causes/syc-20350782. Last accessed 17 Nov 2018.
14. Mayo Clinic, Long QT syndrome homepage. https://www.mayoclinic.org/diseases-conditions/long-qt-syndrome/symptoms-causes/syc-20352518. Last accessed 17 Nov 2018.
15. Mayo Clinic, Brugada syndrome homepage. https://www.mayoclinic.org/diseases-conditions/brugada-syndrome/symptoms-causes/syc-20370489. Last accessed 17 Nov 2018.
16. Mayo Clinic, Hypoplastic left heart syndrome homepage. https://www.mayoclinic.org/diseases-conditions/hypoplastic-left-heart-syndrome/symptoms-causes/syc20350599. Last accessed 17 Nov 2018.
17. Mayo Clinic, Arteriosclerosis/ atherosclerosis homepage. https://www.mayoclinic.org/diseases-conditions/arteriosclerosis-atherosclerosis/symptoms-causes/syc20350569. Last accessed 17 Nov 2018.
18. Mayo Clinic, Tachycardia homepage. https://www.mayoclinic.org/diseases-conditions/tachycardia/symptoms-causes/syc-20355127. Last accessed 17 Nov 2018.
19. Mayo Clinic, Amyloidosis homepage. https://www.mayoclinic.org/diseases-conditions/amyloidosis/symptoms-causes/syc-20353178. Last accessed 17 Nov 2018.
20. Mayo Clinic, Abdominal aortic aneurysm homepage. https://www.mayoclinic.org/diseases-conditions/abdominal-aortic-aneurysm/symptoms-causes/syc20350688. Last accessed 17 Nov 2018.
21. Mayo Clinic, Thoracic aortic aneurysm homepage. https://www.mayoclinic.org/diseases-conditions/thoracic-aortic-aneurysm/symptoms-causes/syc-20350188. Last accessed 17 Nov 2018.

LL(1) as a Property Is not Enough for Obtaining Proper Parsing

Ismail A. Ismail and Nabil A. Ali

Abstract The LL(1) is a top-down technique. This technique manipulates with any set of strings to decide which one belongs to the language that deals with. This technique uses a set of properties of two main functions: first and follow. These properties are the main constrains of using LL(1). This research answers the following question: Is the LL(1) property for a language sufficient for applying parsing method? And if not, what would be the alternative?

Keywords Software LL(1) · First · Follow and LR

1 Introduction

The methods commonly used in compilers can be classified as being either top-down or bottom-up. The top-down methods build parse trees from the top (root) to the bottom (leaves), while the bottom-up methods start from the leaves and work their way up to the root. In either case, the input to the parser is scanned from left to right, one symbol at a time [1]. The aim of the LL(1) technique as a top-down parser is to verify the correctness of programs in terms of the syntax of the programming language. By using the compilation theory, the context-free grammars (CFGs) are used as a precise description of the programming languages [2]. Therefore, the source programs written by users can be considered as sentences of the context-free grammar if they are correctly written [3]. The first "L" in LL(1) means that the input is scanned from left to right, the second "L" means that it uses leftmost derivation for input string, and the "1" in parenthesis stands for using one input symbol of look ahead at each step to predict the parsing process [4]. LL(1)

I. A. Ismail
Department of Computer Science, 6 October University, Cairo, Egypt
e-mail: Amr44_2@hotmail.com

N. A. Ali (✉)
Suez Institute of Management Information Systems, Suez, Egypt
e-mail: nabil92471@gmail.com

A. Z. Ghalwash et al. (eds.), *Internet of Things—Applications and Future*, Lecture Notes in Networks and Systems 114,
https://doi.org/10.1007/978-981-15-3075-3_8

parser starts from a distinguished symbol of the grammar, works down and right toward the right end of a production, step by step, gradually creates the parsing tree with the input string (the source program) being the leaves. In the whole process, any backtracking step is not allowed so that the process of creating the parsing tree is completely deterministic. Obviously the requirement should be that in every step, the creation of the parsing tree is only one way. Therefore the concept of the LL(1) method can be formed as follows:

For a grammar, if for any nonterminal that occurs as left part of more than two productions their derivation sets are disjoint, then the grammar is called LL(1) grammar [3]. There is a number of constraints for any grammar to be LL(1). These constraints reduce the capability of the LL(1) to be applied to most of the numbers of parsing problems.

2 The CFG of Programming Language

The constructs of a programming language are divided into different syntactic categories. A syntactic category is a sub-language that embodies a particular concept. Examples of common syntactic categories in programming languages are iteration statements, conditional statements, expressions, statements and declarations [5] (see Fig. 1).

Any of the iteration statements seems as a general rule in any context [6]. The CFG is a useful way to express the iteration statements: consider for statement of Fig. 1, the expression can be expanded by the following (see Fig. 2).

Now we have the following subset (see Fig. 3):

The CFG of this subset of C-statements can be written as follows (see Fig. 4):

where the lowercase symbols are terminals and the uppercase are nontermials.

Another example is the conditional if statement:

Statement = IF test THEN statement ELSE statement|IF test THEN statement

By using CFGs, the conditional if statement can be written as follows (Fig. 5):

Remark. The expression can be expanded as in the iteration statement.

```
while (expression) statement;
do statement while (expression);
for (expression1;expression2;expression3) statement;
```

Fig. 1 C iteration statements

```
for (Expression;Expression;Expression)Statement;

Expression =Expression-term|term

term= term/factor|factor

factor=(expression)|id
```

Fig. 2 The expression expansion

for (Expression; Expression;Expression)Statement;
Expression =Expression -term|term
term= term / factor|factor
factor=(expression)|id

Fig. 3 For statement expansion

L=f(E;E;E)S;
E = E - T | T
T = T / F| F
F = (E)| id

Fig. 4 The CFGs of iteration statement

S=iEtSeS|iEtS

Fig. 5 The CFGs of if statement

3 The Problems of LL(1)

The LL(1) parser scans input string from left to right and identifies that the derivation is leftmost or rightmost. The LL(1) parser makes use of production rules for choosing the appropriate derivation. The parse tree can be constructed by the LL1 from root to bottom. The choice of productions based on FIRST and FOLLOW sets, LL(1) parser must uniquely choose a production. A grammar that can be parsed using LL(1) parsing is called LL(1) grammar [7].

When a grammar is not LL(1), there are basically two options: Use a stronger parsing method or make the grammar LL(1).

- Using a stronger parsing method is in principle preferable, since it allows us to leave the grammar intact. Two kinds of stronger parsing methods are available: enhanced LL(1) parsers, which are still top-down, and the bottom-up methods LALR(1) and LR(1). The problem with these is that they may not help: the grammar may not be amenable to any deterministic parsing method [8].
- Making a grammar LL(1) means creating a new grammar which generates the same language as the original non-LL(1) grammar and which is LL(1).

In the following subsections, the paper shows the problems that make the grammars not LL(1) grammar and how to make them LL(1) grammars.

E=E+T|T

Fig. 6 Expression with left recursion

E=TE′
E`=+TE′| ε

Fig. 7 Expression without left recursion

3.1 Left Recursion

A grammar is left recursive if it has a nonterminal A such that there is derivation A = A α for some string α. Top-down parsing methods cannot handle left recursive grammars, so a transformation is needed to eliminate left recursion [3]. A left recursive production can be eliminated by rewriting the offending production. We have an expression production as follows (Fig. 6):

By rewriting the two productions to eliminate the left recursion, we get the grammars of Fig. 7 as follows:

3.2 Ambiguity

A grammar that produces more than one parse tree for some sentence is said to be ambiguous. In other words, an ambiguous grammar is one that produces more than one leftmost or rightmost derivation for the same sentence. When ambiguity happens for the derivation we will not be able to decide which derivation we should take for realizing our aim.

3.3 Left Factoring

Left factoring can be applied when two alternatives start directly with the same grammar symbol; consider the production of Fig. 5. It has two problems: it is not LL(1) because the left-hand sides of both productions start with an if token so they can't possibly have disjoint selection sets. Also, it is ambiguous because the same nonterminal appears twice on the right-hand side. The LL(1) can be applied for this production if the problem of left factoring is solved. The problem can be solved by rewriting productions such that the common parts of the two productions must be isolated into a single production [9], (see Fig. 8).

Note that this replacement has also eliminated the ambiguity, because only one parse tree can now be generated from the earlier input.

S=iEtSS′
S′=eS|ε

Fig. 8 The CFGs of if statement

Each of the above problems can be solved by rewriting grammars to be LL1 and the produced new grammars have the properties of LL1 top-down parser. But, is this transformation guarantees the sufficiency of LL1 as a parsing technique? The answer will be gotten from the following section.

4 Parsing Compound Statements Using LL(1)

In this section, the if conditional will be taken as an example of compound statements for parsing using the LL(1) method to show the sufficiency of LL(1) as a parsing technique.

Consider the following if statement grammars (Fig. 9):

First: As the production rule for S needs to be left factored, we will rewrite the grammar as in Fig. 10:

Second: By computing the First and Follow functions for S, S′ and E, we get the First and Follow values (see Fig. 11):

S=iEtSeS|iEtS|a
E=b

Fig. 9 If statement grammars

S=iEtSS′|a
S′=eS|ε
E=b

Fig. 10 The CFGs of if statement

First (S) = {i , a}
First (S′) = {e , ε }
First (E) = {b}
Follow (S) = {e, $}
Follow (S′) = {e, $}
Follow (E) = {t}

Fig. 11 First and follow functions

Table 1 LL(1) parsing table of if statement

	a	b	e	I	t	$
S	S = a			S = iEtSS′		
S′			S′ = ε S′ = eS			S′ = ε
E		E = b				

Third: The predictive parsing table can be constructed according to Table 1 as follows:

Now, we got multiple entries in M[S,e]. To decide whether the grammar of if statement is LL(1) or not, the LL1 table must have a unique value for each entry. If we don't have a unique entry for the LL1 table (as shown in Table 1), the corresponding grammar of if statement is not an LL1 and the LL1 approach must be replaced by LR approach. The if statement grammar is not LL(1) grammar; consequently, LL(1) cannot be applied in the case of if conditional statements.

5 LR Approach for Parsing Conditional Statements

Consider again the following grammar for conditional statements (Fig. 12):

As we noted, this grammar is ambiguous because it does not resolve the dangling-else ambiguity. We solved the problem for using LL1, but the situation is different for LR bottom-up approach. For simplifying, let us consider an abstraction of this grammar, where i stands for if Expr then, e stands for else, and a stands for "all other productions." We can then write the grammar with augmenting production S' = S, (see Fig. 13).

The sets of LR(0) items for grammar of Fig. 13 are shown in Fig. 14. The ambiguity in Fig. 13 gives rise to a shift/reduce conflict in I4. There, = iS•eS calls for a shift of e and, since FOLLOW(S) = {e, $}, item S = iS• calls for reduction by S = iS on input e.

Statement = if Expr then Statement else Statement
| ifExpr then Statement
|other

Fig. 12 If statement grammars

S → S
S → iSeS |iS | a

Fig. 13 The CFGs of if statement

I0: S′ → •S
S → •iSeS
S → •iS
S → •a
I1: goto(I0,S)
S′ → S•
I2: goto(I0,i)
S → i •SeS
S → i • S
S → •iSeS
S → •iS
S → •a
I3: goto(I0,a)
S → a•
I4: goto(I2,S)
S → iS•eS
S → iS•
I5: goto(I4,e)
S → iSe•S
S → •iSeS
S → •iS
S → •a
I6: goto(I5,S)
S → iSeS•

Fig. 14 The LR(0) items

Table 2 SLR parsing table

State	Action				goto
	i	E	a	$	1
0	S2		S3		
1				Accept	
2	S2		S3		4
3		r3		r3	
4		S5 or r2		r2	
5	S2		S3		6
6		rl		r2	

Translating back to the if–then–else terminology, given

FOLLOW(S) = {e, $}. Building the SLR parse table for the above obtained set of items, we get Table 2.

It is clear that, in the above table, there is a shift/reduce conflict at action [S,e]. Logically, we should favor the shift operation as by shifting the else we can

Table 3 SLR parsing table after resolving conflict

State	Action				goto
	i	E	a	$	1
0	S2		S3		
1				Accept	
2	S2		S3		4
3		r3		r3	
4		S5		r2	
5	S2		S3		6
6		rl		r2	

associate it with previous "if Expr then Statement". Therefore, shift/reduce conflict is resolved in favor of shift. Thus, by resolving the conflict we get the SLR parsing table for dangling else problem as in Table 3.

6 Conclusion

We can say that LR(1) is superior in the sense that it need not do any checking before starting any analysis. The LL(1) analysis is different from LR(1), where we have to check whether the grammar is capable of using the LL(1) analysis or not. It checks whether the grammar contains problems (such as left recursion or left factoring or ambiguity), then the LL(1) method cannot be used immediately after they have been resolved. Since LR(1) has no such difficulties, it can to wider categories of grammars and languages. Furthermore, usually there is no need for the transformation of the grammar. Therefore, theoretically LR(1) method is superior than LL(1) [3]. In the sense that, if we have a parsing problem that is not solvable by using the LL(1) method, we already shift to one of the LR classes (SLR parser, LALR parser and canonical LR parser). We choose the most suitable class of solution and apply it to the problem we have.

For citations of references, we prefer the use of square brackets and consecutive numbers. Citations using labels or the author/year convention are also acceptable. The following bibliography provides a sample reference list with entries for journal articles [3], an LNCS chapter [2], a book [5], proceedings without editors [9], as well as a URL [8].

References

1. Su, Yunlin, Song Y. Yan. 2011. *Principles of compilers a new approach to compilers including the algebraic method*. London, New York: Higher Education Press, Beijing and Springer Heidelberg Dordrecht.
2. Ismail, I.A., and N. Amein Ali. 2017. Parsing strings using a subset of the production rules of the C-language. *Egyptian Computer Science Journal* 41(2): 30–35.
3. Aho, Alfred V., Monica S. Lam, and Ravi Sethi. 2007. Compilers—principles, techniques, and tools, 2nd ed. New York: Pearson Addison Wesley.
4. Puntambekar, A.A. 2009. *Compiler design*, 5th ed. Shaniwar, Pune, India: Technical Publication Pune.
5. Ismail, I.A., and N. Amein Ali. 2017. Parallel compilation of a subset of a c-language. In *Future University in Egypt, The 2nd International conference on knowledge engineering and big data analytic 2017*, 78–81. Elsevier.
6. Kernighan, Brian W., and Dennis M. Ritchie. 1988. *The C programming language*. Upper Saddle River, New Jersey: Prentice Hall PTR.
7. Mogensen, Torben Ægidius. 2010. *Basics of compiler design*. Published through lulu.com, University of Copenhagen, Copenhagen, Denmark.
8. Grune, Dick, Kees van Reeuwijk, Henri E. Bal, Ceriel J. H. Jacobs, Koen Langendoen. 2012. *Modern compiler design*, 2nd ed. New York, Heidelberg Dordrecht London: Springer.
9. Holub, Allen I. 1990. *Compiler design in C*. New Jersey: Prentice Hall.

Mixed Reality Applications Powered by IoE and Edge Computing: A Survey

Mohamed Elawady and Amany Sarhan

Abstract IoT field is continuously evolving to cope with the tremendous number of applications that satisfy the user needs. IoT is implemented by physical objects with virtual representations and services. Augmented reality provides the user with a simple interacting interface to applications. The merge between IoT and augmented reality is a realistic consequence of the need to handle the IoT components in non-traditional methods. This merge facilitates the user to interact with the physical components while receiving additional context-aware information about the object. The existing IoT augmented solutions are facing problems in connectivity, security, trustworthiness, latency, throughput, and ultra-reliable, which necessitate using edge computing to assist in solving these problems. In this paper, we shed some light on these topics and the trends of current solutions combining these fields.

Keywords IoT · IoE · Augmented reality · Mixed reality · Edge computing · Cloud computing

1 Introduction

The increasing development of Internet of Things (IoT) technology facilitates connecting various smart devices together in one application. The number of devices belonging to IoT is predicted to grow at an exponential rate predicted to reach around 26 billion devices by the end of the next decade. Internet of

M. Elawady (✉) · A. Sarhan
Computer Engineering and Control Engineering Department,
Faculty of Engineering, Tanta University, Tanta, Egypt
e-mail: elawady5@gmail.com

A. Sarhan
e-mail: amany_sarhan@f-eng.tanta.edu.eg

M. Elawady
Computer Engineering Department, Behera High Institute, Behera, Egypt

A. Z. Ghalwash et al. (eds.), *Internet of Things—Applications and Future*,
Lecture Notes in Networks and Systems 114,
https://doi.org/10.1007/978-981-15-3075-3_9

Everything (IoE) plays an important role in many applications in our lives and offers promising solutions to evolving many industrial systems. IoE connects people, data, processes, and things. Making it easier to construct data flow models between real objects interactions (collected by sensors) and Virtual objects in mixed reality applications (processed by MR devices or any kind of distributed system). Internet of things (IoT), which is a subset of IoE, focuses on the communications between physical objects, managing the devices, and most importantly data gathering. As a result, the market and industry tend to have more focus on building smart devices containing more advanced features (using machine learning algorithms computed on the cloud) to increase the quality of life.

On the other hand, the massive capabilities of hardware used in our daily life like smartphones, tablets and wearable devices, mixed reality applications increase and attract more industries to extended user interface and experience (UI/UX) to new dimensions. Mixed reality aims to create an illusion between the real and digital worlds by creating an entire virtual scene to the user or creating some virtual objects in a real scene. These virtual objects interact in real-time with real objects and user gestures. It's easy now to develop AR and VR applications with free libraries like ARKit [1] and Vuforia [2]. This ease of development leads to the use of mixed reality in more powerful industries and a fast-growing market.

Due to the huge amount of data generated by the exponential increase in the number of IoT devices, the cloud services started to scale much faster in order to manage storing and processing these data. Yet the real challenge here is that the communication is bound by the network itself acting as a bottleneck. In order to solve this issue, fog computing becomes an essential part to decentralize the cloud. As a result, the amount of data sent in the network will be dramatically reduced and the latency of the IoT device's responses will be reduced as well. Fog computing is one of many aspects of edge computing which is the extension of cloud computing where services are brought closer to the end devices. Edge computing is developed to address location awareness, mobility support, and the issues of high latency in delay-sensitive applications.

Edge computing was introduced to improve the performance of latency-sensitive mixed reality applications in some scientific articles. This improvement includes latency reduction and power consumption minimization for MR devices. As a result, MR applications supported by edge computing rely on providing more Quality of Experience (QoE) and supporting more functionalities that need more processing power. There are some solutions introduced to cover a wide area of MR devices using 5G technology. 5G technology offers less latency than 4G and 3G technologies. Edge computers can also improve the rendering scene of MR applications because they have more processing resources than MR devices but need some protocols and standards that control these functions [3–5].

In the past few years, many research efforts have been made on IoE infrastructure, edge computing architecture, and mixed reality. Technology is shaping future smart cities infrastructure and the new generation of applications interacting with the user's environment relying on his activity or his gestures.

This article will survey the recent researches of MR/AR applications with IoE/IoT infrastructure, researches of MR/AR supported by edge computing, and others that mix between these three technologies.

The rest of this paper is organized as follows: Sect. 2 presents the background and current research of IoE, mixed reality, and edge computing. Section 3 provides an in-depth review of the mix between edge computing and IoT introducing how fog computing was deployed to help IoT. Section 4 discusses how mixed reality was used with IoE. Section 5 summarizes how mixed reality benefited from edge computing and how they can merge with IoE. Finally, Sect. 6 concludes the paper.

2 Background and Current Research in IoE, AR, MR, and Edge Computing

2.1 Internet of Everything (IoE)

Internet of Things (IoT) is the network of physical objects accessed through the Internet. These objects contain embedded technology to interact with internal states or the external environment or sending their state through the Internet. Nowadays, we find IoT applications in most of the surrounding fields and the number of things expands exponentially.

IoT Analytics [6] predicts, as shown in Fig. 1a, in 2020 the number of IoT Devices will be 10 Million and double this number in 2025. This leads to a scalability problem in the near future because the amount of data generated from these devices will grow which leads to network congestions. Another challenge facing IoT which is the heterogeneity of communication protocols. IoT devices connect to The Internet through Ethernet, Wi-Fi or cellular networks, but most of the available devices that have non-TCP/IP connection, like Bluetooth, Zigbee or LoRa, are connected through Gateway. Another challenge is the heterogeneity of data models that describe the devices, as shown in Fig. 1b. It can be a simple data model or more complex and have sub devices inside it having their own data models.

Internet of Everything (IoE) has four pillars: people, process, data, and things as shown in Fig. 2. IoT is one pillar of IoE which is things. Besides Machine-to-Machine (M2M) protocols in IoT, IoE includes Machine-to-People (M2P) and technology-assisted people-to-people (P2P) interactions. IoE concerns also to make data into intelligence for better decisions and deliver the right information to the right person or machine at the right time because it relays on metadata for everything.

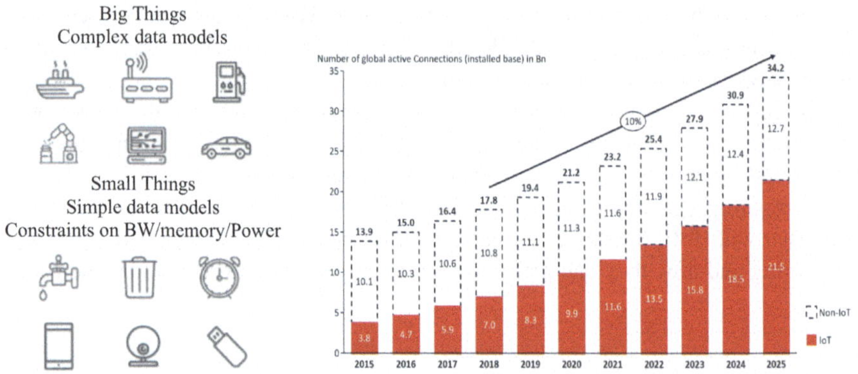

Fig. 1 **a** Total number of devices connections worldwide (right). **b** Types of IoT devices based on data model (left)

Fig. 2 IoE pillars

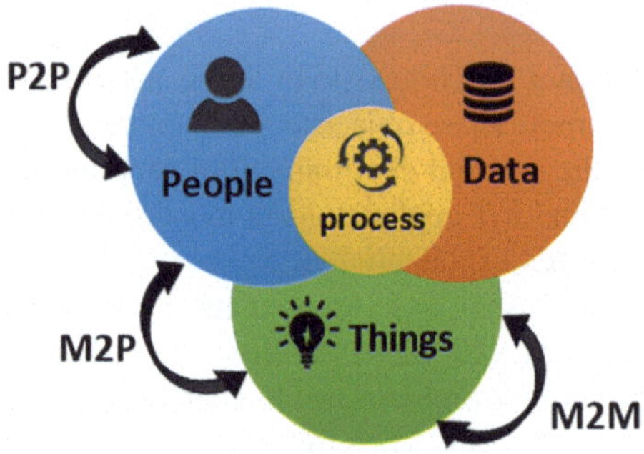

2.2 *Mixed Reality*

Milgram and Kishino [7] introduced the concept of Mixed Reality, which illustrated the merging between real and virtual worlds, and classified interactions between real and virtual elements. As shown in Fig. 3, on the right end is Virtual Reality (VR), where the entire scene is replaced by a computer-generated virtual content. On the opposite left end is the Real Reality (RR) where none of the user's scene is replaced by virtual content. Augmented Virtuality (AV) and Augmented

Fig. 3 Milgram's extended reality (XR) continuum

Reality (AR) is in between. Mixed Reality continuum depends on the rendered scene's objects biased for the real-world or the virtual world.

In the last few years, developers created separate applications that have one of the two technologies: AR or VR. There is also some separated hardware to execute these applications, for example, HTC Vive [8] and Oculus Rift [9] are headsets of VR applications, Microsoft HoloLens [10], and Google glass [11] are headsets for AR applications. Figure 4 shows the four devices mentioned.

Extended reality (XR) or immersive reality applications can mix between VR, AR, and MR, so some scenes immerse users in a fully artificial digital environment like VR and others overlay virtual objects on the real-world environment like AR or MR. In MR the overlays of virtual objects not like AR, but it anchors virtual objects in the real-world seamlessly like real objects and allows the user to interact with them. Figure 5 clarifies the difference between VR, AR, and MR for the same virtual object.

Mixed Reality can improve many Industries, it can be used in the aircraft industry for training engineers involved in repairing engines by wearing special headsets through which they can view the holographic image of engines with various image and voice features and no longer required to extract a real engine out of an aircraft. MR can help senior supervisors in the construction industry or shipyards or any type of construction industries to manage workers and workflow

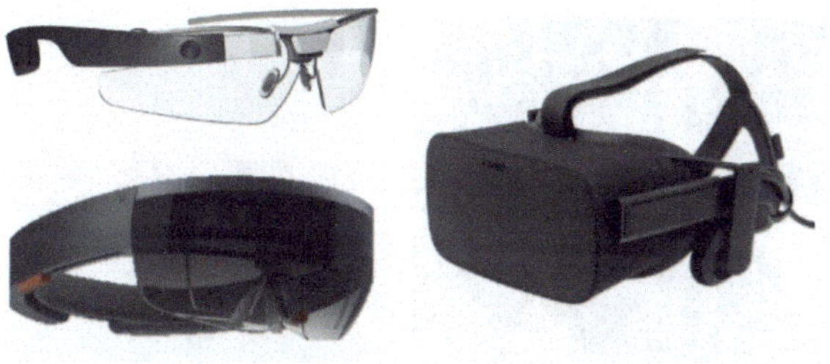

Fig. 4 AR and VR headsets

Fig. 5 The difference between VR, AR, and MR

through visual signals for moving the required equipment or raw materials to the working space. It can also help managers to see real-time information about the overall workflow. MR can also improve the medical industry by helping the medical students to perform virtual surgeries. It can improve their skills or used as imaging layers on a patient for diagnostics. Figure 6 shows some examples of MR applications.

MR applications require the most advanced surface mapping algorithms to create the perfect illusion for users, more quality of graphics, and the new rendering algorithms that interacting with the real-world. Therefore, MR needs more powerful hardware to process these heavy algorithms that include deep learning algorithms and artificial neural networks (NN). Neural Processing Units (NPUs) are designed as hardware accelerators for NN and machine vision applications. Figure 7 contrasts the difference between NPUs and CPUs and demonstrates the time needed to recognize 200 photos on each of them [12]. NPUs can improve the performance of some algorithms of MR applications and it creates a reality that we can make a mixed reality.

Another technology that can help in developing MR applications is 5G networks. They are used to cover a wide area of a vast number of devices with low latency. So, it is building a very reliable network for latency-sensitive MR applications compared to 3G and 4G networks.

Fig. 6 Applications of mixed reality

Fig. 7 Difference between CPU and NPU

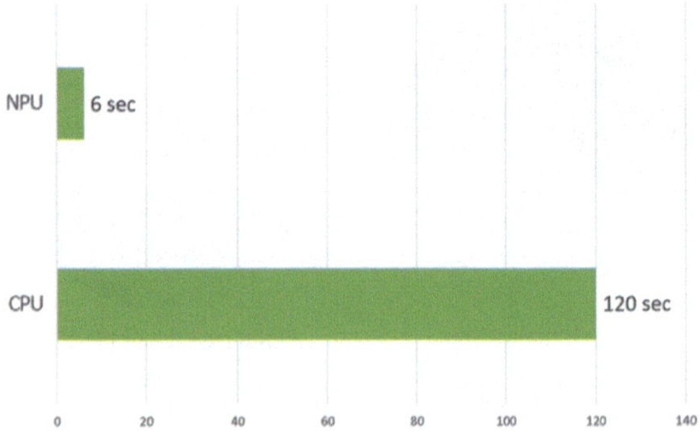

2.3 Edge Computing

Edge computing is an extension of cloud computing that directs computational data and applications away from cloud servers to the edge of a network. It's used for high bandwidth, ultra-low latency, and real-time access to the network information that can be used by several applications. Edge computing minimizes the cloud load by providing resources and services in the edge network and enhances the end user service by reducing the number of multi-hops on their distance on the Internet. Edge computing is location-aware and high mobility support and using a distributed model for server distribution compared to the cloud which uses a centralized model. On the other hand, edge computing has some limitations like the area coverage is limited compared to the globe scope of cloud and has fewer resources than the cloud.

Fog computing, Cloudlets and Mobile Edge Computing (MEC) are the main paradigms of edge computing, all these paradigms not intended to replace cloud computing; they are complementary, Fig. 8 illustrates the difference between cloud computing paradigm vs edge computing paradigms. In [3] classified the Edge computing (Cloudlets, Fog, and MEC) according to the application domain and their areas such as real-time applications, security, resource management, and data analytics, and in [4] surveying architectures of edge computing, applications, challenges facing it, and some paradigms of it like fog, cloudlets, micro data centers, and MEC.

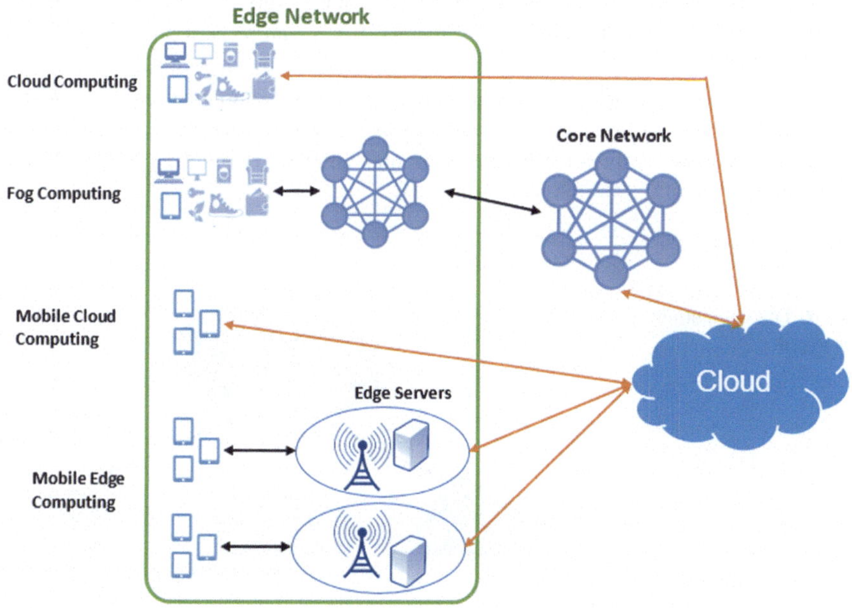

Fig. 8 Cloud computing and edge computing types

MEC, which was coined by ETSI, developed system architectures and standards for a number of APIs, providing capabilities within Radio Access Network (RAN) in close proximity to mobile subscribers like ultra-low latency and high bandwidth, as well as direct access to real-time RAN information, therefore, accelerating services and applications responsiveness from the edge. The deployment of services will not be limited to mobile network operators, but it will also be opened to third party service providers as well. Some of the expected applications include mixed reality, intelligent video acceleration, connected cars, and IoT gateways, among others.

There are some applications that mix the technologies with edge computing, like in [13] which presents a vehicular fog computing for city-wide traffic management system using the integration between cloudlets and fog nodes to reduce the response delay. All the results of the article clarify the huge difference when using edge computing.

3 Fog Computing and IoT

The term of fog computing was coined by "Cisco" [14] in 2012 and has an analogy of real-life fog. As the clouds are said to far above the sky, while the fog is said to be closer to the earth, the same concept is used by fog computing whereas fog computers level is closer to the end devices network, located between the cloud and the users' end devices as shown in Fig. 9. This makes the data transfer easier and faster in wireless and distributed environments. It's meant to be a provision for IoT as a decentralized computing infrastructure can reduce the transferred data to the cloud. Doing so can provide more efficiency, security, and privacy [15]. Some research directions of fog computing development include context-aware resources/ services and how to provide sustainable and reliable computing for real-time applications.

Fig. 9 Fog computing paradigm

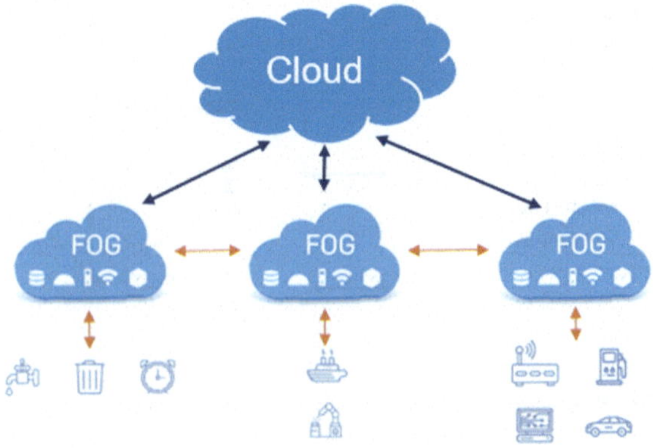

Fog computing has a large storage facility to avoid the need for large-scale storage on cloud, and has processing resources to perform some analytics, control, and management on data received. So, it must support protocols that can handle these requirements. Many application layer communication protocols and their potential of integration in IoT-to-fog-to-cloud system architectures have been proposed including objectives like latency and network throughput [16].

Fog distributed infrastructure for IoT/IoE provides location awareness, low latency, mobility, real-time interactions, interoperability and supporting heterogeneous devices. It can be one layer between cloud and IoT devices, or it can be multiple layers for applications requiring more interoperability between other systems. This realizes a quick response computing with high processing power at neighborhood-wide, community-wide, and city-wide levels, providing the basic infrastructure for high performance and more intelligence in future smart cities. Some of the existing edge IoT solutions and their architecture, functionality, and operational issues are given in [17]. Their demonstration was on healthcare application and the results were clear that bandwidth between IoT cloud systems and local edge devices can be increased and the processing latency can be decreased significantly. In [18], fog computing was directed towards education and the results showed that the amount of traffic was reduced up to five times among all devices in the architecture even using limited resource devices in the fog layer.

4 Mixed Reality and IoE

In Mixed Reality, virtual objects need to be interactive with the real-world and must be rendered to act like real objects. IoE can increase the interactivity between the real-world and the virtual objects, so the data read from the IoT devices can be converted to a process helping rendering scene process. User gestures in MR can also make physical actions in the real-world by IoT actuators.

As in Fig. 10, MR supported by IoE (MRIoE) adds more value to the MR applications because the interactions can flow in both ways; the physical world and the virtual world. IoE models and semantic architecture improve the data flow between the physical objects and the virtual objects and vice versa. IoE also makes it easy to share some virtual objects between multiple users like some objects in a game shared between players, like in Fig. 11, or objects shared between trainer and trainees.

MRIoE can be used in the near future to perform remote surgery that doctors who are trained on MR surgery can operate them online by connecting the doctors with some kind of robots. Another application controls smart devices in homes, production lines or even space by virtual Interactive user interface in the air without any physical switches or controllers.

In [19], a survey on the joint between IoT and AR architecture is given, including the technological aspects of implementation, applications for IoT and AR, which are already developed to its published date. In [20, 21], a generic and scalable

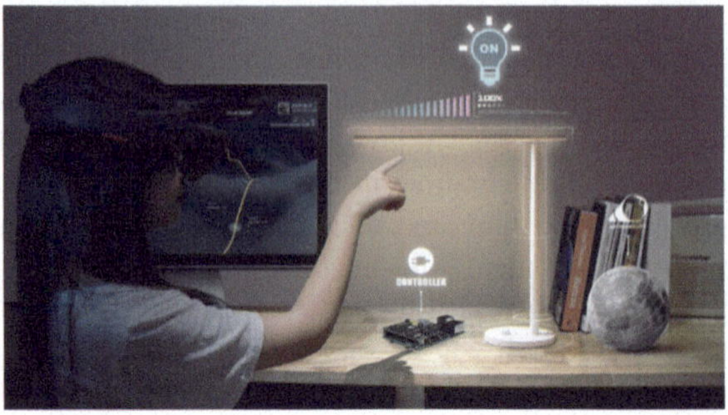

Fig. 10 User gesture change light density

Fig. 11 Game shares the same object between players

augmented reality framework is extended by IoT infrastructure called ARIoT and implementing an application to demonstrate the instructions of Interactions between IoT and AR devices. ARIoT can dynamically identify and filter candidate devices in the AR device perimeter for augmentation then uses a quick guided recognition and tracking objects. It's also concerned about scalability by distributing the information onto some devices and directly communicating with them.

There are several scientific articles that have addressed the issues of interaction between IoE and MR, and their experiments had different scenarios. In [22], AR-middleware is built on IoT gateway of static Wireless Sensor Network (WSN) to enable the relational localization of devices and building the virtual objects based on the calculated location.

Other work used MR and IoT for novel applications, such as in [23] which presents a framework for robot-IoT tasks. These mobile robots serve as binding agents to link static IoT devices and perform a collaborative task. The interface of reprogramming mobile robots is used via AR Interface.

In [24], a prototype for monitoring the current density distribution of a fuel cell in MR that uses MQTT protocol is introduced to connect with the fuel cells which act as IoT devices. In [25], MR is used in the emergency management system by bringing the virtual evacuation simulations into real-world using real-time geometric and semantic data about the built environments, which is provided by the integration between geographic information systems, building information modeling, and IoT.

5 Mixed Reality and Edge Computing

Edge computing can improve MR and solve some challenges facing MR applications including latency reduction, as well as add more processing and storage resources near MR devices for verity usages. In some researches, edge computing is used to solve the latency issues of MR applications. MEC is the most recent paradigm used in the last few years with MR. The work in [26] offers a framework based on MEC for current AR applications to optimize energy efficiency and processing delay. It's integrating the communication, computation, and control functions at the edge layer then taking full advantage of both edge and cloud computing.

In [27], a lightweight and cross-platform solution for AR is proposed which is Web AR using MEC to improve the performance and latency of rendering. It's also discussing the challenges of this solution under 3G, 4G, and 5G networks. In [5], a survey on Web AR was presented and focused on the implementation approaches of mobile AR, challenges of Web AR, and the enabling technologies for reliable Web AR applications including MEC and 5G networks.

In [28], Navantia's Industrial AR (IAR) architecture in shipyards is described which uses cloudlets and fog computing to reduce latency. It was compared with the traditional cloud-based system, and the cloudlets and fog computing were also compared when sending different sizes of payload. In [29], the recent studies that focus on caching and computing at the edge were surveyed, discussing the various types of caching techniques at the network core and network edge, and summarizing the combination of various kinds of caching techniques that reduce the duplicity and eliminate redundant traffic in the network.

On the other hand, Mixed Reality applications that collaborate with IoT devices have a big impact on quality of experience but have one main issue which is latency. As a result, adding fog computing to enhance both IoT layer and MR processing layer became mandatory. Figure 12 shows a fog computing architecture for MR and IoT devices. Some previous work has mentioned this idea in some way or another but not focusing on it directly.

While the work in [28] mentioned that there are Industrial IoT frameworks (IIoT) that can be integrated with IAR in the same level of cloudlets or fog computers. In [30], multi-level cloud service architecture like cloudlets for AR application with IoT Infrastructure was introduced to decrease the number of calls to cloud services. By simulation, modeling and examining three-tire cloud architecture, the results show the average data of delivery time is almost half the time.

In [31], the challenging problem of integrating lightweight virtualization with IoT edge networks was examined by discussing the current issues involving edge computing and IoT network architectures. AR application is one of the scenarios discussed in this article which has all the critical requirements that must be covered such as scalability, multi-tenancy, privacy, security, latency, and extensibility.

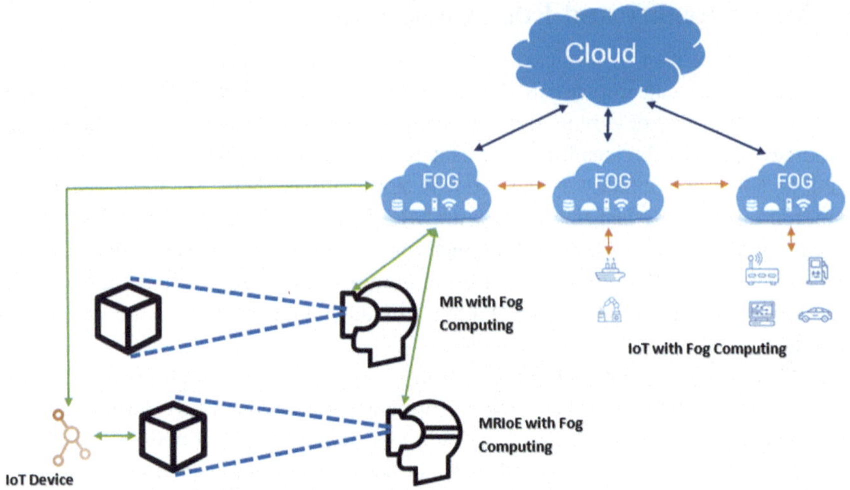

Fig. 12 Fog computing with MR and MRIoE applications

6 Conclusion

In this paper, we have presented the three pillars of three of the most interesting and powerful current technologies. Firstly, the definition of IoE is given and then the difference between IoE and IoT was highlighted. Secondly, the Mixed Reality technology, its applications in different industries, and the challenges facing this technology are summarized. Finally, how Edge computing and Fog computing architectures are used currently to improve IoT and MR. Some of the recent research in IoT, MR, and Edge computing used to improve MR applications are mentioned.

References

1. https://developer.apple.com/augmented-reality/. Accessed 18 June 2019.
2. https://developer.vuforia.com/. Accessed 18 June 2019.
3. Khan, W.Z., E. Ahmed, S. Hakak, et al. 2019. Edge computing: A survey. Future Generation Computer Systems. https://doi.org/10.1016/j.future.2019.02.050.
4. Bilal, Kashif, Osman Khalid, Aiman Erbad, and Samee U. Khan. 2017. Potentials, trends, and prospects in edge technologies: Fog, cloudlet, mobile edge, and micro data centers. Computer Networks. https://doi.org/10.1016/j.comnet.2017.10.002.
5. Qiao, X., P. Ren, S. Dustdar, L. Liu, H. Ma, and J. Chen. 2019. Web AR: A promising future for mobile augmented reality—State of the art, challenges, and insights. Proceedings of the IEEE 107 (4): 651–666. https://doi.org/10.1109/JPROC.2019.2895105.
6. https://iot-analytics.com/state-of-the-iot-update-q1-q2-2018-number-of-iot-devices-now-7b/. Accessed 18 June 2019.

7. Milgram, P., and F. Kishino. 1994. A taxonomy of mixed reality visual displays. *IEICE Transactions on Information and Systems* 77(12):1321–1329.
8. https://www.vive.com/eu/. Accessed 18 June 2019.
9. https://www.oculus.com/rift/. Accessed 18 June 2019.
10. https://www.microsoft.com/en-us/hololens. Accessed 18 June 2019.
11. https://www.google.com/glass/ Accessed 18 June 2019.
12. https://thetechrevolutionist.com/2017/09/huaweis-neural-processing-unit-npu-and.html. Accessed 18 June 2019.
13. Ning, Z., J. Huang, and X. Wang. 2019. Vehicular fog computing: Enabling real-time traffic management for smart cities. IEEE Wireless Communications 26 (1): 87–93. https://doi.org/10.1109/MWC.2019.1700441.
14. https://newsroom.cisco.com/ioe. Accessed 18 June 2019.
15. Mahmud, R., R. Kotagiri, and R. Buyya. 2018. Fog computing: A taxonomy, survey and future directions. In *Internet of Everything. Internet of Things (Technology, communications and computing)*. Singapore: Springer.
16. Dizdarević, Jasenka, Francisco Carpio, Admela Jukan, and Xavi Masip-Bruin. 2019. A survey of communication protocols for Internet of Things and related challenges of fog and cloud computing integration. *ACM Computer Survey* 51 (6).
17. Ray, P.P., D. Dash, and D. De. 2019. Edge computing for Internet of Things: A survey, e-healthcare case study and future direction. *Journal of Network and Computer Applications*. https://doi.org/10.1016/j.jnca.2019.05.005.
18. Akrivopoulos, O., N. Zhu, D. Amaxilatis, C. Tselios, A. Anagnostopoulos, and I. Chatzigiannakis. 2018. A fog computing-oriented, highly scalable IoT framework for monitoring public educational buildings. In *2018 IEEE international conference on communications* (ICC), 1–6. Kansas City, MO. https://doi.org/10.1109/icc.2018.8422489.
19. S. Lanka, S. Ehsan, and A. Ehsan. 2017. A review of research on emerging technologies of the Internet of Things and augmented reality. In *2017 international conference on IoT in social, mobile, analytics and cloud*.
20. Jo, D., and G.J. Kim. 2016. In-situ AR manuals for IoT appliances. In *2016 IEEE international conference on consumer electronics (ICCE)*.
21. Jo, D., and G.J. Kim. 2016. ARIoT: Scalable augmented reality framework for interacting with Internet of Things appliances everywhere. *IEEE Transactions on Consumer Electronics* 62(3):334–410.
22. Baskaran, S., and H.K. Nagabushanam. 2018. Relational localization based augmented reality interface for IOT applications. In *2018 international conference on information and communication technology convergence (ICTC)*, 103–106. Jeju. https://doi.org/10.1109/ictc.2018.8539469.
23. Cao, Yuanzhi, Zhuangying Xu, Fan Li, Wentao Zhong, Ke Huo, and Karthik Ramani. 2019. V. Ra: An in-situ visual authoring system for robot-IoT task planning with augmented reality, 1–6. https://doi.org/10.1145/3290607.3312797.
24. Burkhard Hoppenstedt, Michael Schmid, Kammerer Kammerer, Joachim Scholta, Manfred Reichert, and Rüdiger Pryss. 2019. Analysis of fuel cells utilizing mixed reality and IoT achievements. In *6th international conference on augmented reality, virtual reality and computer graphics (SALENTO AVR 2019)*, June 24–27. Italy: Santa Maria al Bagno.
25. Lochhead, Ian, and Nick Hedley. 2019. Mixed reality emergency management: Bringing virtual evacuation simulations into real-world built environments. *International Journal of Digital Earth* 12 (2): 190–208. https://doi.org/10.1080/17538947.2018.1425489.
26. Ren, Jinke, Yinghui He, Guan Huang, Guanding Yu, Yunlong Cai, and Zhaoyang Zhang. 2019. An edge-computing based architecture for mobile augmented reality. *IEEE Network*: 12–19. https://doi.org/10.1109/mnet.2018.1800132.
27. Qiao, X., P. Ren, S. Dustdar, and J. Chen. 2018. A new era for web AR with mobile edge computing. *IEEE Internet Computing* 22(4):46–55. https://doi.org/10.1109/MIC.2018.043051464.

28. Fernández-Caramés, Tiago, Paula Fraga-Lamas, Manuel Suárez-Albela, and Miguel Vilar-Montesinos. 2018. A fog computing and cloudlet based augmented reality system for the industry 4.0 shipyard. *Sensors* 18. https://doi.org/10.3390/s18061798.
29. Erol Kantarci, Melike, and Sukhmani Sukhmani. 2018. Caching and computing at the edge for mobile augmented reality and virtual reality (AR/VR) in 5G. https://doi.org/10.1007/978-3-319-74439-1_15.
30. Makolkina, M., V.D. Pham, R. Kirichek, A. Gogol, and A. Koucheryavy. 2018. Interaction of AR and IoT applications on the basis of hierarchical cloud services. In *Internet of things, smart spaces, and next generation networks and systems*, ed. O. Galinina, S. Andreev, S. Balandin, Y. Koucheryavy, vol. 11118. NEW2AN 2018, ruSMART 2018. Lecture Notes in Computer Science. Cham: Springer.
31. Morabito, R., V. Cozzolino, A.Y. Ding, N. Beijar, and J. Ott. 2018. Consolidate IoT edge computing with lightweight virtualization. *IEEE Network* 32 (1): 102–111. https://doi.org/10.1109/mnet.2018.1700175.

Computer-Assisted Audit Tools for IS Auditing

A Comparative Study

Sara Kamal, Iman M. A. Helal, Sherif A. Mazen and Sherif Elhennawy

Abstract In a practical sense, Information Systems (IS) have contributed to the success of most of the institutions. The importance of information systems stems from being the main factor in facilitating decision-making and any subsequent emerging problems. Therefore, it is important to verify their efficiency and accuracy by complying with quality standards. An organization can use information systems in several areas with various features; each area needs comprehensive auditing. The IS auditor must perform tasks abiding by existing standards and guidelines. The auditing tasks can be challenging without the use of Computer-Assisted Audit Tools (CAATs). IS auditor uses some of these tools during the audit process. However, these tools do not support all existing IS areas and each tool has its limitations. The aim of this paper is to present a comparative study of the existing information systems' auditing software tools. The results of this study lead to insights into their capabilities and limitations for completing IS auditor's tasks.

Keywords IS auditing · CAATS · Auditing standards · Auditing areas

S. Kamal · I. M. A. Helal (✉) · S. A. Mazen
Faculty of Computers and Artificial Intelligence, Cairo University, Giza, Egypt
e-mail: i.helal@fci-cu.edu.eg

S. Kamal
e-mail: sara.kamal@nub.edu.eg

S. A. Mazen
e-mail: s.mazen@fci-cu.edu.eg

S. Elhennawy
Information Systems Auditing Consultant, Cairo, Egypt
e-mail: selhenawy@gmail.com

A. Z. Ghalwash et al. (eds.), *Internet of Things—Applications and Future*,
Lecture Notes in Networks and Systems 114,
https://doi.org/10.1007/978-981-15-3075-3_10

1 Introduction

Information systems auditing and control mechanisms ensure that an information system is satisfying the needed requirements in different areas [1]. Information systems (IS) auditing contributes to the planning, supervision, and decision-making within the organization. There are several definitions for IS auditing, but the most comprehensive and reliable definition is "an examination of the management controls within an information technology to collect and evaluate evidence of control processes within the organization, which improves the level of services provided by the organization in general" [2].

It is difficult to depend on the human factor to complete the auditing process. Hence, IS auditors use software auditing tools to help in performing all their audit tasks. One of these types that support IS auditors is Computer-Assisted Audit Tools (CAATs). CAATs support and assist IS auditors to complete the audit process within an organization [3]. They allow companies to conduct continuous auditing and monitoring in order to assist business activities and improve the efficiency of the organization's tasks [4].

CAATs can be broadly categorized to cover several domains; such as data analysis, security evaluation, operating systems, database management system, and software and code testing tools [5]. These tools' categorization helps the firms to satisfy their audit requirements. They can perform reliable audit procedures and deliver fast and accurate audit results. Thus, the need for CAATs is increasing, as they allow auditors to execute their review and monitoring tasks effectively and comprehensively.

The previous studies show the importance of using CAATs in achieving quality and the benefits that organizations gain. Yet, there are important questions to consider: what are the auditing areas available in an IT department, what are the available tools in each auditing area, which standards do they follow, are these tools capable of performing all tasks within the IT department within the organization, and finally can the organization rely on only one tool to perform all tasks related to IT department. These questions are the focus of our study and the basic factors affecting the choice of a tool by IS auditor. So, this paper aims to determine the main factors affecting the tool selection, as well as present a comparative study of the existing CAATs with their capabilities and limitations.

The remainder of this paper is organized as follows: an overview of CAATs and their goals in Sect. 2. Section 3 covers a discussion of related work in terms of benefits, limitations, and influence factors for selecting a CAAT. Section 4 categorizes these tools over IS audit areas and presents a comparative study of most of the available CAATs supporting IS auditor. Finally, we conclude the paper in Sect. 5 with an overlook of the future work.

2 Background and Literature Review

CAATs are commonly employed to audit application controls in order to reduce the total audit hours [6]. They enable auditors to test 100% of the population rather than a sample, thereby increasing the reliability of the audit test and conclusions. IS auditor can use CAATs to write a script for automated periodic audits. This automation helps to achieve continuous auditing and monitoring according to management objectives [4]. According to [7], CAATs have several analytic capabilities such as data analysis, applied and managed analytics, and continuous auditing. IS auditors can repeat audit work by executing automatic audits, thereby reducing audit time and costs.

The goal of auditing is to ensure the effectiveness of internal controls, which are designed to facilitate the company's management activities. IS auditors should decide the scope for their projects and develop their CAATs processes based on the audit policies. CAATs performance can be measured using indicators or metrics of team project performance, such as efficiency, completeness, compliance or accommodation with work progress, outcome quality, interaction, and communication [8–10].

Previous studies in [11–15], have found that complete CAATs establishment can conserve auditors' manpower resources, reduce audit costs, reduce the time spent executing audit tasks, increase audit quality, and enable enterprises to improve operating efficiency. CAATs can assist in implementing the Sarbanes–Oxley Act (SOX) requirements, as well as facilitating monitoring activities and reducing their time. Hence, eventually, the use of CAATs can increase enterprise efficiency and overall performance.

In [16], a study introduced some recommendations to increase the efficiency and effectiveness of the available software tools to the auditor. These recommendations include: (a) determine the enterprise's audit mission, objectives and priorities, (b) determine the types and scope of audits, (c) consider the enterprise's technology environment, (d) ensure using the suitable tools, (e) identify the risks, (f) train audit team on the tool usage, and (g) support of periodic review reports.

3 CAATs Benefits, Limitations and Influence Factors

Many organizations have opted to achieve high-quality information systems by developing their business process support, as well as improving their information activities [17]. This increases the need for using CAATs to permit auditors to execute their audit and monitoring tasks successfully. Moreover, it helps auditors to focus on high-risk areas [18]. As a result, CAAT is seen to be a crucial tool to support an internal auditor's work [19]. There are many important and fundamental functions which the auditor considers as the real gain from using CAATs, as follows:

- Replace manual test procedures with automated methods.
- Test the reliability of client software.
- Increase audit tasks precision and efficiency.
- Reduce the auditing time and audits' cost.

Many other functions belong to each field within the organization. The next section addresses two questions: which tools are supporting IS areas and which standards ensure the quality of these tools. Moreover, due to the importance of CAATs, there are some influential factors that affect auditors' decision to adopt CAATs, which need analysis. There are four main factors affecting IS auditor CAATs selection criteria: (1) increasing performance, (2) reducing effort, (3) utilizing social influence, and (4) facilitating both organizational and technical infrastructure conditions to support system use [20]. The authors present a study to check how influential these factors on the internal auditor's decision to use CAATs are. Their study suggests that internal auditors are more willing to utilize CAATs when their usage would improve their job efficiency. Moreover, the two main influential factors are facilitating conditions and increasing performance [21].

Other factors may control the usage and popularity of these tools as in [20]. Examples of these factors are the *cost-benefit* trade-off while using CAATs, audit *firm's size*, audit firm's *top management commitment*, and the *compatibility* between required tasks and the underlying technology used. All these factors can affect the implementation quality of the auditing tasks. They can also affect the efficiency and effectiveness of IS auditor tasks within the organization. Furthermore, using these tools helps to measure the accuracy of audit tests, reduce reviewing time, provide ad hoc reports, and early detection of deviations.

4 Factors Affecting Auditing Tools Selection

The need for CAATs emerged to support IS auditor reviews many areas in an organization. There are some researches investigating the influential factors to use CAATs during the auditing process [1, 20]. In addition, there are other factors that affect the auditor's choice in selecting the most suitable tool to accomplish the required tasks. CAATs have some basic characteristics and factors that affect the auditor decision while searching for a suitable tool [6]:

- Ease of use; a measurement of how easy the tool is to use by its intended users.
- Ease of data extraction; which includes the ability to access a wide variety of data files from different platforms, and to integrate data with different formats.
- Ability to define fields and select from standard formats.
- Menu-driven functionality for processing analysis commands.
- Simplified query building and adjustments.
- Suitable platform and/or operating system for the organization.
- Supporting documentation and periodic audit reports.

There are other items that the IS auditor considers as key items while selecting a comprehensive tool as in [16]. There are some recommendations for increasing the software tools usage efficiency and effectiveness to the auditor. Those recommendations include the following:

- Determine the enterprise's audit mission, objectives, priorities.
- Determine the types and scope of audits.
- Consider the enterprise's technology environment.
- Ensure selecting the right tools.
- Training audit team on the use of the selected tool.
- Beware of the risks.
- Review daily reports.

The following subsections address factors from different perspectives. The discussion starts with the support of existing standards and follows with how the factors are affecting internal auditors' intentions to use and accept CAATs. Due to the huge number of CAATs that assist IS auditors, a sample of tools was collected from websites; e.g. Capterra[1] and Software Advice.[2] These tools cover many IS auditing areas that need further investigation to decide the key factors affecting the tool selection criteria.

4.1 CAATs, Audit Areas and Standards

Nowadays, the information technology (IT) department acts as the backbone of any organization. It supports several areas such as business, database, security, network, governance, risk assessment, etc. Each area has its own features and functions. Therefore, IS auditor needs assistance in deciding which tool suits reviewing the current auditing task and abides by the organization's quality standards. Table 1 addresses the important areas within the IT department and the available tools in each area to the best of our knowledge. It also clarifies the recent ISO standards to be considered during the audit process and sample software tools used in each area.

Table 1 illustrates that the main areas supported by CAATs are risk assessment, security, and general data protection regulation (GDPR), respectively. These tools are mainly supporting ISO 31000:2018 and ISO/IEC 27000:2018 standards. Moreover, there are tools that cover multiple areas but not all of them; such as Janco, Delphix, Debian, Ramce ERP, and ACL. These tools can be very promising to IS auditors due to their area coverage. They can minimize the number of required tools to cover all the areas, as well as minimize the learning curve and training for the personnel and employees.

[1]https://www.capterra.com/audit-software/.

[2]https://www.softwareadvice.com/audit/.

Table 1 Distribution of CAATs supporting Auditing areas (Tools' references see Appendix)

Code	Areas	ISO standard	Tools
a.1	Risk assessment	ISO 31000:2018	ARBUTUS—TeamMate—Tackle—SmartSolve—Symbiant Tracker—R-CAP—Ramce ERP—MKinsight—MetricStream—Isolocity—Qwerks—InfoZoom—DATEV—Debian—Analyzer–Ecomply—TrustArc—Consenteye—BigID—ZenGRC
a.2	Security	ISO/IEC 27000:2018	Onspring—ECAT—Assure—ZenGRC—ManageEngine ADAudit Plus—Debian—Lynis—Janco—Xandria-Onspring-ACL—Delphix—ECAT
a.3	Network security	ISO/IEC 27033-5:2013	WinAudit—Aircrack-ng—cSploit—Open—AudIT—AIDA64—E-Z Audit—Fern Wifi Cracker
a.4	Governance	ISO/IEC TR 38505-2:2018	ACL—Delphix—Collibra
a.5	Hardware	ISO 25119-3:2018	Informer—WinAudit—Belarc Advisor—E-Z Audit—ManageEngine ADAudit Plus
a.6	Software	ISO/IEC/IEEE 24748-8:2019	WinAudit—Belarc Advisor
a.7	Cloud computing	ISO/IEC TR 22678:2019	Skeddly—CloudStack—Netskope Cloud Security Platform—MultCloud—RightScale—Ormuco Stack—Cloud Management—Ramce ERP
a.8	e-Commerce	ISO 10008:2013	DeepCrawl—SEMrush
a.9	Database	ISO 17572-2:2018	Onspring—Form.com—ACL—Active@—IDEA—Xandria—AuditBoard—Delphix
a.10	Sourcing code	ISO 3166	Debian—Clang- Analyzer
a.11	Business continuity	ISO/TS 22318:2015	Janco
a.12	Disaster recovery testing	ISO/IEC 24762:2008	Onspring—Janco—Delphix
a.13	Social media	ISO 26000:2010	NetBase—Tailwind—Clean Cloud
a.14	General data protection regulation (GDPR)	ISO/IEC 27000:2018	Catalystone—Iubenda—Delphix—Cookie Assistant—Ecomply—PYXI—Termly—BigID—consentEye—OneTrust—TrustArc—Quantcast–Consenteye—ACL

4.2 Comparison Between CAATs

Various factors can be categorized into functional and non-functional requirements. In this paper, the selected *functional requirements* are divided into areas of interest (see Table 1) and configurable audit reports. These reports should explain all tasks that were performed and their quality. On the other hand, the *non-functional requirements* cover several factors: easy installation, ease of use, friendly UI, supported operating systems (OS), web interface support, provide a free demo, open-source support, and training support (e.g. offline documentation, online support). Table 2 illustrates a comparison between CAATs based on the specified functional and non-functional requirements.

The use of audit software tools differs from an organization to another. Table 2 presents a comparative study, which investigates the factors affecting the success of each tool to decide the influence factors. This study shows that many tools prefer web interface support than supporting various types of operating systems. This can be due to several reasons, one of them could be due to their ease of use without further installation steps. Another reason could be due to the required time to install and costs to provide support to various operating systems. In addition, not all CAATs provide any type of training, which can be very difficult for the IS auditors. Moreover, several tools do not provide a free demo for the end users testing; which can be an important selling factor for the tool under assessment.

Figure 1 examines the support of influence factors and their coverage in the sample of tools as presented in Table 2. Both configurable audit reports and web support comes on the top of the support list by several CAATs. Moreover, training can be either supported online or through user documentation, but not many CAATs support both. It is notable that more than 85% of these tools support configurable audit reports, see Fig. 2. However, the applicability of CAATs over several operating systems is lacking. Moreover, only ~63% of the tools provide training, this percentage needs more root-cause investigation.

Another perspective of this comparative study is to the discussion of how CAATs can support IS auditors. These tools are designed to help IS auditors to manage all aspects of the audit process. As shown in Table 2, all CAATs support a set of areas (as mentioned in Table 1) to serve a range of audit tasks. Figure 3 analyzes each of these areas and how often they are supported by CAATs. It is evident that the risk assessment area has gained the highest interest of several CAATs, while business continuity area has the least interest within the sample of studied tools. The support of other areas varies, e.g. both governance and disaster recovery testing is supported by 3 tools, while auditing cloud computing and database areas are supported by 8 tools.

There are also tools capable of performing joint tasks between two or more areas. For example, some tools are used in data analysis, task management, interactive audit trail, pivot tables, and graphs, e.g. IDEA—ACL—Delphix, which service each of governance, database, and general data protection regulation areas. We can take one of the recent areas mentioned before such as general data protection regulation (GDPR) and consider how CAATs can help achieve GDPR compliance.

The objective of the GDPR audit is to help management assess how effective it is being governed, monitored, accurately managed. In order to help the assessment and assurance processes, the researchers have categorized GDPR auditing controls.

Table 2 Comparison between CAATs supporting IS auditors

Serial	CAATs	Functional requirements		Non-functional requirements ✓										
		Areas	Config. Audit Reports	Easy install	Ease-of-use	Friendly UI	Windows	Linux	Macintosh	Web Support	Free Demo	Open source	Training offline	Training Online
1	Debian	a.1 a.2	✓	✓	✓	✓	✓	✓	✓	×	×	✓	✓	×
2	AIDA64	a.3	×	✓	✓	✓	✓	✓	✓	✓	✓	×	×	×
3	Lynis	a.2	✓	✓	✓	✓	×	✓	✓	✓	✓	✓	✓	×
4	Fern Wifi Cracker	a.3	✓	✓	✓	✓	✓	✓	✓	✓	×	×	×	×
5	cSploit	a.3	✓	✓	✓	✓	✓	✓	✓	✓	✓	×	×	×
6	TeamMate	a.1	✓	✓	✓	✓	✓	×	×	×	✓	×	✓	✓
7	Onspring	a.2 a.9	✓	✓	✓	✓	×	×	✓	✓	✓	✓	✓	✓
8	ACL	a.4 a.9 a.14	✓	✓	✓	✓	✓	✓	✓	×	✓	×	✓	✓
9	IDEA	a.9	✓	✓	✓	✓	✓	×	×	×	✓	×	×	×
10	Clang	a.10	✓	✓	✓	✓	×	✓	✓	✓	×	✓	×	✓
11	Analyzer	a.1	✓	×	×	×	✓	×	×	✓	×	×	✓	✓
12	E-Z Audit	a.3 a.5	✓	✓	✓	✓	✓	✓	✓	✓	✓	×	✓	✓
13	Janco	a.2 a.11	×	×	✓	✓	✓	×	×	×	×	×	×	✓
14	DATEV	a.1	✓	✓	✓	✓	✓	×	×	×	✓	×	×	✓
15	ARBUTUS	a.1	✓	✓	✓	✓	✓	×	×	×	×	×	✓	×
16	Ecomply	a.1 a.14	✓	✓	✓	✓	✓	×	×	✓	✓	✓	✓	✓
17	Cookie Assistant	a.14	✓	✓	✓	✓	×	✓	✓	✓	✓	×	✓	✓
18	Iubenda	a.14	✓	✓	✓	✓	×	✓	✓	✓	×	×	✓	✓
19	Quant cast	a.14	×	×	✓	✓	×	✓	✓	✓	×	✓	✓	✓
20	TrustArc	a.1 a.14	✓	✓	✓	✓	×	✓	✓	✓	✓	×	✓	✓

(continued)

Table 2 (continued)

21	OneTrust	a.1 a.14	✓	✓	✓	✓	×	✓	✓	✓	✓	×	×	✓
22	Consenteye	a.1 a.14	✓	✓	✓	✓	×	✓	✓	✓	✓	✓	×	✓
23	BigID	a.1 a.14	✓	✓	✓	✓	×	✓	✓	✓	×	✓	×	×
24	Termly	a.14	×	✓	✓	✓	×	✓	✓	✓	×	✓	✓	✓
25	PYXI	a.14	×	✓	×	×	×	×	×	✓	×	×	✓	×
26	NetBase	a.13	✓	✓	✓	✓	×	✓	✓	✓	✓	✓	✓	×
27	Xandria	a.2 a.9	✓	✓	✓	✓	✓	×	×	✓	✓	×	×	✓
28	SEMrush	a.8	✓	✓	✓	✓	×	×	×	✓	✓	✓	×	✓
29	Skeddly	a.7	✓	✓	✓	✓	×	×	×	✓	✓	×	✓	×
30	Ormuco Stack	a.7	✓	✓	✓	✓	×	×	×	✓	✓	×	✓	✓
31	Netskope Cloud	a.7	✓	✓	×	×	×	×	×	✓	✓	×	✓	✓
32	MultCloud	a.7	✓	✓	✓	✓	×	×	×	✓	×	✓	✓	×
33	RightScale	a.7	×	✓	✓	✓	×	✓	✓	✓	×	×	×	×
34	Cloud Management	a.7	×	✓	✓	✓	×	×	×	✓	×	×	×	×
35	Informer	a.5	✓	✓	✓	✓	✓	×	×	✓	✓	✓	✓	✓
36	SmartSolve	a.1	✓	✓	✓	✓	×	✓	✓	✓	×	×	×	×
37	MetricStream	a.1	×	×	×	×	×	✓	✓	✓	✓	×	×	×
38	Assure	a.2	✓	✓	✓	✓	×	×	✓	✓	✓	×	✓	✓
39	ManageEngine ADAudit Plus	a.2	✓	✓	✓	✓	✓	✓	✓	✓	✓	✓	✓	✓
40	Delphix	a.2 a.4 a.9 a.12 a.14	✓	✓	✓	✓	✓	×	✓	✓	✓	✓	✓	✓
41	Catalystone	a.14	✓	×	×	×	✓	✓	✓	✓	✓	×	✓	✓
42	CleanCloud	a.13	✓	✓	✓	✓	✓	✓	✓	✓	✓	×	✓	✓
43	Collibra	a.4	✓	✓	✓	✓	✓	✓	✓	✓	×	×	✓	✓

(continued)

Table 2 (continued)

44	Qwerks	a.1	×	✓	✓	✓	✓	✓	✓	✓	×	×	✓	✓
45	MKinsight	a.1	✓	✓	✓	✓	✓	✓	✓	✓	✓	✓	✓	✓
46	Ramce ERP	a.1-a.7	✓	✓	✓	×	✓	✓	✓	✓	×	×	✓	✓
47	Taskle	a.1	✓	✓	✓	✓	×	✓	✓	✓	✓	×	✓	✓
48	Symbiant Tracker	a.1	✓	✓	✓	✓	×	✓	✓	✓	×	✓	×	✓
49	R-CAP	a.1	✓	✓	✓	✓	×	✓	✓	✓	✓	×	✓	✓
50	Isolocity	a.1	✓	✓	✓	✓	×	✓	✓	✓	✓	×	✓	✓
51	Tailwind	a.13	✓	✓	✓	×	✓	×	×	×	×	×	×	×
52	Active@	a.9	✓	✓	✓	✓	✓	×	×	×	✓	×	×	✓
53	InfoZoom	a.1	✓	✓	✓	✓	✓	×	×	×	✓	×	✓	✓
54	AuditBoard	a.9	✓	✓	✓	✓	✓	×	×	✓	×	×	×	×
55	DeepCrawl	a.8	✓	✓	✓	✓	×	×	×	✓	✓	×	×	×
56	ECAT	a.2	✓	✓	✓	✓	×	×	×	✓	×	×	×	×
57	form.com	a.9	✓	✓	✓	✓	✓	×	✓	×	×	×	✓	×
58	Aircrack-ng	a.3	✓	×	✓	✓	✓	×	×	×	✓	✓	✓	×
59	Belarc Advisor	a.5 a.6	✓	✓	✓	✓	✓	×	×	×	✓	✓	✓	✓
60	WinAudit	a.3 a.5 a.6	✓	✓	✓	✓	✓	×	×	×	×	✓	✓	✓
61	ADAudit Plus	a.2	✓	✓	✓	✓	✓	×	×	×	×	✓	✓	×
62	ZenGRC	a.1 a.2	✓	✓	✓	✓	✓	×	✓	✓	✓	×	×	✓

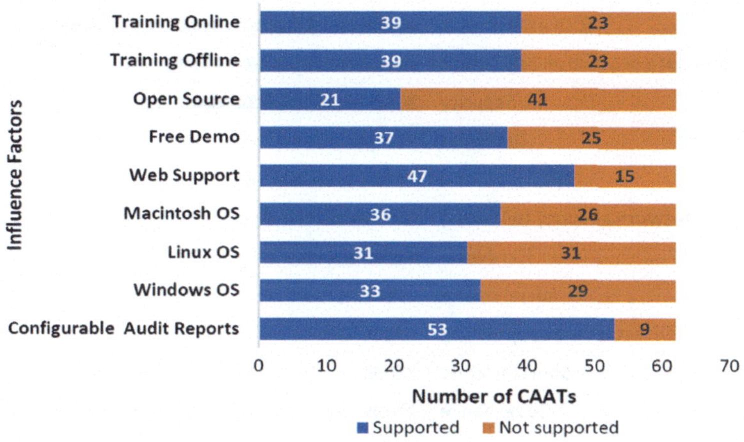

Fig. 1 Influence factors affecting CAATs selection

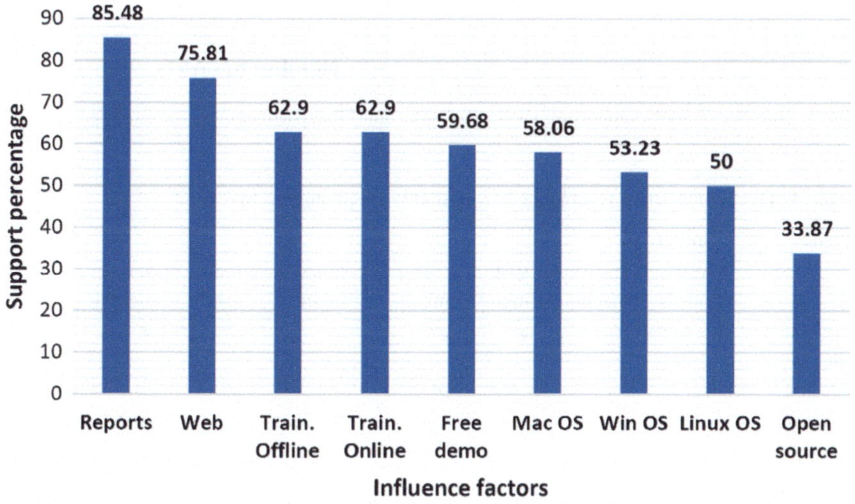

Fig. 2 Percentage of supported influence factors in CAATs

There are basic controls, such as access controls, data mapping, risk management, consent management, incident management, policy management, as well as sensitive data identification. These controls evaluate the effectiveness of GDPR. GDPR is an area (a.14) in Table 1. After studying CAATs, the researchers found that each tool serves a set of features that achieve GDPR compliance, see Table 3. Considering the tools supporting the GDPR areas, there is not a single tool that can fully support all the basic controls.

Fig. 3 Number of tools supporting each area under the IT department

Table 3 Distributed controls which achieve GDPR compliance (a.14)

Tools	Controls						
	Access controls	Data mapping	Risk mgmt.	Consent mgmt.	Incident mgmt.	Policy mgmt.	Sensitive data identification
Catalyst one	✓	✓	✓	✓	✓	✓	✓
Iubenda				✓		✓	
Delphix	✓	✓				✓	✓
Cookie Assistant				✓		✓	✓
Ecomply		✓	✓	✓	✓	✓	
PYXI		✓			✓		✓
Termly				✓		✓	
BigID	✓	✓	✓	✓	✓		✓
consentEye			✓		✓	✓	
OneTrust	✓	✓	✓	✓	✓		

The GDPR area is sensitive to all other areas, e.g. business, cloud computing, social networking, within the IT department and there are many common functions among them. IS auditor need to get a comprehensive report, with the status of all the tasks in IT department areas, to help to review every task efficiently. For example, it is important to review security according to existing standards and guidelines. However, there is not one tool that combines all these controls with various domains. Table 3 shows that Catalyst is the only tool supporting all the seven GDPR controls, while Ecomply, consentEye, and OneTrust support five controls each.

There are some standards such as ISO 27001 and ISO 27002 which help organizations to ensure that they have effective information security programs. ISO 27001 was originally created to help to secure both government services and citizen data at the service provider's side. The use of ISO 27001 ensures the GDPR principles, and the appropriate technological and organizational measures are all preserved to protect information [22]. It helps organizations to define responsibilities, such as who is responsible for information assets and who can authorize access to these data. Also, ISO 27001 provides independent accreditation for information security management systems, while ISO 27002 is a code of practice that is not accredited by external parties. Either standard will help to ensure that an organization has strong aiding controls [23].

Although the factors presented in Table 2 can provide added competitive features, there can be some challenges in software auditing tools, such as

- Lack of compatibility of web application over different browsers.
- The user interfaces need adaptation over different environments.
- Reporting tool needs some improvements and tailored adjustments.
- Online support can be out-of-date and/or incompetent.
- Upgrading the system can be faulty and costly.
- Lack of support forums and community for new CAATs.
- The increased learning curve for advanced features.
- Reports with multimedia charts and tabular information are not easily provided.
- The need for multiple CAATs can be highly expensive.

These challenges can be the reason for most of the prominent defects that have emerged while using CAATs. Yet, one of the most prominent flaws in all CAATs is that there is no single tool capable of covering all the information systems auditing tasks together.

As a result of this study, there are several challenges that have emerged. The main challenge is the difficulty to support all areas in one tool, which leads some organizations to use multiple CAATs. This can be very expensive and complicate the integration of the generated reports from each tool. Consequently, the organization will waste more time and effort into training its personnel. Repeating the implementation of the common controls can generate faulty results. All these obstacles can generate several errors in the final reports and gaps in covering the audit tasks.

5 Conclusion and Future Work

IS auditors use software auditing tools such as CAATs to help in performing all auditing process tasks. The use of these tools helps to measure the accuracy of audit tests, reduce reviewing time, provide ad hoc reports, and early detection of deviations. There are many factors influencing the use of audit software. Recent research finds that the two important factors that affect auditor's decision on whether to use CAATs or not are performance expectancy and facilitating conditions. However, there are other factors that help in determining the appropriate tool to perform the tasks during the auditing process. This paper aims to find these influential factors that help in choosing suitable auditing tools to support the success of the required audit tasks.

In order to achieve this target, the researchers investigate several factors of selecting these tools. Nevertheless, there are also many tools that serve the tasks of IS auditing in all areas. As a result, we found that the most recent auditing tools comply with ISO standards, which provide accurate guidelines to help the auditors achieve high-quality audit results. Each of these CAATs cannot solely support all areas of IS auditing. This can be very challenging to IS auditor to generate a comprehensive and accurate report with the minimum cost and effort. As future work, the researchers aim to create a framework for an integrated IS auditing tasks in one comprehensive tool.

Appendix

See Table 4.

Table 4 Tools with their URL references—last checked on 19th June 2019

Tool name	**Website URL reference**
ACL	https://www.acl.com/
Active@	http://www.lsoft.net/
AIDA64	https://www.aida64.com/products/aida64-network-audit
Aircrack-ng	https://www.aircrack-ng.org
Analyzer	https://clang-analyzer.llvm.org/
ARBUTUS	https://www.arbutussoftware.com/products-solutions/audit-analytics
Assure	https://www.asuresoftware.com/
AuditBoard	https://www.auditboard.com/
Belarc advisor	https://download.cnet.com/Belarc-Advisor/3000-2094_4-10007277.html
BigID	https://bigid.com/

(continued)

Table 4 (continued)

Tool name	**Website URL reference**
Catalystone	https://catalystone.com/gdpr-data-audit-tool/
Clang	www.createaclang.com
CleanCloud	https://cleancloudapp.com/
Cloud management	https://www.virtustream.com/software/xstream/features
CloudStack	https://reviews.financesonline.com/p/apache-cloudstack/
Collibra	https://www.collibra.com/
Consenteye	https://www.consenteye.com/
Cookie assistant	https://www.cookieassistant.com/
cSploit	https://www.apksum.com/app/csploit/org.csploit.android
DATEV	https://www.datev.com/
Debian	https://www.debian.org/security/audit/tools
DeepCrawl	https://www.deepcrawl.com/pain-point/regular-site-audit/
Delphix	https://delphix.github.io/
ECAT	https://ecat-group.com/audit-management-software/?utm_source=capterra-visit-website&utm_medium=referral&utm_campaign=Capterra
Ecomply	https://ecomply.io/
E-z audit	http://www.ezaudit.net/features/
Fern Wifi Cracker	https://n0where.net/fern-wifi-cracker
Form.com	https://form.com/
IDEA	https://www.casewareanalytics.com/products/idea-data-analysis
Informer	https://informer.freshdesk.com/support/solutions/articles/5000665438-auditfile
InfoZoom	https://www.softlakesolutions.com/
Isolocity	https://www.isolocity.com/
Iubenda	https://www.iubenda.com/en/gdpr
Janco	https://sourceforge.net/projects/janco/
Lynis	https://cisofy.com/lynis/
Manage engine ADAUDIT Plus	https://www.manageengine.com/products/active-directory-audit/download.html
MetricStream.	https://www.metricstream.com/solutions/audit-management.htm
MKinsight	http://www.mkinsight.com/functionality.aspx?id=9
MultCloud	https://project-management.com/multcloud-software-review/
NetBase	https://www.netbase.com/
Netskope cloud	https://www.netskope.com/
OneTrust	https://www.onetrust.com/
Onspring	https://www.onspring.com/#difference
Open-Audit	https://opmantek.com/network-discovery-inventory-software/
Ormuco Stack	https://ormuco.com/
PYXI	http://www.pyxi.co.uk/
Quant cast	https://alternativeto.net/software/quantcast/

(continued)

Table 4 (continued)

Tool name	**Website URL reference**
Qwerks	https://getqwerks.com/
Ramce ERP	http://www.ramco.com/
R-CAP	http://www.r-cap.com/
RightScale	https://reviews.financesonline.com/p/rightscale/
SEMrush	https://www.semrush.com/
Skeddly	https://cloudcheckr.com/partners/skeddly/
SmartSolve	https://www.pilgrimquality.com/
Symbiant tracker	https://www.symbiant.co.uk/
Tailwind	https://www.tailwindapp.com/
Taskle	https://www.taskle.com/
Teammate	http://www.teammatesolutions.com/audit-management.aspx.
Termly	https://termly.io/
TrustArc	https://www.trustarc.com/
WinAudit	https://www.techspot.com/downloads/2307-winaudit.html
Xandria	https://www.syslink-xandria.com/en
ZenGRC	http://unbouncepages.com/reciprocity-zengrc-risk-management-gdm/?directory=Risk_Management&source=SoftwareAdvice

References

1. Alcarraz, Gerardo D. and others. 2009. Certified Information Systems Auditor – CISA—Review manual 2009.
2. Romney, M.B., P.J. Steinbart, and B.E. Cushing. 2000. Accounting information systems, vol. 2. Upper Saddle River, NJ: Prentice Hall.
3. Coderre, David, and Royal Canadian Mounted Police. 2005. Global technology audit guide: continuous auditing implications for assurance, monitoring, and risk assessment, 1–34. Lake Mary: The Institute of Internal Auditors.
4. Sun, Chia Ming. 2012. The adaptation and routinization processes of a continuous auditing system implementation..
5. Braun, Robert L., and Harold E. Davis. 2003. Computer-assisted audit tools and techniques: analysis and perspectives. *Managerial Auditing Journal* 18 (9): 725–731.
6. Zhang, Li, Amy R. Pawlicki, Dorothy McQuilken, and William R. Titera. 2012. The AICPA assurance services executive committee emerging assurance technologies task force: the Audit Data Standards (ADS) initiative. *Journal of Information Systems: Spring* 26(1):199–205.
7. ACL. 2011. The ACL audit analytic capability model: navigating the journey from basic data analysis to continuous monitoring. A white paper.
8. Henderson, J.C., and S. Lee. 1992. Managing I/S design teams: a control theories perspective. *Management Science* 38 (6): 757–777.
9. Keil, M., A. Rai, and S. Liu. 2012. How user risk and requirements risk moderate the effects of formal and informal control on the process performance of IT projects. *European Journal of Information Systems* 22 (6): 650–672.
10. Lu, Y., C. Xiang, B. Wang, and X. Wang. 2011. What affects information systems development team performance? An exploratory study from the perspective of combined socio-technical theory and coordination theory. *Computers in Human Behavior* 27 (2): 811–822.

11. Vasarhelyi, M.A., M. Alles, S. Kuenkaikaew, and J. Littley. 2012. The acceptance and adoption of continuous auditing by internal auditors: a micro analysis. *International Journal of Accounting Information Systems* 13:267–281.
12. Gonzalez, G.C., P. N. Sharma, and D.F. Galletta. 2012. The antecedents of the use of continuous auditing in the internal auditing context. *International Journal of Accounting Information Systems* 13(3):248–262.
13. Masli, A., Peters, G. F., Richardson, V. J., & Sanchez. Examining the potential benefits of internal control monitoring technology. The Accounting Review, J. M. (2010). Vol. 85, No.3, pp. 1001-1034.
14. Janvrin, D., J. Bierstaker, and D.J. Lowe. 2008. An examination of audit information technology use and perceived importance. *Accounting Horizons* 22(1):1–21.
15. Rezaee, Z., A. Sharbatoghlie, R. Elam, and P.L. McMickle. 2002. Continuous auditing: building automated auditing capability. *Auditing: A Journal of Practice & Theory* 21(1):147–163.
16. Mahzan, N., and F. Verankutty. 2011. IT auditing activities of public sector auditors in Malaysia. *African Journal of Business Management* 5 (5): 1551–1563.
17. Ramamoorthi. S., M.L. Windermere. 2004. The pervasive impact of information technology on internal auditing, Ch. 9. Lake Mary: Institute of Internal Auditors Inc.
18. Shukarova Savovska, K., and B.A. Sirois. 2018. Audit data analytics: opportunities and tips (English). Centre for financial reporting reform (CFRR). Washington, D.C.: World Bank Group.
19. Mahzan, N., and A. Lymer. 2014. Examining the adoption of computer-assisted audit tools and techniques: cases of generalized audit software use by internal auditors. *Managerial Auditing Journal* 29(4):327–349.
20. Bierstaker, J., D. Janvrin, and D.J. Lowe. 2014. What factors influence auditors' use of computer-assisted audit techniques? *Advances in Accounting* 30(1):67–74.
21. Al-hiyari, A. and E. Hattab. 2019. Factors that influence the use of computer assisted audit techniques (CAATs) by Internal Auditors in Jordan. ISSN: 1096-3685.
22. Calder, A. 2018. EU GDPR: a pocket guide. Cambridgeshire: IT Governance Publishing Ltd.
23. Tzolov, T. 2018. One model for implementation GDPR based on ISO standards. In *2018 International Conference on Information Technologies (InfoTech)*, 1–3. IEEE.

On the Selection of the Best MSR PAPR Reduction Technique for OFDM Based Systems

Mohamed Mounir, Mohamed B. El_Mashade and Gurjot Singh Gaba

Abstract Orthogonal Frequency Division Multiplexing (OFDM) is used in the physical layer of IoT–based networks, due to its ability to mitigate frequency selectivity. However, OFDM suffers from high Peak-to-Average Power Ratio (PAPR) problem, which reduces the Power Amplifier (PA) efficiency or otherwise degrade BER and increases out-of-band (OOB) radiation. This means IoT networks will consume more power. However, IoT devices are expected to have limited power sources. Many PAPR reduction techniques were introduced in the literature. Among them, Multiple Signal Representation (MSR) techniques have high PAPR reduction gain and a small number of Side Information (SI) bits. However, they require high computational complexity to provide high PAPR reduction gain. In this paper, three MSR techniques namely Partial Transmit Sequence (PTS), Selective Mapping (SLM), and interleaving are compared in terms of PAPR reduction gain and computational complexity. Results showed that PTS is the best among others. It provides PAPR reduction performance similar to SLM and better than interleaving with less computational complexity.

Keywords OFDM · PAPR · MSR · PTS · SLM · Interleaving

M. Mounir (✉)
Communication and Electronics Department, El-Gazeera High Institute for Engineering and Technology, Cairo, Egypt
e-mail: Eng_mohamedvip@yahoo.com

M. B. El_Mashade
Electrical Engineering Department, Faculty of Engineering, Al-Azhar University, Cairo, Egypt
e-mail: Elmashade@yahoo.com

G. S. Gaba
School of Electronics and Electrical Engineering, Lovely Professional University, Jalandhar, India
e-mail: Er.gurjotgaba@gmail.com

A. Z. Ghalwash et al. (eds.), *Internet of Things—Applications and Future*, Lecture Notes in Networks and Systems 114,
https://doi.org/10.1007/978-981-15-3075-3_11

1 Introduction

Orthogonal Frequency Division Multiplexing (OFDM) is widely used in IoT–based networks, due to its ability to mitigate frequency selectivity. However, OFDM suffers from a serious disadvantage. OFDM has a high Peak-to-Average Power Ratio (PAPR). High PAPR is a result of the superposition of many narrowband subcarriers. Signals with high PAPR causes BER degradation and out-of-band (OOB) radiation. Otherwise, Power Amplifier (PA) efficiency will be degraded significantly and this will increase the rate of battery failure in IoT devices. Different PAPR reduction techniques were introduced in the literature to enhance PA efficiency without BER degradation or excessive OOB radiation [1]. These PAPR reduction techniques can be classified according to strategy into three categories: adding signal techniques, Multiple Signal Representation (MSR) techniques, and finally coding and precoding techniques [2]. In MSR techniques several statistically independent OFDM symbols are generated using the same data sequence and the one with minimum PAPR is selected for transmission. MSR techniques include Partial transmit sequence (PTS), Selective Mapping (SLM), and interleaving techniques [3]. Theoretically, for the same number of alternatives, each one of the three techniques must provide the same performance, but this requires the alternatives to be statistically independent. On the other hand, the computational complexity of these techniques may be different to provide the same PAPR reduction performance.

In this paper, the three MSR techniques (i.e., PTS, SLM, and interleaving) are compared in terms of PAPR reduction performance and computational complexity.

The rest of this paper is organized as follows: Sect. 2 provides a brief description of the OFDM system model and the PAPR problem. Conventional MSR PAPR reduction techniques are briefly reviewed and analyzed in Sect. 3. In Sect. 4, simulation results in terms of PAPR reduction performance and computational complexity are discussed. Finally, the conclusions are drawn in Sect. 5.

2 System Model and PAPR Description

Consider a sequence of N information symbols A_k (QAM symbol) modulating N subcarriers in the frequency domain. Then, an IFFT is used to convert the frequency domain vector $\mathbf{A} = [\,A_1 \quad A_2 \quad \cdots \quad A_{N-1}\,]$ to a time-domain vector $\mathbf{x}$ (OFDM symbol), conventional written as $\mathbf{x} = \mathbf{IFFT}(\mathbf{A})$. Time-domain samples $x[n]$ of time-domain vector $\mathbf{x}$, are given by

$$x[n] = \frac{1}{\sqrt{N}} \sum_{k=0}^{N-1} A_k e^{j\frac{2\pi nk}{N}}, 0 \le n \le N-1 \tag{1}$$

For large number of subcarriers N, real and imaginary parts of the time-domain samples x[n] of complex OFDM symbols (which are the summation of uniformly distributed frequency domain symbols A_k) will be Gaussian distributed due to central limit theorem. Assume statistical independence among frequency domain symbols A_k. Then, the amplitude of the OFDM signal will be Rayleigh distributed suffering from high PAPR.

The PAPR of one OFDM symbol is given by

$$\text{PAPR} = \frac{\text{Max}_{n\in[0,N]}|x[n]|^2}{E\left\{|x[n]|^2\right\}} \tag{2}$$

where $E\{\cdot\}$ is the expectation operator. The PAPR is usually described by means of statistical distribution as PAPR itself seems to be a random variable. Thus, the probability of OFDM symbol's PAPR equal to or larger than the threshold PAPR_o is given by Complementary Cumulative Distributive Function (CCDF) as follows:

$$\text{CCDF} = \text{prob}\{\text{PAPR} > \text{PAPR}_o\} \tag{3}$$

Practically, PAPR reduction capability is measured by the amount of CCDF reduction achieved [4]. Reference [5] introduced this empirical approximation for the CCDF of PAPR of the oversampled signal;

$$CCDF_c = Pr\{\text{PAPR} > \xi_0\} = 1 - exp\left(-Ne^{-\xi_0}\sqrt{\frac{\pi}{3}}\log N\right) \tag{4}$$

Where ξ_0 is the given threshold PAPR_o. Oversampling by factor L is used to accurately estimate the peak of the continuous OFDM signal.

3 MSR PAPR Reduction Techniques

In MSR techniques several statistically independent OFDM symbols are generated using the same data sequence and the one with minimum PAPR is selected for transmission. Side information (SI) may be required at the receiver for correct detection [3]. Assuming a set of *M* statistically independent OFDM symbols are generated the probability that PAPR of one OFDM symbol exceeds certain threshold ξ_0 is given by [6];

$$CCDF_c = Pr\{\text{PAPR} > \xi_0\} = \left(1 - exp\left(-Ne^{-\xi_0}\sqrt{\frac{\pi}{3}}\log N\right)\right)^M \tag{5}$$

3.1 Partial Transmit Sequence (PTS)

PTS technique was presented in [7] by Muller and Huber. In this technique, the OFDM symbol $\mathbf{A} = [\, A_0 \quad A_1 \quad \cdots \quad A_{N-1} \,]$ is partitioned into V disjoint sub-Blocks $\mathbf{A}^v = \left[A_0^v, A_1^v, \ldots, A_{N-1}^v\right]$ where $1 \le v \le V$. All subcarrier positions that already represented in a sub-Block are set to zeros in all other sub-Blocks, so that $\mathbf{A} = \sum_{v=1}^{V} \mathbf{A}^v$. Thus, all sub-Blocks have the same size, that equal to the original block size. Scrambling (rotating phase independently) in the PTS technique is applied to each sub-block. PTS in the OFDM transmitter is shown in Fig. 1.

Then the time-domain version of the v_{th} sub-block $\mathbf{x}^v$ is obtained by tacking the IFFT of A^v. Summation of time-domain versions of all sub-blocks produces the time-domain version of the original OFDM symbol. Due to the linearity property of IFFT, which ensure the following

$$\mathbf{x} = \sum_{v=1}^{V} \mathbf{x}^v = \mathbf{IFFT}(\mathbf{A}) = \mathbf{IFFT}\left(\sum_{v=1}^{V} A^v\right) = \sum_{v=1}^{V} \mathbf{IFFT}(A^v) \tag{6}$$

Reduction of the PAPR of the OFDM symbol takes place by multiplying each sub-Block $\mathbf{x}^v$ by an optimum phase factor $\hat{b}^v$ before combining these sub-Blocks or Partial Transmit Sequences where $|b^v| = 1, b^v = e^{j\phi^v}$ and $\phi^v \in [0, 2\pi]$. The optimum phase vector $\hat{\boldsymbol{b}}$ achieves the global minimum as per (7) so that the PAPR of $\mathbf{x}$ can be minimized.

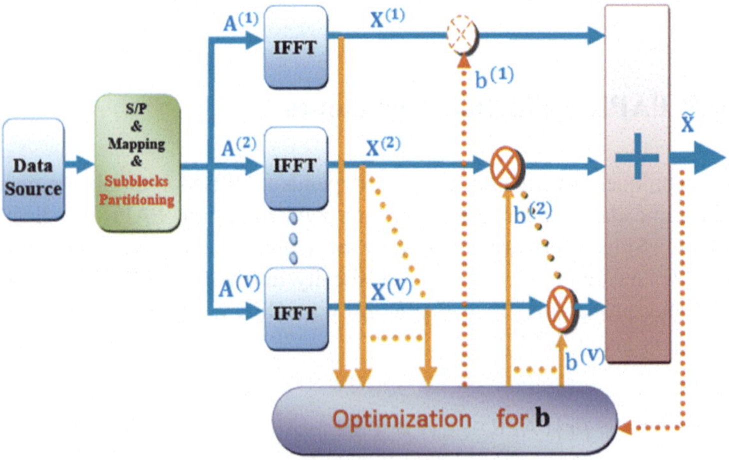

Fig. 1 Block diagram of PTS technique

$$\hat{\boldsymbol{b}} = \left[\hat{b}^1 \cdots \hat{b}^v \cdots \hat{b}^V\right] = \underset{\left[b^1 \quad \cdots \quad b^V\right]}{argmin} \left(\frac{Max_{n\in[0,N]} \left|\sum_{v=1}^{V} b^v \mathrm{x}^v[\mathrm{n}]\right|^2}{E\left\{\left|\sum_{v=1}^{V} b^v \mathrm{x}^v[\mathrm{n}]\right|^2\right\}} \right) \tag{7}$$

The time-domain signal after optimization and combining of partial transmit sequences is given by

$$\hat{\mathbf{x}} = \sum_{v=1}^{\mathrm{V}} \hat{b}^v \mathbf{x}^v = \sum_{v=1}^{\mathrm{V}} \hat{b}^v \mathbf{IFFT}(\boldsymbol{A}^v) \tag{8}$$

Where, $\hat{\mathbf{x}}$ is the optimum sequence that has the lowest PAPR among all other alternative transmit sequences.

In general, the selection of the phase factors $b^v|_{v=1}^{\mathrm{V}}$ are limited to a set with a finite number of elements to reduce the search complexity. The set of allowed phase factors is given by

$$\phi^v \in \left.\frac{2\pi l}{W}\right|_{\substack{0\leq l\leq W-1 \\ 1\leq v\leq V}} \tag{9}$$

where W is the number of allowed phase factor. Usually number of allowed phase factor W is set to 2 or 4, i.e., $b^v \in \{\pm 1\}$ or $b^v \in \{\pm 1, \pm \mathrm{j}\}$, respectively, to get a reduction in the required computational complexity, as no actual multiplications will be needed in (8). PTS technique requires $\log_2 W^{(\mathrm{V}-1)}$ side information bits if $b^1 = 1$ i.e., $\phi^1 = 0$ to reduce the computational complexity without loss of generality [8]. The amount of PAPR Reduction of PTS depends on

1. Number of Sub-Blocks V
2. Number of allowed phase factors W
3. Sub-Block Partitioning Scheme (SPS)

In literature, there are three SPS for PTS, namely Adjacent, Interleaved, and Pseudo-Random. As illustrated in Fig. 2. Although Pseudo-Random SPS has been stated to be the best SPS scheme for PTS, Adjacent SPS does not require any change in the already found standards (such as 802.16) if it used in their transmitters, only it requires at least one pilot in each sub-Block to allow the receiver to equalize the phase rotation of each sub-Block (as if it was a channel effect) rather than side information (i.e., ensure downward compatibility). The computational effort of PTS consists of the $\mathrm{IFFT_s}$ of the V partial sequences, the $W^{(V-1)}$ superpositions of all partial sequences, and the calculation of the PAPRs (metric) for selection. The computational complexity of PTS is then given by

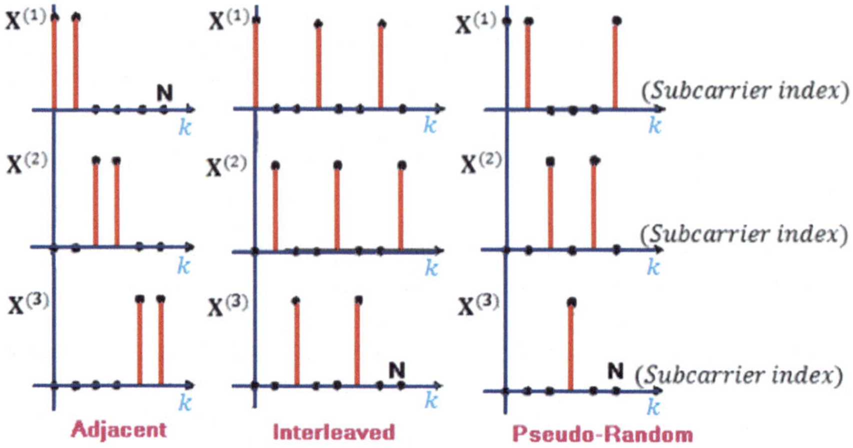

Fig. 2 Different Subblock Partitioning Schemes (SPS) of PTS

$$c_{PTS} = V \cdot c_{FFT} + W^{(V-1)} \cdot \left(c_{sp} + c_{met}\right) \tag{10}$$

The three components of the computational effort of PTS (*i.e.*, c_{FFT}, c_{sp}, and c_{met}) are given in Table 1 in terms of real additions and multiplications [4].

3.2 *Selective Mapping (SLM) Technique*

Selective Mapping (SLM) technique was introduced in [9] by Bauml, Fischer, and Huber. In this technique, a set of several alternative multicarrier signals for the same OFDM symbol are generated by modifying the phase of input symbol, then the signal with the lowest PAPR will be selected for transmission. The OFDM transmitter applying SLM is shown in Fig. 3. The input data block A is multiplied by a set of U different distinct phase rotation vectors, which are known at both transmitter and receiver. The u_{th} phase rotation vector $B^{(u)}$ has the same size as A i.e., $B^{(u)} = \left[B_0^{(u)}, B_1^{(u)}, \ldots, B_{N-1}^{(u)}\right]$, where $B_k^{(u)} = e^{j\phi_k^{(u)}}$ and $\phi_k^{(u)} \in [0, 2\pi)$. In general, $B_k^{(u)}$ are unit magnitude complex number selected from a binary or quaternary elements sets that is $\{\pm 1\}$or $\{\pm 1, \pm j\}$. $B^{(1)}$ is an all one vector to include the unmodified data block in the set of modified data blocks i.e., $x^{(1)} = x$. After multiplication we get a set of U alternative data blocks. The u_{th} alternative is given by $A^{(u)} = \left[A_0^{(u)}, A_1^{(u)}, \ldots, A_{N-1}^{(u)}\right]$ where $A_k^{(u)} = A_k B_k^{(u)}, 1 \le u \le U$, and $0 \le k \le N-1$. Then, all the U modified subcarrier vectors are transformed into time-domain to get $x^{(u)} = \text{IFFT}\left(B^{(u)} A^{(u)}\right)$. Finally the vector with the lowest

Table 1 Computational complexities of MSR techniques

	Real multiplication			Real addition			SI
	c_{FFT}	c_{sp}	c_{met}	c_{FFT}	c_{sp}	c_{met}	
PTS	$2NV \log_2 N$	0	$W^{(V-1)}2N$	$3NV \log_2 N$	$2N(V-1)W^{(V-1)}$	$W^{(V-1)}N$	$(V-1) \log_2 W$
SLM	$2NU \log_2 N$	–	2NU	$3NU \log_2 N$	–	NU	$\log_2 U$
RI	$2NK \log_2 N$	–	2NK	$3NK \log_2 N$	–	NK	$\log_2 K$

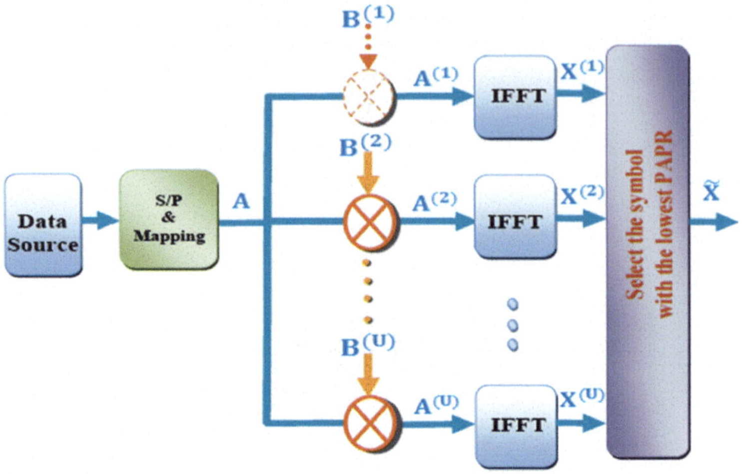

Fig. 3 Block diagram of SLM technique

PAPR $x^{(\hat{u})}$ is chosen to be transmitted where

$$\hat{u} = \underset{1 \le u \le U}{\text{argmin}} \left(\frac{\text{Max}_{n \in [0, N-1]} \left| x^{(u)}[n] \right|^2}{\boldsymbol{E}\left[\left| x^{(u)}[n] \right|^2 \right]} \right) \tag{11}$$

The amount of PAPR reduction achieved by SLM depends on

1. Number of phase sequences U.
2. Design of phase sequences.

SLM technique requires U IFFT operation per alternative symbol, and a $\lfloor \log_2 U \rfloor$ side information bits, to allow the receiver to know the index of the phase sequence used in the transmitter [8].

The computational effort of SLM consists of U calls of the IFFT and U calculations of the PAPRs (metric). In this case, the complexity is given by [10]

$$c_{SLM} = U \cdot (c_{FFT} + c_{met}) \tag{12}$$

where c_{FFT} and c_{met} are given in Table 1.

3.3 Interleaving Techniques

While it is well known that interleaving can be used to combat the effect of burst noise in error correction systems, Jayalath and Tellambura used interleaving to reduce the PAPR. By using $U - 1$ Interleavers, U alternative signal representations can be generated. The one with the lowest PAPR is selected for transmission. In the receiver, $\log_2 U$ SI bits are required to recover the data. There are two types of Interleaver that can be used in PAPR reduction, Random Interleaver (RI) and Periodic Interleaver, each one of them can be used in symbol level or bit level, as shown in Figs. 4 and 5.

In Random Interleaving a block of N symbols or bits is permuted in a pseudo-random order, while in Periodic Interleaving with period C a block of N symbols or bits is stored into the matrix $(C \times R)$ column by column and then read as row by row, where $R = N/C$. The computational effort of interleaving is similar to SLM, as shown in Table 1. In addition to that, bit level interleaver requires extra U mappers [11].

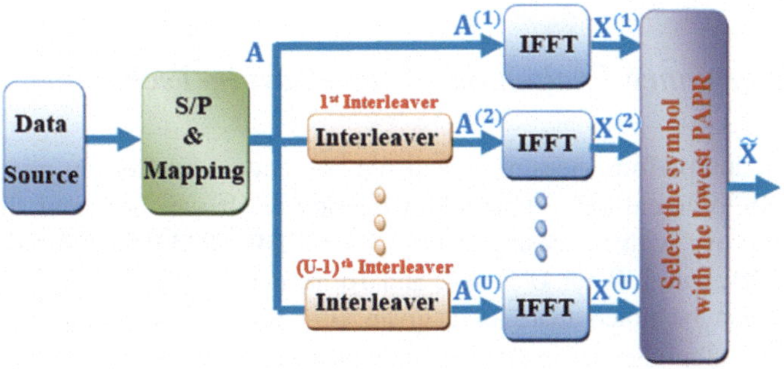

Fig. 4 Block diagram of symbol level interleaver

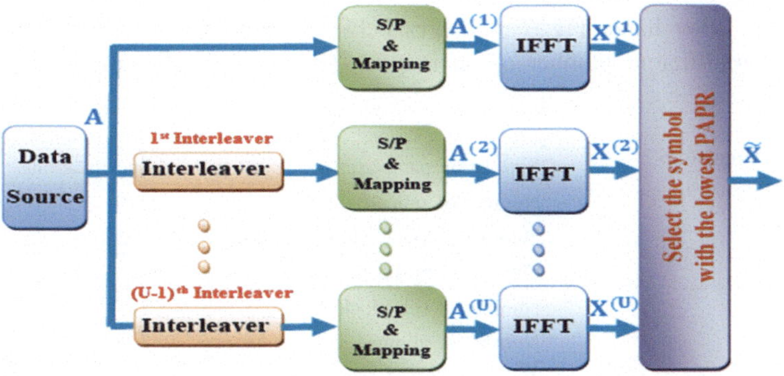

Fig. 5 Block diagram of bit level interleaver

4 Simulation and Results

In this section, computer simulations are performed to evaluate and compare the performance of the three MSR techniques.

Firstly, PAPR reduction performance of interleaving techniques is evaluated, and then it will be compared with the PAPR reduction performance of SLM, after that performance of PTS is evaluated in terms of different values for the different parameters of PTS. Finally, SLM and PTS are compared in terms of PAPR reduction performance and computational complexity.

The performance of PAPR reduction techniques is usually measured by means of CCDF versus PAPR in dB. The CCDF of all proposed schemes are compared with the case of no PAPR reduction technique is used. Simulation parameters are listed in Table 2.

4.1 Performance Evaluation of Interleaving Techniques

PAPR reduction performances of bit and symbol random interleavers are compared in Fig. 6 for different (U – 1) random interleaver. It can be noted that PAPR statistics improve with increasing U (at 0.1% $CCDF_c$, PAPR is reduced by about 1.2dB, 2.2dB, and 3dB for U = 2, 4, and 16, respectively).

Also, it can be noted that, for a given U, both bit interleaving and symbol interleaving offer the same PAPR reduction in the region $CCDF_c > 10^{-3}$. However, symbol interleaving doesn't improve the PAPR statistics after $CCDF_c = 10^{-3}$ while bit interleaving performs well in the region $CCDF_c < 10^{-3}$. This means that the number of statistically independent permutations is limited when symbol interleaving is performed.

On the other hand, Fig. 7 compares between random and periodic bit interleavers for different number of alternatives $(U = 1, 3, 7)$. Results indicate that

Table 2 Simulation parameters

Simulation parameters	Specifications
Number of subcarriers $\boldsymbol{N}$	256
Number of Data subcarriers N_{DSC}	192
Oversampling Factor $\boldsymbol{L}$	4
Number of sub-blocks $\boldsymbol{V}$	2 *to* 10
Number of allowed rotation factors $\boldsymbol{W}$	2 or 4
Phase rotation factors $(\mathbf{b}^{v})$	$\in \{\pm 1\}\, or \{\pm 1, \pm j\}$
Number of alternatives $\boldsymbol{U}$	2 *to* 512
Phase rotation vectors $\mathbf{B}^{(u)}$	$\in \{\pm 1, \pm j\}$
Modulation scheme	16-QAM

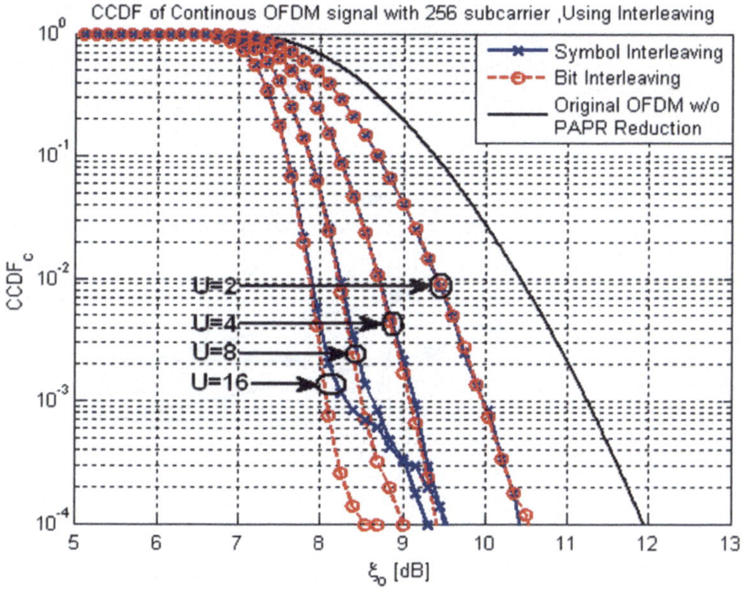

Fig. 6 Comparison between PAPR reduction performance of symbol interleaving and bit interleaving for different numbers of random interleavers

periodic interleaver does not give satisfactory PAPR reduction performance as random interleaver.

4.2 Performance Evaluation of SLM Technique

PAPR reduction Performance of SLM technique is shown in Fig. 8, as can be seen, increasing numbers of phase sequences U (i.e., increasing number of alternatives of OFDM symbol) will enhance the PAPR reduction performance of SLM technique. However, an excessive increase in the number of alternatives is not efficient as the enhancement in PAPR reduction gain becomes small and not comparable with the increase in the computational complexity and numbers of SI bits.

A comparison between SLM and random bit interleaving for the same number of IFFT U (i.e., the same complexity if SLM restrict their phase shifts to $\{\pm 1, \pm j\}$ so that no additional multiplication is required) and the same amount of redundancy bits is shown in Fig. 9 which implies that performance of SLM and Random bit interleaving is coincident in the region $CCDF_c > 10^{-3}$. While SLM performs better than random bit interleaving in the region $CCDF_c < 10^{-3}$. Where the maximum PAPR reduction of random bit interleaving occurs at 0.1% $CCDF_c$ (i.e., interleaving technique cannot reduce high PAPR values). Although random bit

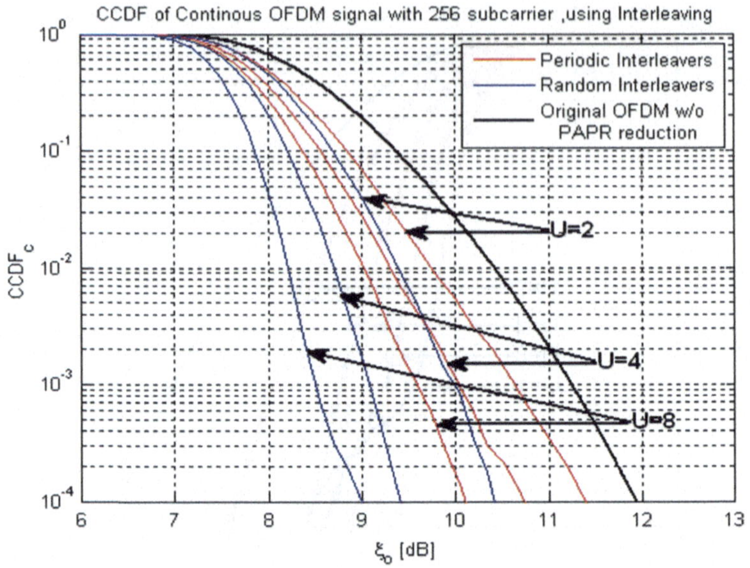

Fig. 7 Comparison between PAPR reduction performance of Periodic interleaving and random interleaving for different numbers of (K-1) interleavers

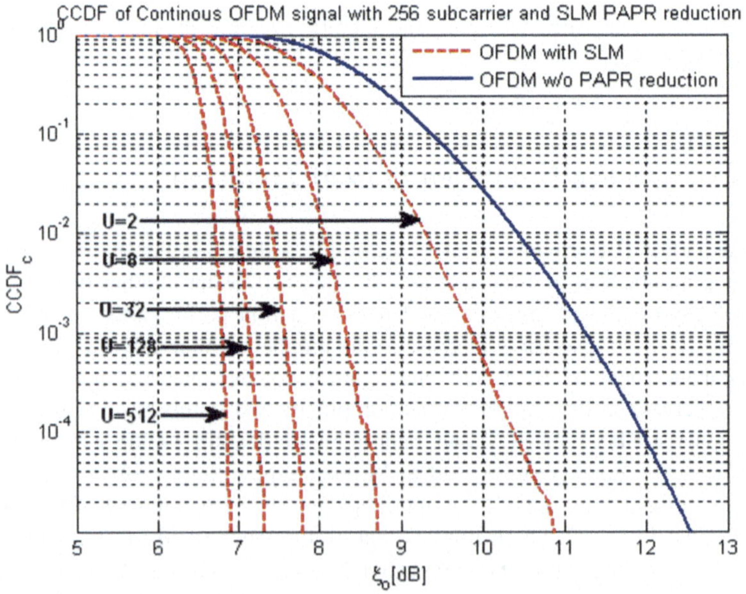

Fig. 8 PAPR reduction performance of SLM technique, with different number of phase sequences

interleaving requires extra U − 1 mappers and serial to parallel converters compared with SLM.

4.3 Performance Evaluation of PTS Technique

Performance of PTS depends on three parameters; the number of sub-blocks (V), number of allowed phases (W), and SPS. Changing V or W can be used to obtain the same number of alternatives (I). But, what is the best, increasing (V) or (W)? Increasing (V) is better than increasing (W) as shown in Fig. 10. On the other hand, Fig. 11 shows that pseudo-random SPS is the best SPS among others. Increasing number of sub-blocks (V) will enhance the PAPR reduction performance of PTS technique as shown in Fig. 12. However, increasing number of sub-blocks (V) excessively is not efficient, as the enhancement in PAPR reduction gain becomes small and not comparable with the increase in the computational complexity and numbers of SI bits.

For the same number of alternatives SLM and PTS (with pseudo-random SPS) perform the same as shown in Fig. 13. But, which is the best? Figure 14 compares the computational complexity of SLM and PTS for the same number of alternatives, based on (10) and (12). It can be noted that for the same number of alternatives SLM is more complex (in terms of required numbers of additions and

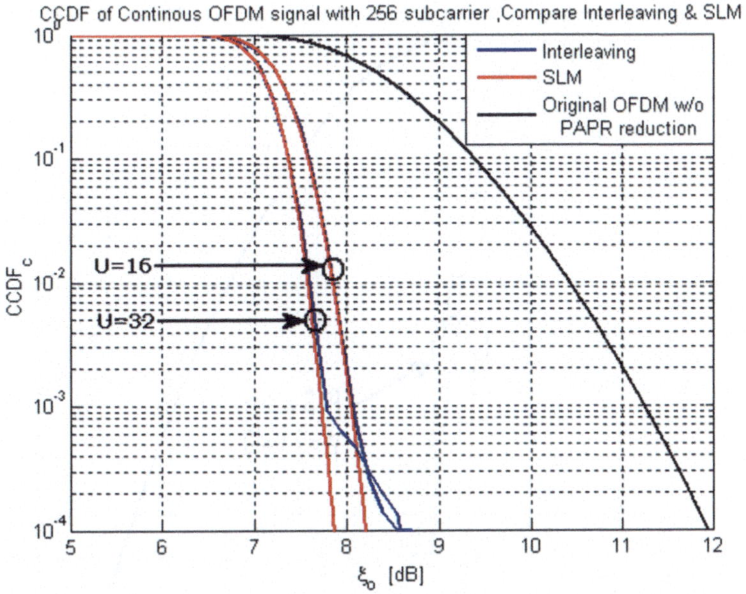

Fig. 9 Comparison between PAPR reduction performance of SLM and random interleaving (bit level), for the same number of alternatives

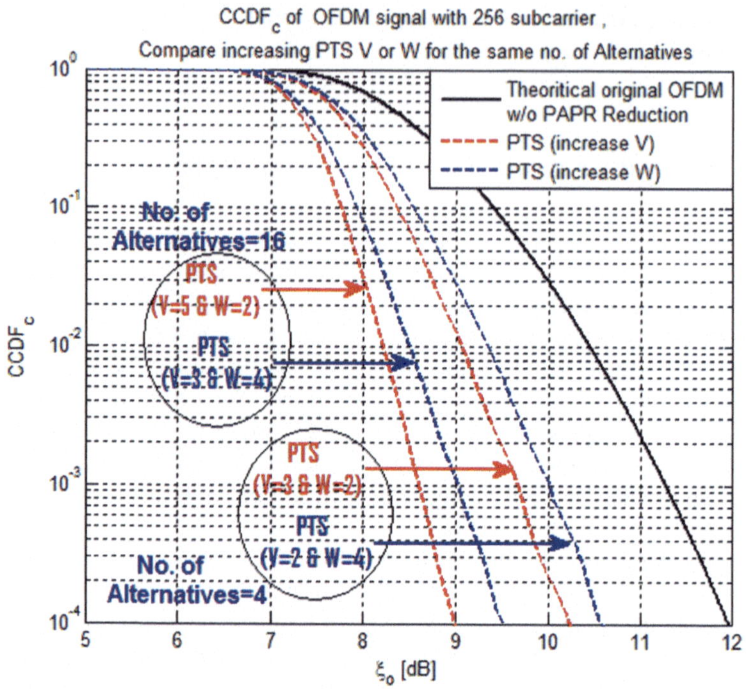

Fig. 10 Increase V or W for the same number of alternatives?

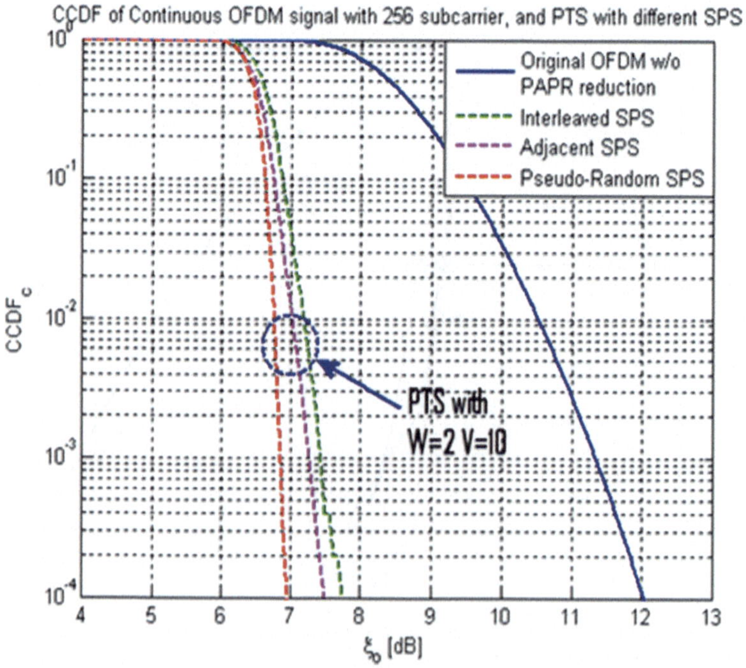

Fig. 11 PAPR reduction performance of different SPS of PTS

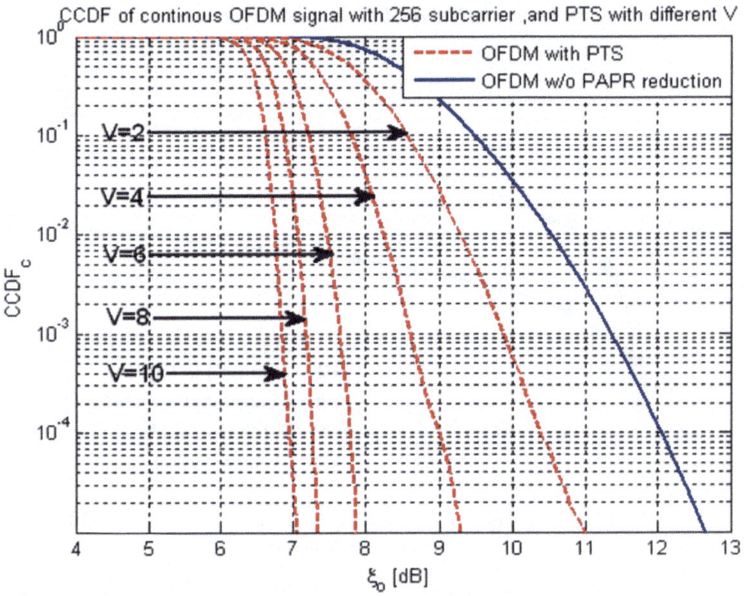

Fig. 12 Comparison between PAPR reduction performance of SLM and random interleaving (bit level), for the same number of alternatives

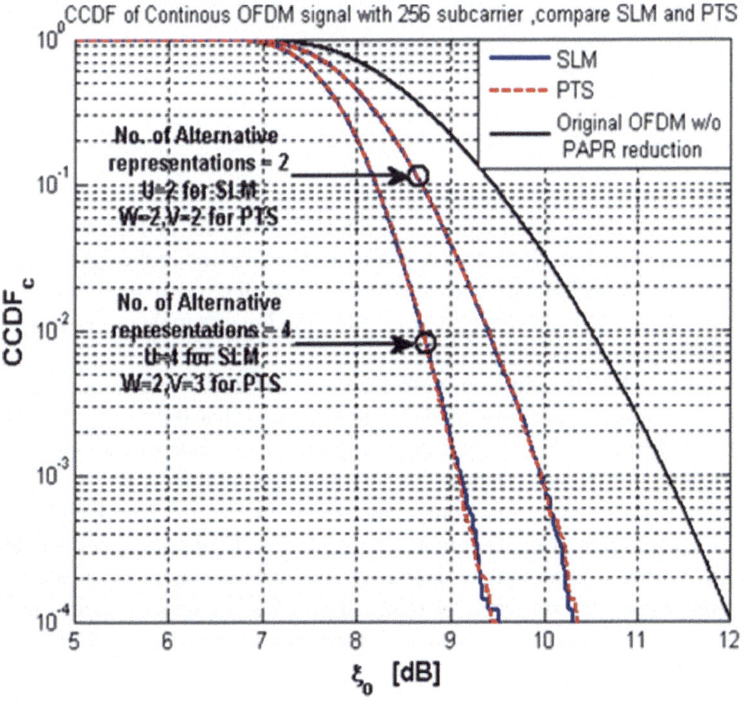

Fig. 13 Comparison between PAPR reduction performance of SLM and PTS, for the same number of alternatives

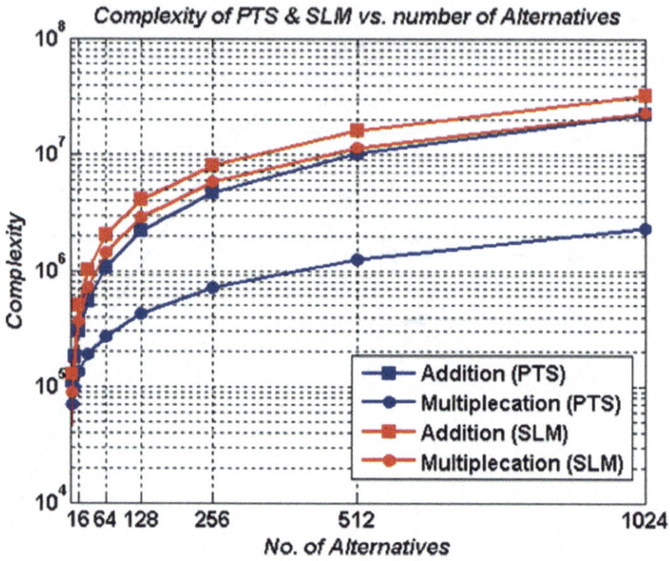

Fig. 14 Comparison between the computational complexity of SLM and PTS, for the same number of alternatives

multiplications) than PTS. Finally, it can be said PTS is the best MSR technique among others.

5 Conclusions

OFDM is an attractive technique for IoT-based networks in frequency selective channels. However, high PAPR is its main drawback. Among the PAPR reduction techniques, MSR techniques provide high PAPR reduction gain with a small number of SI bits. But, they require high computational complexity to provide this high PAPR reduction gain.

In this paper, PAPR reduction performance of three MSR techniques (i.e., PTS, SLM, and interleaving) is compared. Results showed that for the same number of alternatives SLM provides PAPR reduction performance better than that of random bit interleaving. However, random bit interleaving requires extra U − 1 mappers and serial to parallel converters compared with SLM. On the other hand, SLM and PTS with pseudo-random SPS provide the same PAPR reduction performance for the same number of alternatives. However, for the same number of alternatives, SLM is more complex (in terms of required numbers of additions and multiplications) than PTS. Finally, it can be concluded that PTS is the best MSR technique.

References

1. Cho, L., X., Yu, C., Chen, C., Hsu. 2018. Green OFDM for IoT minimizing IBO subject to a spectral mask. In *4th international conference on applied system invention (ICASI)*, 5–8. Chiba: IEEE. https://doi.org/10.1109/icasi.2018.8394252.
2. Langlais, C., S., Haddad, Y., Louet, N., Mazouz. 2011. Clipping noise mitigation with capacity approaching FEC codes for PAPR reduction of OFDM signals. In *8th international workshop on multi-carrier systems & solutions*, 1–5. Herrsching: IEEE. https://doi.org/10.1109/mc-ss.2011.5910733.
3. Jayalath, A.D.S., C., Tellambura. 2002. Peak-to-average power ratio of orthogonal frequency division multiplexing (Technical report 2002/116). School of computer science and software engineering, Monash University.
4. Youssef, M. I., I.F., Tarrad, M., Mounir. 2016. Performance evaluation of hybrid ACE-PTS PAPR reduction techniques. In *11th international conference on computer engineering & systems (ICCES)*, 407–413. Cairo: IEEE. https://doi.org/10.1109/icces.2016.7822039.
5. Wei, S., D. L., Goeckel, P. E., Kelly. 2002. A modern extreme value theory approach to calculating the distribution of the peak-to-average power ratio in OFDM systems. In *2002 IEEE international conference on communications (ICC 2002)*, vol. 3, 1686–1690. New York: IEEE. https://doi.org/10.1109/ICC.2002.997136.
6. Tellambura, C., M., Friese. 2006. Peak power reduction techniques. In *Orthogonal frequency division multiplexing for wireless communications*, eds. Li, Y., G. L., Stuber, 199–224. Berlin: Springer. https://doi.org/10.1007/0-387-30235-2_6.
7. Müller, S. H., J. B., Huber. 1997. OFDM with reduced peak–to–average power ratio by opti-mum combination of partial transmit sequences. *Electronics Letters* 33(5), 368–369. https://doi.org/10.1049/el:19970266.
8. Zahra, M. M., I. F., Tarrad, M. Mounir. 2014. Hybrid SLM-PTS for PAPR reduction in MIMO-OFDM. *International Journal of Advanced Research in Electrical, Electronics and Instrumentation Engineering* 3(5), 9313–9324.
9. Bauml, R. W., R. F. H., Fischer, J. B., Huber. 1996. Reducing the peak to average power ratio of multicarrier modulation by selected mapping. *Electronics Letters* 32(22), 2056–2057. https://doi.org/10.1049/el:19961384.
10. Siegl, C., R. F., Fischer. 2008. Partial transmit sequences for peak-to-average power ratio reduction in multiantenna OFDM. *EURASIP Journal on Wireless Communications and Networking* 1–11. https://doi.org/10.1155/2008/325829.
11. Jayalath, A. D. S., C., Tellambura. 2000. Reducing the peak-to-average power ratio of orthog-onal frequency division multiplexing signal through Bit or symbol interleaving. *Electronics Letters* 36(13), 1161–1163. https://doi.org/10.1049/el:20000822.

A Framework for Stroke Prevention Using IoT Healthcare Sensors

Noha MM. AbdElnapi, Nahla F. Omran, Abdelmageid A. Ali and Fatma A. Omara

Abstract Stress and depression are negative emotions which could have impact on the persons' life when repeated for long time. There are several systematic studies showing that these emotions promote chronic diseases, such as stroke, hypertension, heart attack, high blood pressure (BP), digestive diseases, and cardiovascular diseases, reduce immune functions, and so on. Therefore, emotions analysis and management represent most attractive sides of recent emerging of internet of medical things (IoMT) applications and smart wearable sensors. The work in this paper aims to use IoMT technology to recognize stress level using measured data from wearable sensors and show how stress may pose risks on the human health and may lead to chronic diseases such as stroke. Moreover, an effective system called—EmoStrokeSys—has been proposed for health monitoring that combines three wearable sensors: Glucometer sensor to measure glucose level, iHealth wireless blood pressure to measure BP and heart rate, and GSR sensor to detect stress level. The main goals of the proposed EmoStrokeSys system are to control all outcomes of the mentioned parameters under stress situation and minimize the risk of failure by closely monitoring an individual level with regular checkups for these parameters. The measured data from wearable sensors has found significant effects of continuous chronic stress which could reverse changes in BP, blood glucose, and heart rate and may lead to stroke.

Keywords Bioinformatics · Healthcare · Human emotion · Internet of things (IoT) · Mobile and wireless health · Stress · Stroke

N. MM. AbdElnapi (✉)
Faculty Computer Science, Nahda University, Beni Suef, Egypt
e-mail: Noha.mohamed@nub.edu.eg

N. F. Omran
Department of Mathematics, Faculty of Science, South Valley University, Qena, Egypt

A. A. Ali
Faculty of Computers and Information, Minia University, Minia, Egypt

F. A. Omara
Faculty of Computers and Information, Cairo University, Cairo, Egypt

A. Z. Ghalwash et al. (eds.), *Internet of Things—Applications and Future*,
Lecture Notes in Networks and Systems 114,
https://doi.org/10.1007/978-981-15-3075-3_12

1 Introduction

IoT technologies and its applications are considered the most attractive aspects for the healthcare. It plays a dynamic role in serving patients by making healthcare more inexpensive, accessible, and available [1]. Moreover, it has the potential to give rise to many medical applications such as remote health monitoring to prevent chronic diseases, fitness programs, and emotion recognition, as shown in Fig. 1 [2]. The emergency of IoT in the healthcare field produces new terminology by integrating the internet and smart medical devices. This new technology is known as internet of medical things (IoMT), which is used to collect and share health data between themselves and cloud [3, 4].

A stroke is a medical definition of "brain attack," which means a sudden death of brain cells due to lack of oxygen which causes blockage of blood flow or ruptures an artery to the brain. It may also occur when a blood vessel bursts and leaks blood into the brain, causing damage. In addition, lack of blood pumped to the brain due to heart attack may also cause a stroke [5, 6]. There are two categories of stroke: ischemic stroke and hemorrhagic stroke. Ischemic stroke is considered the most common type, where 80% of strokes types are ischemic. In this category, a part of brain cells begins to die due to the stopping of oxygen and nutrients from getting to brain because a blood clot blocks a blood vessel in the brain. Hemorrhagic stroke causes bleeding into or around the brain (see Fig. 2a, b) [7].

Although hemorrhagic stroke is considered less common, but it is considered more deadly than the ischemic stroke. It is more difficult to treat hemorrhagic stroke

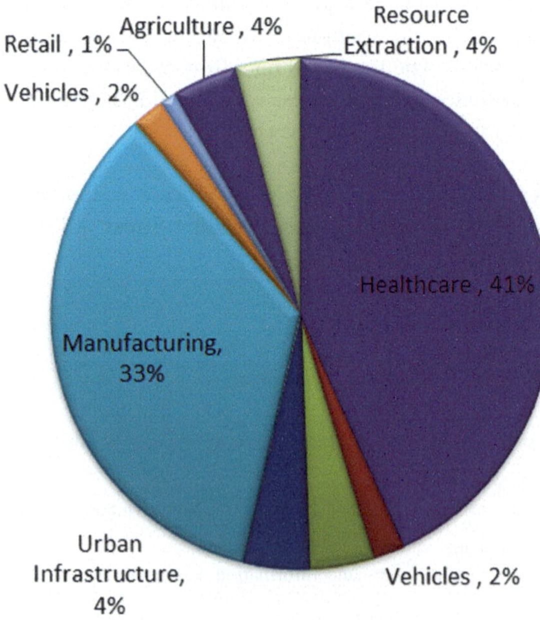

Fig. 1 Projected market share of dominant IoT applications [2]

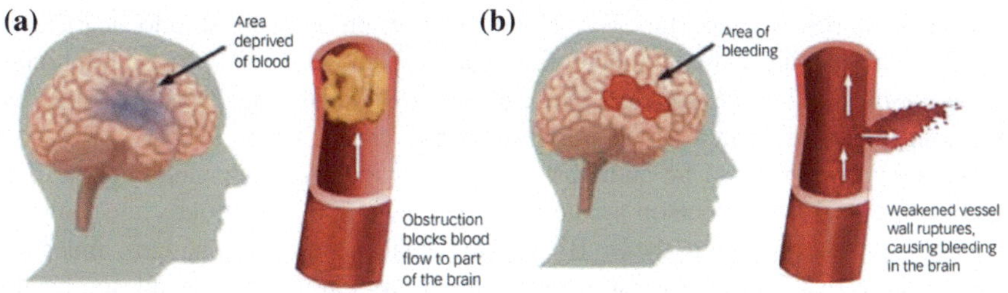

Fig. 2 The two types of stroke: **a** ischemic stroke; **b** hemorrhagic stroke [7]

because the treatment may be by surgery to stop bleeding or lower pressure in the brain. Sometimes, medications are used to control blood pressure and brain swelling [8, 9].

On the other hand, stroke is considered the second leading cause of death and occurs at any time and at any age. It is associated with some modifiable and non-modifiable risk factors (see Fig. 3) [8]. The modifiable risk factors are the most common reasons for stroke. These factors can be controlled through pharmacological or surgical interventions and lifestyle changes [5]. Therefore, several IoT healthcare systems and applications have been developed to recognize human emotions and bio-signals [10]. During the few last years, scientific community and industry has been giving a great attention to designing and development of wearable biosensors and smart medical devices [11, 12].

The work in this paper aims to extract the relationship between stress and other health parameters, such as high BP (> 20/80 mmHg), high blood glucose (>125, HBA1C > 5.6), and accelerated pulse which may lead to stroke using IoT technologies in healthcare. Therefore, the concept of extending the self-reporting

Fig. 3 List of modifiable and non-modifiable risk factors [8]

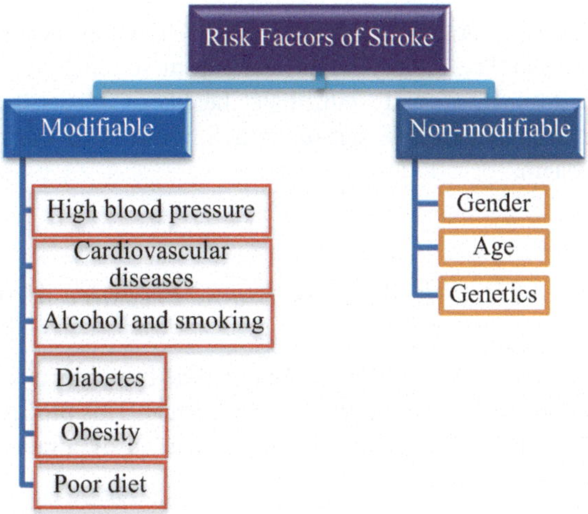

control of the bio-signals (i.e., BP, blood glucose, emotion level, and heart rate) considering how individuals and professionals use modern IoT with smartphones and wearable medical devices to capture these bio-signals. It has been presented by proposing a system called EmoStrokeSys for preventing stroke.

Stroke prevention can be summarized by knowing and understanding the risk factors that increase a person's likelihood of developing a stroke, and how to manage or possibly avoid them. By making healthy lifestyle changes and controlling these risk factors, a person can avoid stroke. On the other hand, the health specialists classify stroke prevention into two types; (1) *primary prevention*; it is the prevention of a first-ever stroke and (2) *secondary prevention*; it applies to individuals who have had (or survived) a stroke, thus preventing future attacks [19].

According to this work the proposed framework had designed for the first type "*primary prevention*."

The proposed system collects and stores data from three wearable sensors: iHealth smart glucometer to measure glucose level in blood, iHealth wireless blood pressure monitor to measure BP and heart rate, and GSR to detect stress level. One of the main goals of this system is to provide maximum convenience to the user or patient during sensors measurements, especially for prolonged use.

The rest of this paper is organized as follows: A brief overview of the most important smart medical devices and sensors related to this work is presented in Sect. 2. Section 3 discusses the related studies. The important points and some mathematical relations between the measured parameters are presented in Sect. 4. The structure of the proposed EmoStrokeSys system and the experimental measurements are demonstrated in detail in Sect. 5. Finally, the paper is concluded and future work is presented in Sect. 6.

2 Smart Medical IoT Devices

The most important advantages of IoT within healthcare systems are availability, accessibility, ability to provide a more "individual" system, and high-quality cost-effective, and immediate healthcare. This section summarizes the three important wearable sensors which are used to form the proposed EmoStrokeSys system.

2.1 IHealth Wireless Blood Pressure Monitor

Blood pressure is the pressure outcome from blood pumping by the heart around the whole body in the arteries, when it beats. This process produces a force that creates pressure on the arteries and it is recorded by a ratio of two numbers [13]:

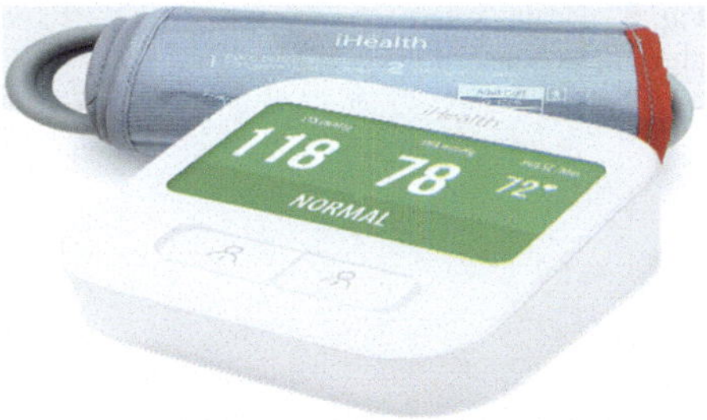

Fig. 4 iHealth wireless blood pressure monitor sensor [13]

- ***Systolic*** (the upper number)—represents the value of blood pressure against the walls of the arteries. It is recorded during heartbeat (when the heart muscle contracts) and can be measured as mL per beat. The optimal value is 120.
- ***Diastolic*** (the lower number)—represents the value of blood pressure against the walls of the arteries while the heart rests between the pulses. It is recorded between the heart rate (when the heart stops and is filled with blood). The optimal value is 80.

In this work, iHealth wireless blood pressure monitor has been used to measure live heart rate data and BP data. It composes a simple smart sensor to measure heart rate and medical device for measuring blood pressure (see Fig. 4) [14].

2.2 IHealth Smart Glucometer Sensor

The key element for energy production in the people's life is blood glucose. In the first few days of person's life during neonatal and postnatal periods, the concentrations of blood glucose are kept within the range of 3.5–5.5 mmol/L. After that, hormones which control glucose production and utilization try to keep its concentrations within this normal range by a complex interplay [15]. These hormones include insulin, glucagon, cortisol, epinephrine, norepinephrine, and growth hormone. Monitoring glucose level in the body is one of the important parameters in the proposed system. It can be done using iHealth smart glucometer sensor (see Fig. 5) [16].

2.3 Galvanic Skin Response (GSR) Sensor

GSR is used to measure and detect the different conductance of the skin when a person is under stress or in normal case. It sends different data from the two electrodes which are placed on the fingers to a coordinator via ZigBee and, then sends this data to a mobile application [17]. The main idea of GSR sensor is that strong stress emotions can cause stimulus to sympathetic nervous system, so sweat

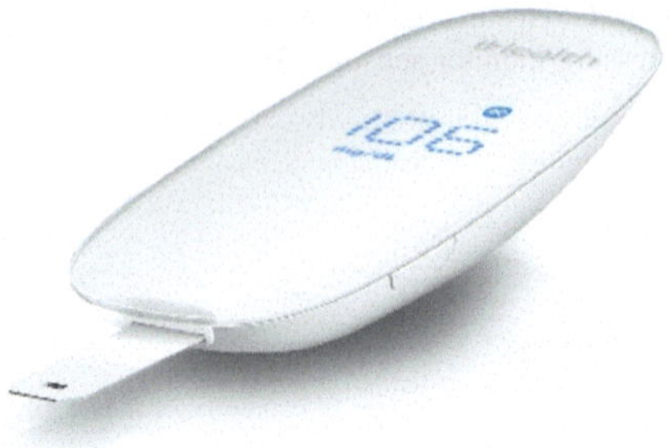

Fig. 5 iHealth smart glucometer sensor [13]

Fig. 6 Galvanic skin response (GSR) sensor [17]

glands secrete more sweat and make change in the electrical properties of the skin. So it allows us to spot such strong emotions by simply attaching the two electrodes to two fingers on one hand (see Fig. 6) [18, 19].

3 Related Works

A lot of studies have been done to show the relationship between IoT and healthcare to improve quality of health.

Villarejo et al. [20] have designed and built a stress sensor to detect and control different emotions in diabetes. They used ZigBee as a bridge between the two electrodes which are placed on the fingers and a coordinator to share different emotional data. The measurement results have demonstrated that increasing blood glucose level will increase GSR resistance value. Picano et al. [18] have presented prospective research study of stress echocardiography (SE) applications, in and beyond coronary artery disease (CAD) by considering a variety of signs in addition to regional wall motion abnormalities. Mohanraj and Narayanan [19] have proposed a method that can be used as a diagnostic tool in the detection of diabetic autonomic neuropathy by comparing the sympathetic skin response (SSR) and galvanic skin resistance (GSR) values in males with type 2 diabetes. The nervous system and the autonomic nervous system are affecting with changes in various metabolic pathways, and causes a metabolic disorder, which is known as diabetes mellitus (DM). Diabetes is categorized into two main types, namely, Type 1 DM and Type 2 DM. Type 1 DM occurs more predominantly during childhood and Type 2 DM is majorly found within the people with the age group of above 35 [21]. Esler [22] has presented a review about the relationship between sympathetic nervous stress responses with blood pressure and cardiovascular disease. Goshvarpour et al. [23] have introduced a system based on electrocardiogram (ECG) and galvanic skin response (GSR) to examine the effectiveness of matching pursuit (MP) algorithm in emotion recognition.

4 Mathematical Relations Between the Measured Bio-Signals

According to the work in this paper, it is found that significant effects of emotional state, especially continuous chronic stress could reverse changes in BP and heart rate which may lead to stroke. This can be explained by some psychological theories and concepts. This section discusses the obtained results from the proposed sensors using theories and mathematical relations. There is a simple mathematical remodel primarily based on BP that explains how BP is affected by heart rate, stroke quantity, and total blood vessel resistance [24, 25].

The equation which explains the relation between BP, blood flow (Bf), and blood vessel resistance (R) is as follows [24]:

$$\mathrm{BF} = \mathrm{BP}/\mathrm{R} \tag{1}$$

and the equation which explains the relation between cardiac output (CO), stroke volume (SV), and heart rate (HR) is as follows:

$$\mathrm{CO} = \mathrm{SV} \times \mathrm{HR} \tag{2}$$

Equating (1) and (2) we get

$$\mathrm{BP} = \mathrm{SV} \times \mathrm{HR} \times \mathrm{R} \tag{3}$$

According to (3), it is found that when the blood vessel resistance (R) is high, BP will increase, and when the heart rate (HR) is high, BP will increase. Moreover, stroke volume will be increased when BP is high [26].

5 The Proposed EmoStrokeSys System

Cloud computing is a kind of internet-based computing, where shared data is provided with computers and other devices on-demand [27–30]. The proposed EmoStrokeSys system combines the internet and mobile applications, and offers opportunities to enhance disease prevention and management by expanding health fast interventions beyond conventional care. It generates various streams of health bio-signals which locally analyze to build a user profile and send to private cloud (see Fig. 7).

To evaluate the performance of the proposed EmoStrokeSys system, a dataset has been measured using the system from 10 healthy volunteers (four men, six women) of various ages (min = 19; max = 53; mean = 36) during stress and relax situations in their home and work. Relevant personal habits like smoking, alcoholism, details about medication, demographic details, and basal vital parameters have been taken into consideration. The collected data about main modifiable factors of stroke (i.e., BP, HR, glucose level, emotion level) in the case of relax situation is listed in Table 1 and illustrated in Fig. 8. In the case of stress situation, the measured data is listed in Table 2 and illustrated in Fig. 9.

Standard deviation (SD) is a statistical measure of how spread out values are, its symbol is σ (Sigma), and its formula is [31, 32]:

$$\sigma = \sqrt{\frac{\Sigma(X-\mu)^2}{N}}$$

where X denotes each separate number, μ denotes the mean overall numbers, N is the number of measurements, and $\sum$ denotes a sum.

BP, emotion, glucose level, and heart rate values have been analyzed for each volunteer. Besides presenting the measurements in the two situations (i.e., stress and relaxation), the SD for each volunteer is calculated to describe differences between the measurements results in relaxation and stress in order to show the effect of stress on the factors of stroke. According to the measurements in Tables 1 and 2, it is clear that volunteers 1 and 8 are more expected to have a stroke if chronic stress

Fig. 7 Architecture of the proposed platform based on IoT technologies

Table 1 Relax (low arousal) situation

ID	Age	Gender	BP (mm Hg)		HR (bpm)	GSR	Glucose level (mg/dl)	SD
			Upper	Lower				
1	53	M	120	77	84	1.611	126	44.43198275
2	35	F	120	80	95	0.823	106	41.87918973
3	30	M	107	72	101	1.621	97	38.92262918
4	27	F	118	78	119	1.406	98	43.42467103
5	26	M	116	77	150	0.793	115	51.01455769
6	22	F	120	80	110	1.557	101	42.57409682
7	21	M	121	82	130	0.958	100	46.04781995
8	21	F	120	80	127	0.876	120	47.35226521
9	21	F	119	80	109	1.138	91	41.71860792
10	19	F	115	79	115	1.001	74	41.65728844

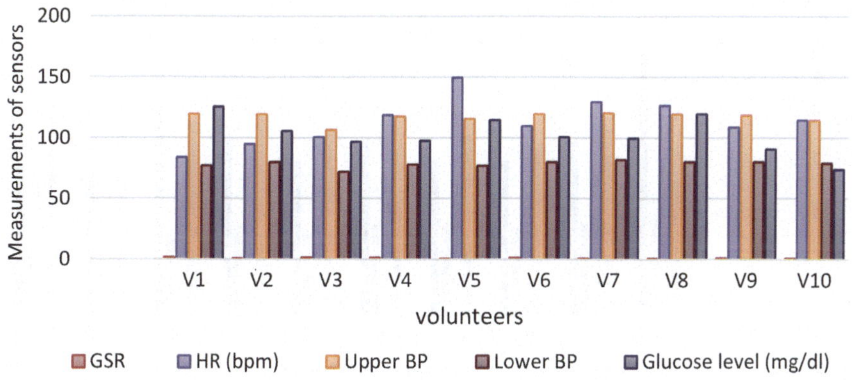

Fig. 8 Relax (low arousal) situation

continues for long periods of time (see Fig. 10). So, the medical specialists will send back an alarm to these volunteers (i.e., V1, V8) immediately.

According to the measured results in Tables 1 and 2, it is noticed that the GSR values were higher in the stress situation which implies that the skin resistance is high during stress.

Therefore, the proposed system can play an important role in immediately detecting stroke pathological signatures and control the risk of these factors.

Table 2 Stress (high arousal) situation

ID	Age	Gender	BP (mm Hg)		HR (bpm)	GSR	Glucose level (mg/dl)	SD
			Upper	Lower				
1	53	M	145	95	130	2.829	250	79.84335989
2	35	F	137	89	183	1.515	167	65.36907553
3	30	M	130	87	160	3.63	150	57.04127895
4	27	F	135	88	173	1.018	150	60.91518343
5	26	M	138	88	110	1.052	170	57.21671183
6	22	F	125	79	181	2.143	149	62.19763526
7	21	M	139	89	190	2.112	169	67.07737448
8	21	F	128	78	187	3.123	200	72.62813078
9	21	F	126	83	191	2.369	149	64.26672129
10	19	F	127	79	200	3.097	145	66.33233645

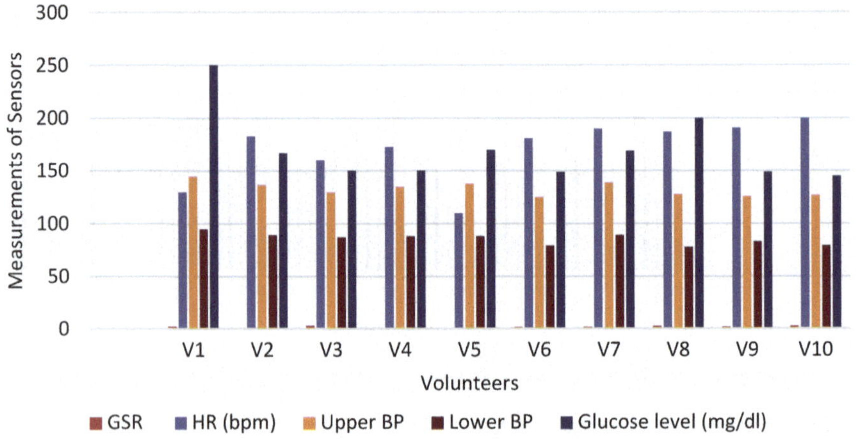

Fig. 9 Stress (high arousal) situation

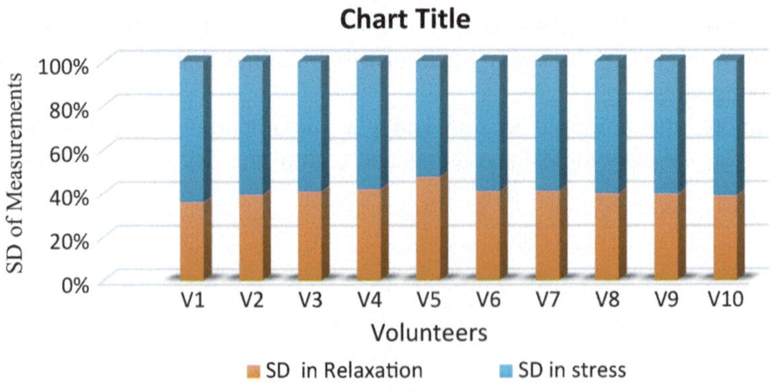

Fig. 10 The SD between the relax and stress situations

6 Conclusion

IoT healthcare systems and applications have been developed to detect and monitor bio-signals. The work in this paper presents a system consisting of three medical devices which can be used to monitor some of these bio-signals. Remote home monitoring after stress situation prevent the rise of the mentioned risk factors (i.e., BP, blood glucose, stress level, and heart rate) and better control of them by increasing measurements and awareness and uncovering undetected high risk factors, which will prevent strokes happens. In spite of these results, we discover some noise associated with the emotional signal generated by GSR sensor. Moreover, this sensor doesn't work well with some people because of the different nature of their skin resistance and conductance. In the future, we will include family history of cardiovascular disease to the proposed work and change the GSR sensor to get accurate results. In addition, there are different open source frameworks that introduce virtual health data such as mHealth platform, and we will try to compare the obtained results of the proposed EmoStrokeSys system and virtual data.

Acknowledgments The authors would like to thank the volunteers and members of the Smart Portable Clinical group for providing their sensors.

References

1. Siegel, J.E., S. Kumar, and S.E. Sarma. 2018. The future internet of things: secure, efficient, and model-based. *IEEE Internet of Things Journal* 5 (4): 2386–2398.
2. AbdElnapi, N.MM., N.F. Omran, A.A. Ali, and F.A. Omara. 2018. A survey of internet of things technologies and projects for healthcare services. In *Innovative Trends in Computer Engineering (ITCE), 2018 International Conference on*, 48–55. IEEE.
3. Lin, J., W. Yu, N. Zhang, X. Yang, H. Zhang, and W. Zhao. 2017. A survey on internet of things: Architecture, enabling technologies, security and privacy, and applications. *IEEE Internet of Things Journal* 4 (5): 1125–1142.
4. Zhao, P., W. Yu, X. Yang, D. Meng, and L. Wang. 2017. Buffer data-driven adaptation of mobile video streaming over heterogeneous wireless networks. *IEEE Internet of Things Journal*.
5. Vijayan, M., and P.H. Reddy. 2016. Stroke, vascular dementia, and Alzheimer's disease: molecular links. *Journal of Alzheimer's Disease* 54 (2): 427–443.
6. Warlow, C., et al., 2002. *Stroke: a practical guide to management*. Blackwell Science.
7. Wright, J.H., G.K. Brown, M.E. Thase, and M.R. Basco. 2017. *Learning cognitive-behavior therapy: An illustrated guide*. American Psychiatric Pub.
8. Billinger, S.A., et al. 2014. Physical activity and exercise recommendations for stroke survivors: a statement for healthcare professionals from the American Heart Association/ American Stroke Association. *Stroke* 45 (8): 2532–2553.
9. Bhogeshwar, S.S., M. Soni, D. Bansal. 2019. Study of structural complexity of optimal order digital filters for de-noising ECG signal. *International Journal of Biomedical Engineering and Technology* 29(2): 101–133.
10. López, J.M., R. Gil, R. García, I. Cearreta, and N. Garay. 2008. Towards an ontology for describing emotions. In *World summit on knowledge society*, 96–104. Springer.

11. Gatzoulis, L., and I. Iakovidis. 2007. Wearable and portable eHealth systems. *IEEE Engineering in Medicine and Biology Magazine* 26 (5): 51–56.
12. Pantelopoulos, A., and N.G. Bourbakis. 2010. A survey on wearable sensor-based systems for health monitoring and prognosis. *IEEE Transactions on Systems, Man, and Cybernetics Part C (Applications and Reviews)* 40 (1): 1–12.
13. http://www.heart.org/HEARTORG/Encyclopedia/Heart-Encyclopedia_UCM_445084_Encyclopedia.jsp?title=blood%20pressure.
14. https://thewirecutter.com/reviews/best-blood-pressure-monitors-for-home-use/.
15. Sonksen, P., and J. Sonksen. 2000. Insulin: understanding its action in health and disease. *British Journal of Anaesthesia* 85 (1): 69–79.
16. Güemes, M., S.A. Rahman, and K. Hussain. 2016. What is a normal blood glucose? *Archives of Disease in Childhood* 101 (6): 569–574.
17. Sharma, M., S. Kacker, and M. Sharma. 2016. A brief introduction and review on galvanic skin response. *International Journal of Medical Research Professionals* 2: 13–17.
18. Picano, E., et al. 2017. Stress echo 2020: the international stress echo study in ischemic and non-ischemic heart disease. *Cardiovascular ultrasound* 15 (1): 3.
19. Mohanraj, S., and K. Narayanan. 2016. Sympathetic skin response and galvanic skin resistance in males with type 2 diabetes mellitus. *Journal of Evidence Based Medicine and Healthcare* 3: 2544.
20. Villarejo, M.V., B.G. Zapirain, and A.M. Zorrilla. 2012. A stress sensor based on Galvanic Skin Response (GSR) controlled by ZigBee. *Sensors* 12 (5): 6075–6101.
21. Snekhalatha, U., T. Rajalakshmi, C. Vinitha Sri, G. Balachander, and K. Shankar. 2018. Non-invasive blood glucose analysis based on galvanic skin response for diabetic patients. *Biomedical Engineering: Applications, Basis and Communications* 30(02): 1850009.
22. Esler, M. 2017. Mental stress and human cardiovascular disease. *Neuroscience and Biobehavioral Reviews* 74: 269–276.
23. Goshvarpour, A., A. Abbasi, and A. Goshvarpour. 2017. An accurate emotion recognition system using ECG and GSR signals and matching pursuit method. *Biomedical journal* 40 (6): 355–368.
24. Schoenthaler, A.M., and D.M. Rosenthal. 2018. Stress and hypertension. In *Disorders of blood pressure regulation: phenotypes, mechanisms, therapeutic options*, ed. A.E. Berbari and G. Mancia, 289–305. Cham: Springer International Publishing.
25. R. Ferdiana, and O. Hoseanto. 2018. The implementation of computer based test on BYOD and cloud computing environment. *International Journal of Advanced Computer Science and Applications* 9(8): 121–124.
26. Van Drumpt, A., et al. 2017. The value of arterial pressure waveform cardiac output measurements in the radial and femoral artery in major cardiac surgery patients. *BMC anesthesiology* 17 (1): 42.
27. Hong, H.-J., C.-L. Fan, Y.-C. Lin, and C.-H. Hsu. 2016. Optimizing cloud-based video crowdsensing. *IEEE Internet of Things Journal* 3 (3): 299–313.
28. Noha, F. Omara, and N. Omran. 2016. *A hybrid hashing security algorithm for data storage on cloud computing*.
29. Alworafi, M., A. Al-Hashmi, A. Dhari, Suresha, and A.B. Darem. 2018. *Task-scheduling in cloud computing environment: cost priority approach*, 99–108.
30. Rani, A.A.V., E. Baburaj. 2019. Secure and intelligent architecture for cloud-based healthcare applications in wireless body sensor networks. *International Journal of Biomedical Engineering and Technology* 29(2): 186–199.
31. Piętka, E., J. Kawa, and W. Wieclawek. 2014. *Information technologies in biomedicine*. Springer.
32. Nagavci, D., M. Hamiti, B. Selimi. 2018. Review of prediction of disease trends using big data analytics. *Optimization* 9(8): 46–50.

An Improved Compression Method for 3D Photogrammetry Scanned High Polygon Models for Virtual Reality, Augmented Reality, and 3D Printing Demanded Applications

Mohamed Samir Hassan, Hossam-Eldeen M. Shamardan and Rowayda A. Sadek

Abstract The creation of three-dimensional (3D) games, augmented reality (AR), virtual reality (VR), and 3D printing high polygon demanded applications mainly depend on the number of 3D meshes. 3D photogrammetry scanned models are usually constituted from millions of meshes in order to capture all details, which requires high processing and large memory requirements. Consequently, it's too difficult to use 3D photogrammetry scanned models in its original state with its high number of meshes and the huge sizes because it will severely impact the needed computational cost and storage. This paper proposes an efficient compression method that provides the 3D photogrammetry scanned model that can be efficiently worked on these demanded applications by reducing the number of meshes with post-processing for good visual quality. The proposed method improved the results significantly compared to the existed method by achieving a higher compression ratio with satisfactory quality.

Keywords 3D mesh compression · Laplacian smoothing · Vertex deletion · VR · AR · 3D games · Virtual reality · Augmented reality · 3D printing

M. S. Hassan (✉) · H.-E. M. Shamardan · R. A. Sadek
Faculty of Computers and Information, Helwan University, Helwan, Egypt
e-mail: mohamedsamir@fci.helwan.edu.eg

H.-E. M. Shamardan
e-mail: hossam@fci.helwan.edu.eg

R. A. Sadek
e-mail: rowayda_sadek@yahoo.com

A. Z. Ghalwash et al. (eds.), *Internet of Things—Applications and Future*, Lecture Notes in Networks and Systems 114,
https://doi.org/10.1007/978-981-15-3075-3_13

1 Introduction

The 3D models represented as 3D meshes. The 3D meshes can be created by a lot of mesh generation methods such as the cad programs, 3D photogrammetry scan, etc., the meshes can be classified to structured meshes and unstructured meshes. Structured mesh has the same faces topology and it is created manually in most cases because their creation is very difficult on the generation step. It can be described as better assemblage and higher resolution than unstructured meshes for the complex models. Generation of structured meshes is divided into two types, elementary approaches (hand-generated) and complex approaches (algebraic). Unstructured meshes are characterized as the meshes in which polygons have different topologies for the complex models. It's easy to generate unstructured meshes. Moreover, it needs fewer components as compared to the structured mesh for the same domain as they grade in size rapidly. Generation of unstructured meshes can be done by several approaches, namely, advancing front, quadtree decomposition, and Delaunay triangulation. The meshes with a low number of polygons do not constitute harm for memory resources. The problem with the meshes that contain a large number of vertices and faces require large space (memory/hard disk) and bandwidth. The 3D mesh compression can be classified into two main techniques. First is the single-rate and second is a progressive approach [1]. Those techniques depends on whether the model is decoded during, or only after, the transmission. Single-rate is lossless coding work by removing the redundancy in the original 3D model. The lossy single-rate can be achieved by modifying the 3D model data without losing too much information. Progressive compression aims to achieve the best trade between data size and approximation accuracy. These techniques are called re-meshing [2]. So in the case of storing 3D data, it can be chosen the single-rate technique, it's ensuring to get a high compression ratio for the 3D mesh file. However, the reconstructed mesh is only available when all data are decoded at the decompression stage. Progressive mesh compression is highly related to mesh simplification because the progressive mesh ha simplified the meshes in every level of progressive mesh [3]. 3D meshes are being complex to describe the high level of 3D model details. mesh simplificationworks by reducing the number of meshes to solve the complexity, rendering speed, and file size [4]. The existing progressive compression algorithm rarely considers the attribute information [5]. The progressive mesh compression is approached by Hoppe, 1997, Hoppe and Wa, 1998 [6, 7]. The applications for progressive representations are very useful when data is transmitted over the internet, the sender can control the quality of the mesh details equal size. The single-rate is more efficient on this point because it can compress the 3D model file with a very high compression ratio. Then decompression it with full quality on the side of the receiver. The single-rate compression for 3D meshes is very useful in the part of the size and send mesh files through the internet. It depends on algorithms that can compress the code of V and F. This kind of compression is near to encryption because there is no way to preview the compressed file without the decompression process. The main problem of this technique is that most 3D applications have a number of standard methods able to read the 3D files so the 3D mesh app is not able to read the compressed files with the single-rate.

2 Background

(A) **3D Scan Using Photogrammetry**

Photogrammetry is one of the best 3D scan techniques. The photos that are captured by the photogrammetry is able to build multi-scales of 3D model details. The photos that using to build the 3D models photogrammetry can be captured using digital cameras, satellites, sensors or etc. Then processed following the typical photogrammetric pipeline based on sensor calibration, image orientation, surface measurement, feature extraction, and Orthophoto generation [8]. Photogrammetry is used to capture the real world to be a part of the digital world. This method is very useful on 3D multi-fields to create a virtual world but it still has a bad effect on the part of storage and complexity.

(B) **Progressive decimation (mesh simplification)**

Progressive Decimation Algorithm [9] is reducing vertices (V) and faces (F) numbers. This algorithm by working deletes some unnecessary vertices along with associated triangles. First, it Calculates the number of all V and F in the original mesh and selects some of it by a certain criterion and demonstrates the deletion priorities of vertices, then delete vertices. And re-triangulation meshes to close the empty holes caused by the deletion of vertices. After the mesh simplification process, the V and F numbers of the simplified 3D model are less than the original model. This algorithm is very useful to reduce the number of meshes that are meaning decreases the complexity and increases the compression ratio. On the 3D meshes field, the high number of V and F always represent the full mesh details so, when the 3D model loses a high number of V and F it means losing the details and mesh quality.

The meshes of the 3D photogrammetry scanned model will lose a lot of their details and that can be acceptable on the 3D world (3D games, VR, AR, and 3D printing) At the expense of Speed, reduce complexity and size. But the distortion that may occur is unacceptable Fig. 2a. A progressive mesh is a series of different sizes, polygons, and vertices numbers of meshes.

$$\begin{aligned} &M^l = M^l(V,K) \quad \text{Related by the operations} \\ &\left(\hat{M} = M^n\right) \to M^{n-1} \to \cdots M^1 \to M^0 \end{aligned} \tag{1}$$

where $\hat{M}$ and M^n represent the full resolution scanned Model, And M^0 is a simplified 3D model. (V) is the vertex R3 and K is the mesh topology (connectivity of the mesh triangles).

M^0 is the simplified mesh in every new time.

$$M^0 \rightarrow M^1 \rightarrow \cdots \rightarrow M^{n-1} \rightarrow M^n \quad (2)$$

To achieve reduction levels every new level must be less than M^0 of the previous process [9].

(C) **Curvature and Laplacian algorithm**

Laplacian smoothing algorithm [10] is used to smooth F and V in the 3D model. The meshes smoothing can be done by choosing a new position for each F and V on the mesh based on the neighbor's position. The Laplacian coordinate (LC) is a difference between the V position and it is the neighbor's position. It is converting between absolute and intrinsic representations by solving a sparse linear system. LC is based on translation, but can't effort on scaling and rotation. The Laplacian smoothing algorithm is solving the distortion that happens after vertex deletion algorithm.

The meshes will become smooth by using the Laplacian smoothing algorithm, which can implement a quick smoothing by explicit integration. If the integration methods equal zero, it will use the inverse distance between vertices as weights. If equal one it will use the normalized curvature operator as weights. Laplacian smoothing is saving the original ratio in length between edges and protecting the mesh topology [10].

$$(\bar{\kappa}\mathbf{n})_{normalized} = \frac{1}{\sum_j \left(\cot \alpha_j^l + \cot \alpha_j^r\right)} \sum_j \left(\cot \alpha_j^l + \cot \alpha_j^r\right)(X_i - X_j)$$

where $\bar{\kappa}$ represents the matrix of the curvature normal, cot is cotangent, α_j is the angle opposite to the edge and X_j are the neighbors of the vertex X_i.

3 Previous Work

Authors [5] proposed a progressive compression algorithm using the color attribute and material attribute. The 3D mesh model is simplified with geometric coding and attributes coding during the compression. The authors proposed an algorithm that can achieve a high compression ratio of 1:50 on the basis of maintaining the original appearance and decrease the processing speed of the model. It has reduced the size and complexity of the 3D model. This paper cares about the attribute information on the progressive compression process. The color and materials of 3D models are very important to present a realistic view of the 3D models. In this paper, the author improved Lee [11] algorithm. Lee uses two conquests to remove a set of vertices, generating different levels of details; decimation and cleansing conquests. If a vertex is less than or equal to 6, the vertex is removed [12].

This paper introduced a new method to simplify the complex 3D mesh by progressive mesh compression based on a reverse interpolation loop subdivision (RILSP) by proposing a butterfly subdivision through the treatment of extraordinary vertex the algorithm can interpolate the subdivision by the loop subdivision mask. And that can improve the traditional loop algorithm. The authors used the reverse operation to simplify the high polygons models through the progressive mesh by initial mesh and a series of vertex offsets. The algorithm reduces the regular point relative reverse butterfly subdivision of compensation operation and greatly reduces simplification and reconstruction. Likewise, the reverse butterfly subdivision algorithm reduces the regular point compensation operation and greatly reduces simplification and reconstruction compared with the existing reverse loop subdivision by considering more control vertices. Furthermore, the edge point with respect to the center point is compensated by sacrificing a small amount of time to calculate the smaller vertex offset, this method gives high transmission speed and low offset. RLSP is a lossless simplification and it works to reconstruct the vertices. Luo's Reverse Butterfly Subdivision Process (RBSP), RILSP cancels the offsets compensation or even vertex in the update process of vertices and significantly increases the speed of mesh simplification.

The authors [13] proposed an improved deformation algorithm of Laplacian based on the mesh model simplification. The simplified method of vertex deletion is used to simplify the original 3D mesh model and the Laplacian deformation technique is adopted in the simplified mesh module. The deformation efficiency of the model is improved while the rotation invariant feature is guaranteed. The result of the experiment also verifies the effectiveness of this algorithm.

Authors [14] proposed a mesh simplification method using the salient feature and preserving efficiency. An energy-operator is used to calculate the complexity of triangles, the authors simplified the dragon Stanford model and decreased the number of polygons from 42,000 polygons to 22,300 polygons [15]. The authors proposed a partial differential equation (PDE) working by progressive mesh decimation and solve noise problems based on spectral methods. In this method, the arbitrary mesh model is partitioned into patches, each of which can be represented by a small set of coefficients of its PDE spectral solution.

4 Proposed Method

The proposed method is based on three algorithms, the high level of mesh decimation [9] by vertex progressive decimation algorithm to decimate the number of V and F, Laplacian smoothing algorithm [10] to improve the quality of meshes and Variable precision arithmetic to delete the redundancy on the 3D mesh file Fig. 1.

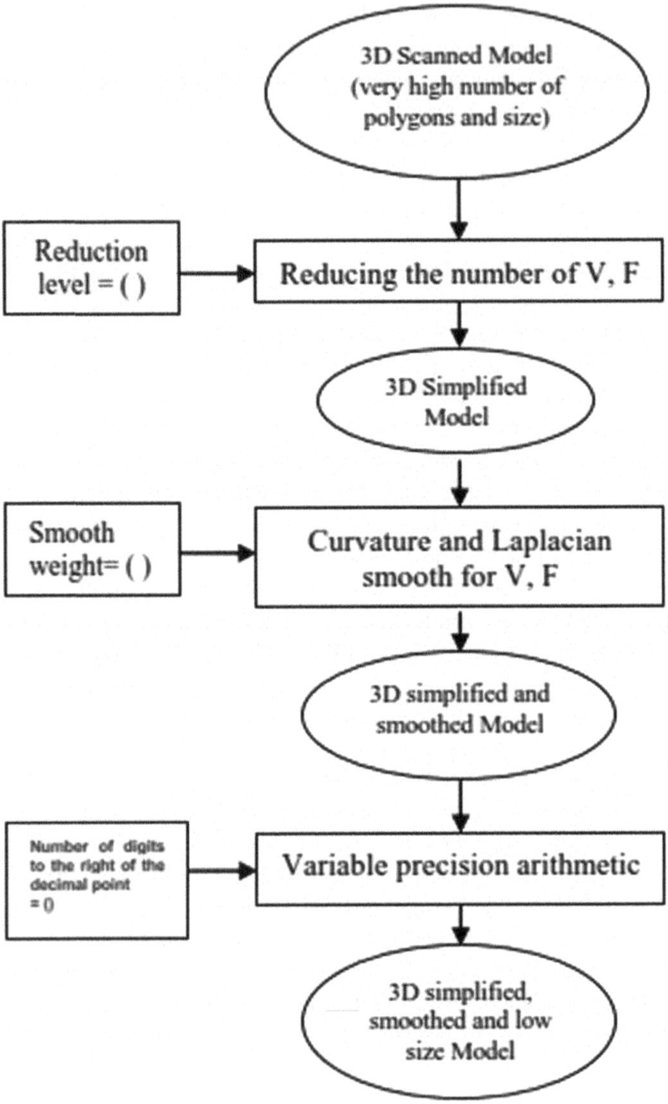

Fig. 1 Flow graph of the proposed method

(A) **3D photogrammetry scanned model (input)**

The 3D photogrammetry scanned model is containing a very high number of polygons, vertices by default, that is meaning high size. The default 3D photogrammetry scanned model is the perfect input of the proposed method. Because it contains all 3D model details.

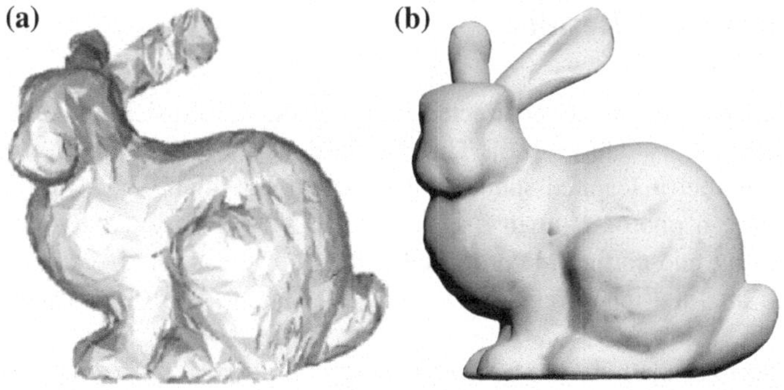

Fig. 2 The original Bunny scanned by Stanford **a** is simplified bunny by the author [14] (with 7795 polygons) and **b** is simplified by the proposed method (with 7795 polygons and 232 KB) the proposed simplified bunny is ready for games, virtual reality, augmented reality, and 3D printing without any distortion

(B) **Reducing the number of V, F**

Reduce the number of vertices (V) and faces (F) at mesh based on reduction level (R). R is the number of vertices or percentage to reduce the number of faces.

The quality of the simplified meshes is always based on the number of faces, this is meaning the high number of faces equal high quality. Fig. 2 is the simplified model by the author [14]. It's the lowest model in his implementation contain a number of faces and vertices and by visual measure their quality is very bad comparing with our method b in the same figure. But in the step of reducing the number of vertices, we get the same bad result.

(C) **Curvature and Laplacian smooth for V, F**

Curvature and Laplacian smooth for V, F implement by smooth the polygons and vertices of the 3D model. The algorithm chooses a new position for every vertex and face on the mesh based on the position of neighbors.

$$\text{Weight (w), W} = 1/(\text{inverse distance} + \text{sigma})(\text{default sigma} = 1)$$

(D) **Variable precision arithmetic**

The 3D model contains unnecessary of a long length number in the part of storing the vertices. The proposed method solved this problem through Variable precision arithmetic by decrease the number of digits to the right of the decimal point.

(E) **3D simplified, smoothed, and low size Model (output)**

The output of the proposed method is a simplified 3D model with high quality and low size. The proposed method improved the use of the 3D photogrammetry models in 3D games, augmented reality, virtual reality, and 3D printing Fig. 1.

5 Experimental Results

The experimental results are conducted by using the Stanford 3D photogrammetry Scanning Repository dataset [16]. The dataset consists of a bunny, a dragon, and 3D photogrammetry scanned statue of the ancient Egyptian God Amun Ra Luxor museum using a photogrammetry scan by [17]. Matlab 2016a software has been used for testing and coding the proposed method. The case study environment is a laptop with an Intel(R) Core(TM) i7-2820QM CPU @2.30 GHz, installed RAM 8 GB.

The proposed method implemented by the high level of vertex decimation algorithm [9] and Laplacian smooth [10], the goal of this method is to make the 3D photogrammetry scanned models for games, virtual reality, augmented reality, and 3D printing. The previous works of authors [5] and [14] have a bad effect on the 3D model meshes as bad triangulations and distortion that happened on the high level of vertex deletion. In this paper, our method had solved this problem and proposed a high level of decimation with high quality of meshes to be compared with authors [5] and [14]. In particular the part of the quality of mesh after high level of vertex decimation (Figs. 2 and 3).

The proposed method improved the results significantly compared to the author [14] by achieving a higher compression ratio on all level of details Table 1. In addition to very high mesh quality Fig. 2 (Fig. 4).

By comparing between write.obj file on the proposed method (step (D) Fig. 1) and 3D studio max software write.obj [18] with the same number of polygons and vertices Fig. 5 this results for bunny model scanned by Stanford (Table 2 and Fig. 5).

To achieve the intended purpose of this proposed method, it must be implemented in a realistic situation so the next implementation is done using a realistic data set with a very high size. This dataset scanned by [17] (Figs. 6 and 7).

(D) The 3D photogrammetry scanned Amun prepared for virtual reality, augmented reality, and 3D printing Fig. 8

The proposed compressed 3D model of Amun is very useful for 3D games, virtual reality, and augmented reality inspite of the bad effect of the face details but this can be solved easily by adding some textures as a material. It's very good because the proposed method saved the topology and achieved a very good compression ratio with high accuracy for the faces.

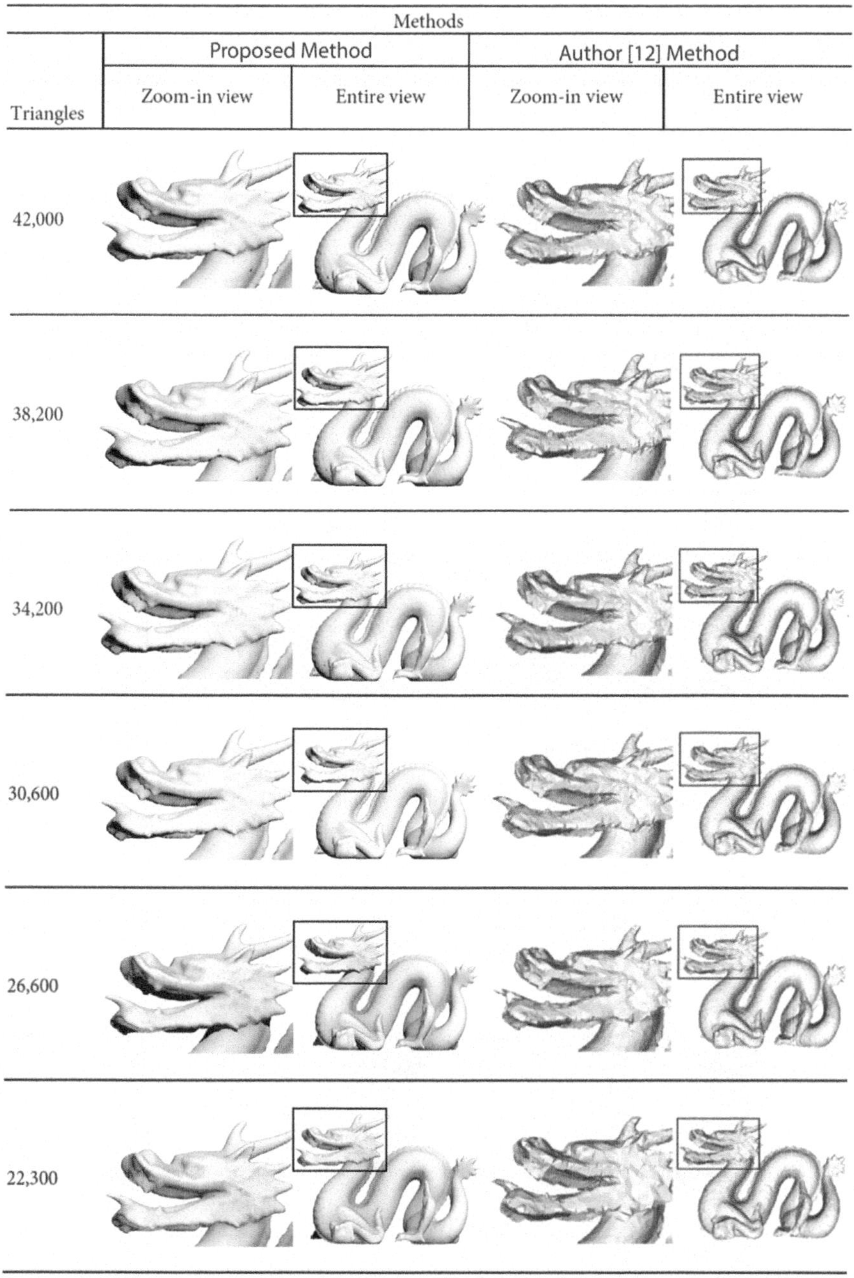

Fig. 3 Simplified dragon models with different levels of detail by the author [14] and proposed method Table 1

Table 1 Statistics of geometrical information of dragon models simplified by the author [14] and the proposed method on the part of size, vertices and triangles remained Fig. 2

Triangles remained	The proposed method		Author [14] method		The proposed method **Step (B) only** Fig. 1	
	Vertices remained	Size (MB)	Vertices remained	Size (MB)	Vertices remained	Size (MB)
42,000	20,261	**1.27**	20,340	**1.38**	20,261	**1.74**
38,200	18,471	**1.15**	18,425	**1.24**	18,471	**1.58**
34,200	16,578	**1.03**	16,455	**1.11**	16,578	**1.41**
30,600	14,869	**0.940**	14,737	**0.991**	14,869	**1.26**
26,600	12,956	**0.809**	12,777	**0.855**	12,956	**1.08**
22,300	10,890	**0.668**	10,719	**0.702**	10,890	**0.925**

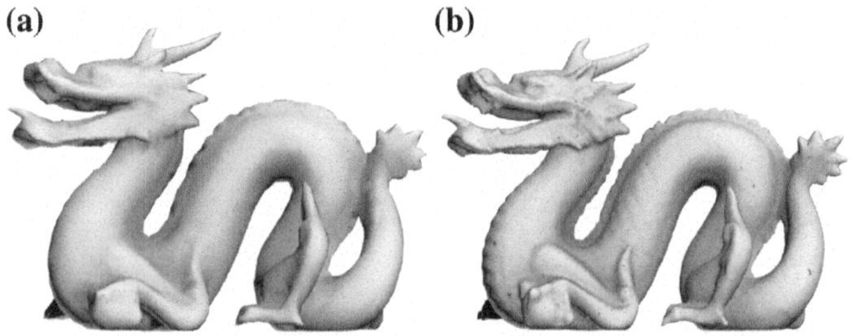

Fig. 4 **a** Is the very high level of simplification simplified by proposed method (with 5000 polygons and 295 K only) ready for games, virtual reality, augmented reality, and 3D printing without any distortion and **b** is the original Dragon scanned by Stanford (52,794 polygons and 6296 KB)

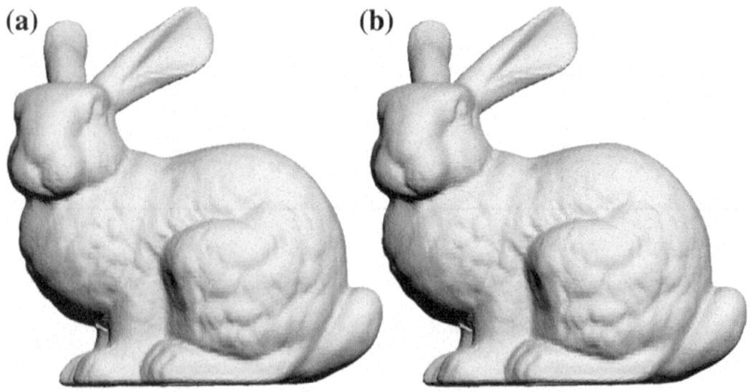

Fig. 5 Bunny **a** has been written by the proposed method (step (D) Fig. 1) and **b** written by 3D Studio Max with the same number of triangles and vertices

Table 2 .

Triangles remained	The proposed method (step (D) Fig. 1) (A)		3D studio max software (B)	
	Vertices remained	Size (MB)	Vertices remained	Size (MB)
69,451	34,834	2.16	34,834	9

Table 3 Statistics of Amun size and vertices numbers before and after implement the proposed method

Model status	Size (MB)	Number of vertices
Original Amun [17]	429	1,499,980
The proposed method	7.21	224,977

Table 4 Statistics of Amun size and vertices numbers before and after implementing the proposed method to prepare Amun for games, virtual reality, augmented reality and 3D printing

Model status	Size	Number of vertices
Original Amun [17]	429 MB	1,499,980
Simplified Amun (0.003% of the original model) by the proposed method	441 KB	14,987

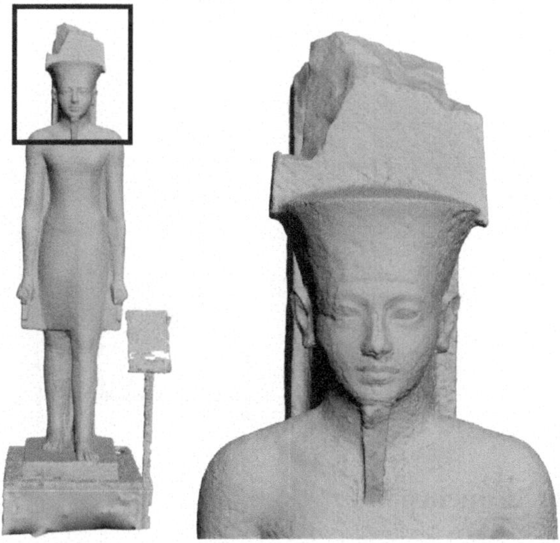

Fig. 6 The original 3D photogrammetry scanned statue of ancient Egyptian God Amun Ra Luxor museum with a very high size and details using photogrammetry scan by [17]

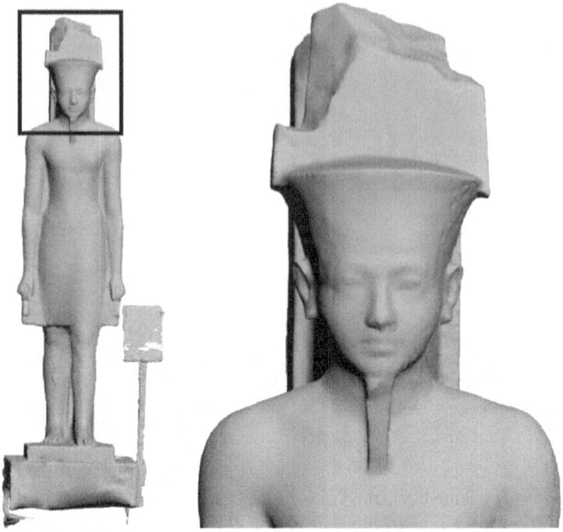

Fig. 7 3D photogrammetry scanned statue of ancient Egyptian God Amun Ra after solve distortion by laplacian smooth algorithm (Table 3)

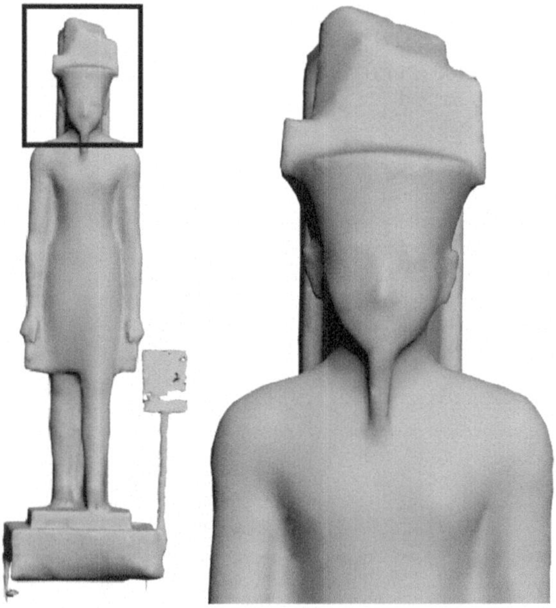

Fig. 8 3D photogrammetry scanned statue of ancient Egyptian God Amun Ra after very high level of simplification (0.003% of the original model), smooth the distortion and decrease the file size by proposed method (Table 4)

6 Conclusion

The proposed method aims to achieve a high compression ratio for 3D meshes and high level of meshes quality with avoidance or resolve the distortion that may happen from mesh decimation algorithms. The proposed method is very useful for preparing the 3D photogrammetry scanned models for games, virtual reality, augmented reality, and 3D printing Results show the superiority of the proposed work in most of the cases compared to [14] and [18]. The method can work with any kind of data set that contains meshes. The implementation succeeded to compress a very high number of faces at the 3D photogrammetry scanned of king Amun [17] that was contain (1,499,980 face and 429 MB) with resolve distortion and make it ready to merge easily to games, virtual reality, augmented reality, and 3D printing with very low size and number of faces.

7 Future Work

The proposed results are very useful on the part of 3D mesh quality after reducing the number of faces but there are many developments that need to be achieved on in the future as

1. The quality of mesh decimation algorithms, almost a high number of algorithms work with the same technique by just deleting a number of vertices or faces, these algorithms need to focus on how to detect important areas and care about all vertices and faces as the same technique. It must save and protect the important areas to achieve better quality.
2. There are not enough researchers that work on the point of 3D models attributes, the texture of the 3D model is very important and with some changes it can build fantastic results and add a lot of details on the 3D model without high size or complexity.
3. The artificial intelligence (AI) has achieved great results in recent years with the 2D fields. AI can work side by side with image processing on the part of 3D. this is the next step to achieve better results in all 3D fields.
4. The quality measures of 3D mesh decimation are very limited because most researchers depend on vision quality of the 3D model to decide which compression ratio or reducing number is useful for these cases, we will work on the future to resolve this point.

References

1. Siddeq, Mohammed Mustafa. 2017. Novel methods of image compression for 3D reconstruction. Sheffield Hallam University, Doctoral.
2. Alliez, P., and C. Gotsman. 2005. Recent advances in compression of 3D meshes. In *Advances in multi resolution for geometric modelling*, ed. by N.A. Dodgson, M.S. Floater, and M.A. Sabin. Berlin, Heidelberg: Mathematics and visualization Springer Journal.
3. Peng, Jingliang, Chang-Su Kim, and Jay Kuo. 2005. Technologies for 3D mesh compression. *Journal of Visual Communication and Image Representation* 16: 688–733 (Elsevier Inc.).

4. Tamie, Abderazzak, Abderrahim Saaidi, and Khalid Satori. 2016. *Comparative study of mesh simplification*. Springer International Publishing Switzerland.
5. Dong, Yuanyuan, Xiaoqing Yu, and Pengfei Li. 2018. 3D model progressive compression algorithm using attributes. In *International conference on smart and sustainable city*. IEEE.
6. Hoppe, H. 1997. View-dependent refinement of progressive meshes. In *Proceedings of the 24th annual conference on computer graphics and interactive techniques, SIGGRAPH*, 189–198. New York, NY, USA: ACM Press/Addison Wesley Publishing Co.
7. Hoppe, H. 1998. Smooth view-dependent level-of-detail control and its application to terrain rendering. In *Proceedings of the conference on visualization*, 35–42. Los Alamitos, CA, USA: IEEE Computer Society Press.
8. Remondino, Fabio. 2011. Heritage recording and 3D modeling with photogrammetry and 3D scanning. 3D Optical Metrology (3DOM) Research Unit, Bruno Kessler Foundation (FBK), 38122 Trento, Italy.
9. Schroeder, W.J. 1997. A topology modifying progressive decimation algorithm. In *IEEE visualization', Proceedings*, 205–212. Phoenix, AZ, USA.
10. Desbrun, Mathieu, Mark Meyer, Peter Schroder, Alan H. Barr. 1999. Implicit fairing of irregular meshes using diffusion and curvature flow. In *Proceedings of the 26th annual conference on computer graphics and interactive techniques*, 317–324. New York, NY, USA: ACM Press/Addison-Wesley Publishing Co.
11. Lee, Ho, Guillaume Lavoue, and Florent Dupont. 2012. Rate-distortion optimization for progressive compression of 3D mesh with color attributes. *The Visual Computer*: 137–153.
12. Shi, Zhou, Qian Kong, Ke Yu, and Xiaonan Luo. 2018. A new triangle mesh simplification method with sharp feature. In *7th international conference on digital home (ICDH)*. IEEE.
13. Sun, J., Y. Ding, Z. Huang, N. Wang, X. Zhu, and J. Xi. 2018. Laplacian deformation algorithm based on mesh model simplification. In *IEEE 3rd international conference on image, vision and computing (ICIVC)*.
14. Wang, Yongzhi, Jianwen Zheng, and Hui Wang. 2019. Fast mesh simplification method for three-dimensional geometric models with feature-preserving efficiency. Hindawi Scientific Programming Volume Article ID 4926190.
15. Shen, Qiqi, Yun Sheng, Congkun Chen, Guixu Zhang, and Hassan Ugail. 2017. A PDE patch-based spectral method for progressive mesh. Germany: ©Springer-Verlag GmbH.
16. Stanford dataset. http://graphics.stanford.edu/data/3Dscanrep/. Last visit June 13, 2019.
17. 3D scanned statue of ancient Egyptian God Amun Ra Luxor museum using photogrammetry scan by Ibrahim Mustafa Ibrahim. https://sketchfab.com/imhotep25. Last visit June 13, 2019.
18. D Studio Max software. https://www.autodesk.com/products/3Ds-max/overview. Last visit June 16, 2019.
19. Cignoni, P., C. Montani, and R. Scopigno. 1998. A comparison of mesh simplification algorithms. *Computer & Graphics* (Elsevier Science Ltd.).
20. Sorkine, O., D. Co, Y. Lipman, M. Alexa, C. Rössl, and H. Seidel. 2004. Laplacian surface editing. In *Proceedings of the Euro graphics ACM SIGGRAPH symposium on geometry processing. SGP '04*, 175–184. Nice, France: ACM.
21. Sun, J., Z. Huang, X. Zhu, L. Zeng, Y. Liu, and J. Ding. 2017. Deformation corrected workflow for maxillofacial prosthesis modeling. *Advances in Mechanical Engineering* 9: 1–10.

Data Quality Dimensions

Mona Nasr, Essam Shaaban and Menna Ibrahim Gabr

Abstract Data quality dimension is a term to identify quality measure that is related to many data elements including attribute, record, table, system or more abstract groupings such as business unit, company or product range. This paper presents a thorough analysis of three data quality dimensions which are completeness, relevance, and duplication. Besides; it covers all commonly used techniques for each dimension. Regarding completeness; Predictive value imputation, distribution-based imputation, KNN, and more methods are investigated. Moreover; relevance dimension is explored via filter and wrapper approach, rough set theory, hybrid feature selection, and other techniques. Duplication is investigated throughout many techniques such as; K-medoids, standard duplicate elimination algorithm, online record matching, and sorted blocks.

Keywords Missing values · Feature selection · Duplication · Data quality issues · Quality dimensions

M. Nasr
Department of Information Systems, Faculty of Computers & Information, Helwan University, Helwan, Egypt
e-mail: m.nasr@helwan.edu.eg

E. Shaaban
Department of Information Systems, Faculty of Computers & Information, Beni Suef University, Beni Suef, Egypt
e-mail: essam.shaban@fcis.bsu.edu.eg; essam_shaaban@cic-cairo.com

E. Shaaban
Canadian International Collège CIC, Zayed Campus, Zayed, Egypt

M. I. Gabr (✉)
Faculty of Commerce & Business Administration, BIS, Helwan University, Helwan, Egypt
e-mail: menna.ibrahim@commerce.helwan.edu.eg

A. Z. Ghalwash et al. (eds.), *Internet of Things—Applications and Future*,
Lecture Notes in Networks and Systems 114,
https://doi.org/10.1007/978-981-15-3075-3_14

1 Introduction

As information is a vital asset for any organization it's important to make sure that this information is at a high level of quality. The concept of data quality may differ from one to another, but they all agree that the satisfaction of customer needs means quality. Data quality can be fully described and illustrated through its dimensions or parameters. All dimensions are complementary dimensions they're supporting each other, but each dimension has its own definition and techniques through which it can be analyzed and tested. By checking quality dimensions on the information we can assure that we have information quality [1, 2]. For the purpose of this paper, we present three data quality dimensions (Completeness, Relevance, and Duplication dimensions). A detailed coverage for each dimension is presented next during the state-of-the-art section.

2 The State of Art

During this section, significant prior research and a thorough analysis of all used techniques to solve each data quality problem are presented. This section is organized into three parts as follows:

2.1 Completeness/Missing Values Dimension

"Completeness means the extent to which data is not missing and is of sufficient breadth and depth for the task at hand."

In 2007, Maytal Saar-Tsechansky compares different methods for dealing with missing data. Predictive value imputation, distribution-based imputation used by C4.5, and reduced models are compared to apply classification trees to instances with missing values [3].

In 2008, Xiaoyuan Su concludes that the accuracy of classifiers produced by machine learning algorithms generally deteriorates if the training data is incomplete, and preprocessing this data using simple imputation methods, such as mean imputation (MEI), does not generally produce much better classifiers [4].

In 2009, Shariq Bashir and Saad Razzaq Introduced a Hybrid Missing values Imputation Technique (HMiT) using association rules and hybrid combination of k-nearest neighbor approach. HMiT gives good results in terms of accuracy and processing time [5].

In 2009, B. Mehala et al. analyze the behavior and efficiency of missing data treatment. C4.5 algorithm is used to treat missing data and K-means for missing data imputation. For the data sets Bupa, Breast Cancer, and Pima the K-means imputation provides good results in most cases [6].

In 2011, R. S. Somasundaram and R. Nedunchezhian implement three missing value imputation methods namely (1) Constant substitution, (2) Mean attribute value substitution, and (3) Random attribute value substitution. Along with Three different clustering algorithms namely k-means, fuzzy c-means, and SOM based clustering algorithm. Attribute mean based missing value imputation method reconstructed the missing values in a better manner where k-means and fuzzy c-means algorithm were almost performed equally when the percentage of missing values is lower [7].

In 2012, Ms. R. Malarvizhi and Dr. Antony conclude that K-Means and KNN methods provide fast and accurate ways of estimating missing values. While KNN-based imputations provides a robust and sensitive approach to estimate missing data [8].

In 2013, Rahul Singhai compares between five methods namely (1) Mean Method of Imputation (2) Ratio Method of Imputation (3) Compromised Method of Imputation (4) Ahmed Methods of Imputation (5) Factor Type Methods of Imputation. The Ratio method of imputation and Factor type method provides very good results, even if the training sets have a large amount of missing data [9].

In 2013, Luciano C. Blomberg et al. show the effect of missing values at different levels ranging from 1 to 50% on the performance and accuracy of five different classifiers (J48, IBZ, MLP, SMO, and Naïve Bayes) according to MCAR mechanism (Missing Completely At Random). Classification performance decreases with the increase of missing values in data sets. Naïve Bayes is the least influenced by missing data, SMO is the next, J48 is the next, MLP is the next, and IBK is the most influenced presenting the lowest accuracy [10].

In 2013, Geaur Rahman and Zahidul Islam present two novel techniques (EMI & DMI) for the imputation of both categorical and numerical missing values. The techniques used a decision tree and forests to identify segments where records belong to a segment having higher similarity and attribute correlation. Results show superiority to the two techniques [11].

In 2014, Gimpy et al. present five methods for estimating missing values. (1) Litwise Deletion (2) Mean/Mode Imputation (MMI) (3) K-Nearest Neighbor Imputation (KNN) (4) Case Deletion (CD) (5) Hot and Cold Deck Imputation [12].

In 2014, Minakshi et al. compared between Mean/Mode Imputation (MMI), K-Nearest Neighbor Imputation (KNN) and Lit wise Deletion techniques. KNN was able to correctly classify 149 instances out of 200. So the accuracy of KNN is 74.5% which is greater than the other compared techniques [13].

In 2015, Geaur Rahman and Zahidul Islam present a novel technique called A Fuzzy Expectation Maximization and Fuzzy Clustering-based Missing Value Imputation Framework for Data Pre-processing (FEMI). FEMI performance is compared with the performance of five high-quality existing techniques, namely EMI, GkNN, FKMI, SVR, and IBLLS. The results indicate that FEMI performs significantly better than EMI, GkNN, FKMI, SVR, and IBLLS [14].

In 2015, Archana Purwar and Sandeep Kumar present a novel Hybrid Prediction Model with missing value imputation (HPMMI) that analyzes various imputation techniques using Simple K-means clustering and apply the best one to a data set.

The proposed system has significantly improved data quality by use of the best imputation technique after a quantitative analysis of eleven imputation approaches. Results are best in comparison with existing methods where the accuracy rate was over 99% [15].

In 2015, Emily Grace et al. compare four methods of imputation namely (median, kNN, ½ minimum, and zero) with respect to revealing their effects on the normality and variance of data, on statistical significance and on the approximation of a suitable threshold to accept missing data as truly missing. KNN imputation was the best method for restoring data statistics relative to the full-data control [16].

In 2015, Sameh Shohdy et al. focus on speeding up the missing values imputation process using the bitmap indexing technique. The bitmap indexing is used to directly access the required records for the imputation method (i.e., Direct Access Imputation (DAI)). Then, the bitmap indexing technique is used for missing value estimation using the pre-generated bitmap indexing vectors (i.e., Bitmap-Based Imputation (BBI)). bitmap-based methods give satisfying results by accelerating data mining classification of incomplete data while also maintaining precision [17].

In 2015, Xiaobo Yan et al. propose three new models to solve missing data problems, and they are model of missing value imputation based on context and linear mean (MCL), model of missing value imputation based on binary search (MBS), and model of missing value imputation based on Gaussian mixture model (MGI). Experimental results showed that the three models can improve the accuracy, reliability, and stability of missing value imputation greatly and effectively [18].

In 2017, Amit Paul et al. propose an integrated framework which first imputes missing values (using the concept of fuzzy similarity measure) and then in order to achieve maximum accuracy in classifying patient a classifier (using optimized fuzzy rule generation) has been designed. Analysis of the results reveals that the proposed framework is able to classify the diseased and normal patients with improved accuracy [19].

In 2017, Alma B Pedersen et al. provide insights on the type of missing data, traditional methods (complete-case analyses, missing indicator method, single value imputation, and sensitivity analyses incorporating worst- and best-case scenarios), and multiple imputations as alternative methods to deal with missing data. Multiple imputations provide unbiased and valid estimates of associations based on information from the available data under MAR, MCAR, and MNAR assumptions, unlike other techniques [20].

In 2017, B. Dalvand et al. apply the data envelopment analysis DEA-base clustering method to know the cluster that has missing units, and then the prediction of the missing value is done by training the neural network algorithm with this cluster. Then the results are compared using the EM algorithm and Monte Carlo simulation [21].

It is important to fill the missing data before any processing steps. There are many missing value imputation techniques that have been discussed in the literature. Among these techniques are KNN imputation, mean/median/mode imputation, lit wise deletion, and other novel techniques that proved its efficiency during the experiments.

2.2 Relevance/Feature Selection Dimension

"Relevance means the extent to which data is applicable and relevant for the task at hand."

In 2007, Helyane and Júlio apply selection attributes algorithms in two gene expression databases. The Used evaluation measures were based on the filter and wrapper approach. Naive Bayes (NB), C4.5 (C4.5), SVM (SVM) and the KNN algorithms were selected and used as evaluation measures. Comparing the results obtained fromthe two attribute selection approaches, the wrapper approach with sequential search produces the best results [22].

In 2007, WU Ming and YAN Puliu proposed an improved DS based feature selection algorithm which takes both the difference and similitude into account. The algorithm employs the function values to evaluate the importance of each attribute. The results showed that the algorithm is not superior to the compared algorithm. But the experiments showed that it is an effective algorithm, especially for large-scale databases [23].

In 2007, Yvan Saey et al. provide a basic taxonomy of feature selection techniques, and discussing their use, variety, and potential in a number of both common, as well as upcoming bioinformatics applications. These techniques are Wrapper, Filter, and Embedded techniques [24].

In 2009, Y.Y. Yao and Y. Zhao used rough set theory to deal with uncertainty and vagueness of the decision system and it has been used to identify the reduct set of the set of all attributes of the decision system [25].

In 2009, Thangavel, K., and Pethalakshmi propose Rough set and neural network based reduction to bring out the most relevant attributes [26].

In 2009, Wa'el M. Mahmud et al. proposed a new hybrid model RSC-PGA (Rough Set Classification Parallel Genetic Algorithm) presented to address the problem of identifying important features in building an intrusion detection system [27].

In 2010, Patricia et al. proposed a new feature selection algorithm that incorporates domain specific definition of low, medium, and high correlations. The decision rule-based feature selection enables to incorporate domain specific definition of feature relevance into the feature selection process. Which gives a high level of predictive classification performance [28].

In 2010, Huan Liu et al. give a brief introduction about the key components of feature selection and review its developments with the growth of data mining. Then overview FSDM and the papers of FSDM. Then nascent demands in data-intensive applications have been examined [29].

In 2011, Hui-Huang Hsu et al. introduce a hybrid feature selection method which combines two feature selection methods—the filters and the wrappers. Candidate feature is first selected from the original feature set via computationally efficient filters. The candidate feature set is further refined by more accurate wrappers. The results showed that equal or better prediction accuracy can be achieved with a smaller feature set. This feature set can be obtained in a reasonable time period [30].

In 2011, K. Manimala et al. propose two novel wrapper-based hybrid soft computing techniques (genetic algorithm "GA" & simulated annealing "SA") for feature selection and parameter optimization. The proposed algorithm significantly improved the classification accuracy rate by eliminating relatively useless features and proper parameter selection for the classifier [31].

In 2011, Huihui Zhao et al. presented a novel computational strategy to select biomarkers as few as possible for the disease. A comparison between three types of feature selection based data mining methods, i.e., Filter, Wrapper, and Embedded methods is made and used 3 fold cross validation to evaluate computational performances. The results show that Wrapper-based feature selection methods perform better than Filter and Embedded methods. Based on this, a novel feature selection by combing t-test and classification data mining method to select 6 proteins with 96% prediction accuracy and 5 metabolites with 81% prediction accuracy is presented [32].

In 2011, P. Ravisankar et al. use t statistic technique for feature selection and compared the performance of many data mining techniques before and after doing feature selection. PNN outperformed all the techniques without feature selection and, GP and PNN outperform all the techniques with feature selection with equal accuracies [33].

In 2011, Jasmina NOVAKOVIĆ et al. presented a comparison between several feature ranking methods. Six ranking methods (namely Information Gain (IG) attribute evaluation, Gain Ratio (GR) attribute evaluation, Symmetrical Uncertainty (SU) attribute evaluation, Relief-F (RF) attribute evaluation, One-R (OR) attribute evaluation, and Chi-Squared (CS) attribute evaluation) that can be divided into two broad categories: statistical and entropy-based. They conclude that there is no best ranking index, the only way to be sure that the highest accuracy is obtained in practical problems requires testing a given classifier on a number of feature subsets, obtained from different ranking indices [34].

In 2012, Shan Suthaharan and Tejaswi Panchagnula suggest an approach which analyzes the intrusion datasets, evaluates the features for its relevance to a specific attack, determines the level of contribution of feature, and eliminates it from the dataset automatically by using Rough Set Theory (RST) based approach and select relevance features using multidimensional scatter-plot [35].

In 2012, Sunita Beniwal and Jitender Arora provide a detailed introductory paper on different techniques used for classification and feature selection. Filtering is done using different feature selection techniques like wrapper, filter, and embedded technique [36].

In 2012, Anand Kannan et al. Proposed a genetic-based feature selection algorithm for cloud networks. Experimental results showed that it improves the detection accuracy of the fuzzy SVM classifier by providing minimum and required number of features. This helps to reduce the classification time in addition to the increase in classification accuracy [37].

In 2013, Newton Spolaor and Everton Alvares compare four multi-label feature selection methods that use the filter approach. Two standard multi-label feature selection approaches (Binary Relevance (BR) and Label Powerset (LP)). Besides these two problem transformation approaches, it uses ReliefF and Information Gain to measure the goodness of features. Methods that use ReliefF are superior to the ones that use Information Gain [38].

In 2013, B. Azhagusundari and Antony Selvadoss Thanamani introduce a new algorithm based on the discernibility matrix and Information gain to reduce attributes. The selection method using Information Gain and Discernibility shows better results in terms of number of features selected and accuracy than applying methods individually [39].

In 2013, Verónica Bolón et al. A review of 11 feature selection methods applied over 11 synthetic datasets and 2 real datasets was presented aimed at studying their performance with respect to several situations that can hinder the process of feature selection. Three major approaches were evaluated: filters, wrappers, and embedded methods besides, four classifiers were selected to measure the effectiveness of the selected features and to check if the true model was also unique [40].

In 2014, Hassan Rabeti and Martin Burtscher introduce a new Tree Search of Correlation-Adjusted Class Distances (TS-CACD). A comparison is made between TS-CACD feature selection algorithm to the L0, mRMR, InfoGain, ReliefF, and CfsSubset feature selection methods. The results showed that TSCACD is the best or close to the best accuracy [41].

In 2014, Jiaqiang Chen et al. propose a novel two-stage data preprocessing approach, which incorporates both feature selection and instance reduction. In particular, in the feature selection stage, they propose a new algorithm using both feature selection and threshold-based clustering which contains both relevance analysis and redundancy control. The final results demonstrate the effectiveness of the two-stage data preprocessing approach which can greatly reduce both the number of features and the number of instances of the original dataset, which then simplify the training process of the classification models [42].

In 2014, Adel Sabry Eesa et al. present a new feature selection approach based on cuttlefish optimization algorithm (CFA) which is used for intrusion detection systems. The proposed model uses CFA as a search strategy to find the optimal subset of features and decision tree algorithm as a judgment on the selected features produced by CFA. The results revealed that CFA gives a higher detection rate and accuracy rate with lower false alarm when compared with the results obtained from all features [43].

In 2015, Divya Tomar and Sonali Agarwal propose a hybrid feature selection (HFS) based on an efficient disease diagnostic model for Breast Cancer, Hepatitis, and Diabetes. It combines the positive aspects of both Filter and Wrapper FS approaches, adopts weighted least squares twin support vector machine (WLSTSVM) as a classification approach, sequential forward selection (SFS) as a search strategy, and correlation feature selection (CFS) to evaluate the importance of each feature. It is not only useful for selecting significant features, but it also handles the class imbalance problem. Experimental results show the effectiveness of the proposed disease diagnosis model. The predictive accuracy for Pima Diabetes dataset is 89.71% which is far better than other existing approaches. For Hepatitis and Breast Cancer diseases, the predictive accuracy is 87.50% and 98.55%, respectively, which is also comparable with other existing models [44].

In 2015, E. Emary et al. A system for feature selection based on an intelligent search of gray wolf optimization has been proposed. Compared with PSO and GA over a set of UCI machine learning data repository, GWO proves much better performance in both classification accuracy and feature size reduction, as well as it proves much robustness and fast convergence speed. Moreover, the gray wolf optimization approach proves much robustness against initialization in comparison with PSO and GA optimizers [45].

In 2015, H. Hannah Inbarani et al. present a novel feature selection approach to deal with issues of high dimensionality in the medical dataset. It is a supervised feature selection method based on Rough Set Quick Reduct hybridized with an Improved Harmony Search algorithm. The quality of the reduced data is measured by the classification performance. The proposed algorithm reveals more than 90% classification accuracy in most of the cases and the time taken to reduct the dataset also decreased than the existing methods. The experimental result demonstrates the efficiency and effectiveness of the proposed algorithm [46].

In 2015, Zhongyi Hua et al. propose a hybrid filter–wrapper approach for short-term load forecasting (STLF) feature selection. This proposed approach, which is believed to have taken full advantage of the strengths of both filter and wrapper methods, first uses the Partial Mutual Information based filter method to filter out most of the irrelevant and redundant features, and subsequently applies a wrapper method, implemented via a firefly algorithm, to further reduce the redundant features without degrading the forecasting accuracy. The results confirmed that the forecasting performance of the proposed hybrid filter–wrapper-based model is superior to other well-established counterparts. Therefore, the hybrid filter–wrapper method can be a good alternative for feature selection in STLF [47].

High dimensionality data takes too much time during processing which affects the quality of the results. It is better to remove these useless dimensions before processing. Different feature selection techniques in different areas have been highlighted in the literature, and each technique has proved his effectiveness within certain circumstances.

2.3 Duplication Dimension

"Duplication means a measure of unwanted duplication existing within or across systems for a particular field, record, or data set."

In 2007, Ahmed K. Elmagarmid et al. present a thorough analysis of the literature on duplicate record detection. They cover similarity metrics that are commonly used to detect similar field entries and present an extensive set of duplicate detection algorithms that can detect approximately duplicate records in a database.

They also cover multiple techniques for improving the efficiency and scalability of approximate duplicate detection algorithms. We conclude with the coverage of existing tools [48].

In 2008, Li Huang et al. propose an approach, named length filtering and dynamic weighting (LFDW) for duplicate records cleansing. There are three steps in LFDW. The first step is length filtering. Secondly, this approach detects duplicate records using dynamic weighting properties. Finally, to improve the performance of duplicate detection, a dynamic sliding-window algorithm has been adopted. The result indicates that the time, recall, and precision of LFDW are better than traditional ones [49].

In 2009, Ying Pei et al. proposed an improved K-medoids clustering algorithm (IKMC) to resolve the problem of detecting the near-duplicated records. It uses edit distance method and the weights of attributes to get similarity value among records, then detect duplicated records by clustering these similarity values. This algorithm can automatically adjust the number of clusters by comparing the similarity value with the preset similarity threshold. Through the experiment, this algorithm is proved to have good detection accuracy and high availability [50].

In 2009, Mariam Rehman and Vatcharapon Esichaikul compare among standard duplicate elimination algorithm (SDE), sorted neighborhood algorithm (SNA), duplicate elimination sorted neighborhood algorithm (DE-SNA), and adaptive duplicate detection algorithm (ADD). A prototype is also developed which shows that adaptive duplicate detection algorithm is the optimal solution for the problem of duplicate record detection. For approximate matching, string matching algorithms (recursive algorithm with word base and recursive algorithm with character base) have been implemented and it is concluded that the results are much better with recursive algorithm with word base [51].

In 2009, Xiao Mansheng et al. a sub-fuzzy clustering property optimization method based on grouping is proposed. First, the properties of group record are processed to reduce the dimension of the property effectively and obtain the representation of the group, and then a similarity comparison method is used to detect approximately duplicated records in groups. It is shown in theoretical analysis and experiment, this method has higher detection accuracy and efficiency, and could better solve the recognition problems of approximately duplicated records in large dataset [52].

In 2009, Qin Kan et al. present a synthetic approach for recognizing clusters of approximate duplicate records of multi-language data. The experiments using real data sets showed that this solution has very good efficiency and accuracy. In addition, it is very extendable, since many existing techniques can be used [53].

In 2010, Hussein Issa presents the problem of duplicate records and their detection, and addresses the issue of one type of records in specific which is of great interest in the business world: that of duplicate payments [54].

In 2010, Liu Zhe and Zhao Zhi-gang proposed a segment strategy based on character features and the algorithm of edit distance with variable weight. The experiment results indicate that the segment time is small along with data scale growing. It means that the total running time of detecting duplicate records is not influenced by segment in the big data scale. The experiment results also indicate that the algorithm running efficiency and detect precision can be improved [55].

In 2010, Weifeng Su et al. present an unsupervised, online record matching method, UDD, which, for a given query, can effectively identify duplicates from the query result records of multiple Web databases. After removal of the same source duplicates, the "presumed" non-duplicate records from the same source can be used as training examples alleviating the burden of users having to manually label training examples. Two cooperating classifiers, a weighted component similarity summing classifier and an SVM classifier are used to iteratively identify duplicates in the query results from multiple Web databases. Experimental results show that UDD works well for the Web database scenario [56].

In 2010, Mohamed A. Soliman et al. propose ProbClean, a system that treats duplicate detection procedures as data processing tasks with uncertain outcomes. They use a novel uncertainty model that compactly encodes the space of possible repairs corresponding to different parameter settings. ProbClean efficiently supports relational queries and allows new types of queries against a set of possible repairs [57].

In 2010, Felix Naumann and Melanie Herschel examine closely two main components: (i) Similarity measures are used to automatically identify duplicates when comparing two records. (ii) Algorithms that are developed to perform on very large volumes of data in search of duplicates. Finally, they discuss methods to evaluate the success of duplicate detection [58].

In 2010, Quanping Hua et al. proposed an optimal feature selection method based on fuzzy clustering in groups. First, it deals with attributes of records in groups so as to reduce dimensions of attributes recorded effectively and obtain representative records in groups. Then it detects approximately duplicate records in groups by a computing method which compares with similarity. The results show that identification accuracy and detection efficiency of this method is higher and it can better solve the recognition problem of approximately duplicate records in a large number of datasets [59].

In 2011, Hussein Issa and Miklos A. Vasarhelyi discussed a framework and methodology used in duplicates detection. Exact matching and fuzzy (or near-identical matching) methods are presented [60].

In 2011, Uwe Draisbach and Felix Naumann present a new algorithm called Sorted Blocks in several variants, which generalizes blocking and windowing approaches. Experiments with several real-world datasets show that Sorted Blocks outperforms the two other approaches. A challenge for Sorted Blocks is finding the right configuration settings, as it has more parameters than the other two approaches [61].

In 2011, Mohammadreza Ektefa et al. present a threshold-based method which takes into account both string and semantic similarity measures for comparing record pairs. Experimental results indicate, that the proposed similarity method which is based on the combination of string and semantic similarity measures outperforms the individual similarity measures with the F-measure of 99.1% in Restaurant dataset [62].

In 2012, Subramaniyaswamy and Pandian Chenthur presented a thorough analysis of duplicate record detection and coved almost all the matrices that are commonly used to detect similar entities and a set of duplicate detection algorithms [63].

In 2012, Uwe Draisbach et al. propose with the Duplicate Count Strategy (DCS) a variation of SNM that uses a varying window size. Next to the basic variant of DCS, they also propose and thoroughly evaluate a variant called DCS++ which is probably better than the original SNM in terms of efficiency. They have proven that with a proper threshold, DCS++ is more efficient than SNM without loss of effectiveness [64].

In 2014, Thorsten Papenbrock et al. present two novel, progressive duplicate detection algorithms that significantly increase the efficiency of finding duplicates if the execution time is limited. They maximize the gain of the overall process within the time available by reporting most results much earlier than traditional approaches. Comprehensive experiments show that the progressive algorithms can double the efficiency over time of traditional duplicate detection [65].

In 2015, Arfa Skandar et al. review, analyze, and compare algorithms lying under the empirical technique in order to suggest the most effective algorithm in terms of efficiency and accuracy. The comparison was performed in order to come up with the best algorithm, i.e., DCS++. The selected algorithm was critically analyzed in order to improve its working. On the basis of limitations of the selected algorithm, variation in the algorithm was proposed and validated by the developed prototype. After the implementation of existing DCS++ and its proposed variation, it was found that the proposed variation in DCS++ is producing better results in terms of efficiency and accuracy [66].

In 2016, Dhivyabharathi and Kumaresan present a thorough analysis of the literature survey on adaptive and progressive duplicate detection techniques. They concluded that SNM identifies the duplicates efficiently within the window size, but it is very expensive. Markov Clustering algorithm does not produce efficient results in duplicate detection. An advantage of Sorted Blocks in comparison to the Sorted Neighborhood Method is the variable partition size instead of a fixed size window. This allows more comparisons if several records have similar values. DCS++ (Duplicate count strategy) dynamically improves the efficiency of Duplicate detection but they need to run for a certain period of time and cannot maximize the efficiency for any given time slot. PSNM and PB dynamically adjust the comparison order at runtime based on intermediate results. For a given time slot it can find more duplicate pairs [67].

Duplicates cost money, confuse sales, and break the business automation process. It is best to catch duplicates immediately and sometimes deleting this duplicate record before it has a history. A summary of all covered studies is presented in Table 1.

Table 1 Summary of existing studies

Dimension	Year	Technique (method)	Findings
Completeness	2007	• Predictive value imputation, distribution-based imputation used by C4.5, and reduced models [3]	• Reduced-feature models are preferable both to C4.5's distribution-based imputation and to predictive value imputation
	2008	• Mean imputation (MEI) [4]	• Simple imputation methods, such as mean imputation (MEI), does not generally produce much better classifiers
	2009	• Hybrid Missing values Imputation Technique (HMiT) [5] • K–means [6]	• Good results • Good results in most cases
	2011	• Constant substitution, Mean attribute value substitution and Random attribute value substitution [7]	• Attribute mean based missing value imputation behaved in a good manner
	2012	• K-Means and KNN methods [8]	• Fast and accurate ways of estimating missing values
	2013	• (1) Mean method of imputation (2) Ratio method of imputation (3) Compromised method of imputation (4) Ahmed methods of imputation (5) Factor type methods of imputation [9] • Two novel techniques (EMI & DMI) [11]	• The ratio method of imputation and factor type method provide very good results • Results show superiority to the two techniques
	2014	• Mean/Mode Imputation (MMI), K-Nearest Neighbor Imputation (KNN) and Lit wise Deletion techniques [13]	• The accuracy of Knn is greater than the other compared techniques
	2015	• A Fuzzy Expectation Maximization and Fuzzy Clustering-based Missing Value Imputation Framework for Data Pre-processing (FEMI) [14] • Hybrid Prediction Model with missing value imputation (HPMMI) [15] • Median, kNN, ½ minimum, and zero imputation [16] • Bitmap-Based Imputation (BBI) [17] • Model of missing value imputation based on context and linear mean (MCL), model of missing value imputation based on binary search (MBS), and model of missing value imputation based on Gaussian mixture model (MGI) [18]	• FEMI performs significantly great • Significantly improved data quality • KNN imputation was the best method • Bitmap-based methods give satisfied results • The three models can improve the accuracy, reliability, and stability of missing value imputation greatly and effectively
	2017	• Integrated framework (fuzzy approach) [19] • Insights on traditional methods and multiple imputation methods [20]	• Proposed framework able to classify the diseased and normal patient with improved accuracy

(continued)

Table 1 (continued)

Dimension	Year	Technique (method)	Findings
		• Data envelopment analysis DEA-base clustering method and neural network algorithm [21]	• Multiple imputation provides unbiased and valid estimates • results are compared using EM algorithm and Monte Carlo simulation
Feature - selection	2007	• Filter and wrapper approach [22] • Improved DS based feature selection algorithm [23]	• The wrapper approach with sequential search produce best results • Effective algorithm
	2009	• Rough set theory to identify the reduct set [25] • RSC-PGA (Rough Set Classification Parallel Genetic Algorithm) [27]	• Good results • RSC-PGA proves it effectiveness
	2010	• New algorithm which incorporates domain specific definition of low, medium and high correlations [28]	• High level of predictive classification performance
	2011	• Hybrid feature selection method which combine two feature selection methods—the filters and the wrappers [30] • Two novel wrapper-based hybrid soft computing techniques (genetic algorithm "GA" & simulated annealing "SA") [31] • Filter, Wrapper and Embedded methods [32] • T statistic technique for feature selection [33] • Information Gain (IG) attribute evaluation, Gain Ratio (GR) attribute evaluation, Symmetrical Uncertainty (SU) attribute evaluation, Relief-F (RF) attribute evaluation, One-R (OR) attribute evaluation, and Chi-Squared (CS) attribute evaluation [34]	• The candidate feature set is further refined by more accurate wrappers • Significantly improved the classification accuracy rate • Wrapper-based feature selection methods perform better than Filter and Embedded methods • GP and PNN outperform all the techniques with feature selection • There is no best ranking index
	2012	• Rough Set Theory (RST) based approach and select relevance features using multidimensional scatter-Plot [35] • Genetic based feature selection algorithm [37]	• Better results. • Improves the detection accuracy of the fuzzy SVM classifier by providing minimum and required number of features
	2013	• Multi-label feature selection methods which use the filter approach [38] • New algorithm based on discernibility matrix and Information gain [39]	• Relief are superior to the ones that use Information Gain • Better results in terms of number of features selected and accuracy
	2014	• Tree Search of Correlation-Adjusted Class Distances (TS-CACD) [41]	• TSCACD is the best or close to the best accuracy

(continued)

Table 1 (continued)

Dimension	Year	Technique (method)	Findings
		• Novel two-stage data preprocessing approach, which incorporates both feature selection and instance reduction [42] • New feature selection approach based on cuttlefish optimization algorithm(CFA) [43]	• Effectiveness of the two-stage data preprocessing approach • Higher detection rate and accuracy rate
	2015	• Hybrid feature selection (HFS) [44] • Feature selection based on intelligent search of gray wolf optimization [45] • Novel feature selection approach [46] • Hybrid filter–wrapper approach [47]	• Effectiveness of the proposed disease diagnosis model • Much robustness against initialization in comparison with PSO and GA optimizers • Efficiency and effectiveness of the proposed algorithm • It superior to other well-established counterparts
Duplication	2008	• Length filtering and dynamic weighting (LFDW) for duplicate records cleansing [49]	• Time, recall and precision of LFDW are better than traditional ones
	2009	• Improved K-medoids clustering algorithm (IKMC) [50] • Standard duplicate elimination algorithm (SDE), sorted neighborhood algorithm (SNA), duplicate elimination sorted neighborhood algorithm (DE-SNA), and adaptive duplicate detection algorithm (ADD) [51] • Sub-fuzzy clustering property optimization method based on grouping [52]	• Good detection accuracy and high availability • Much better results with recursive algorithm with word base • Higher detection accuracy and efficiency
	2010	• Segment strategy based on character features and the algorithm of edit distance with variable weight [55] • Online record matching method, UDD [56] • Optimal feature selection method based on fuzzy clustering in groups [59]	• The running efficiency and detect precision can be improved • UDD works well • The identification accuracy and detection efficiency of this method is higher and it can solve the recognition problem of approximately duplicate records
	2011	• Sorted Blocks in several variants, which generalizes blocking and windowing approaches [61] • A method which takes into account both string and semantic similarity measures [62]	• Sorted Blocks outperforms the two other approaches • Outperforms the individual similarity measures
	2012	• Duplicate Count Strategy (DCS) and variation of SNM [64]	• DCS++ is more efficient than SNM without loss of effectiveness
	2014	• Progressive duplicate detection algorithms [65].	• It can double the efficiency over time of traditional duplicate detection
	2015	• Existing DCS++ and its variation [66]	

(continued)

Table 1 (continued)

Dimension	Year	Technique (method)	Findings
			• Variation in DCS++ producing better results in term of efficiency and accuracy
	2016	• Thorough analysis of literature survey on adaptive and progressive duplicate detection techniques [67]	• SNM identifies the duplicates efficiently within the window size but it is very expensive. DCS+ + improve the efficiency of Duplicate detection but they need to run for certain period of time and cannot maximize the efficiency for any given time slot

3 Conclusion

In this survey, we have discussed various adaptive techniques used to deal with data quality problems. As data preprocessing is an essential step to catch data quality defects, we present various existing approaches to catch and remove certain quality issues related to three data quality dimensions. Finally, we can conclude that there are many available techniques that can be assigned to each data quality problem and their effectiveness may vary from one case to another based on the surrounding case study.

References

1. Sidi, F., P. Hassany, S. Panahy, L.S. Affendey, M.A. Jabar, H. Ibrahim, and A. Mustapha. 2012. Data quality: A survey of data quality dimensions, 300–304.
2. Vasile, G., and O. Mirela. 2008. *Data quality in business intelligence applications*. Oradea: Analele Universitatii, 1364–1369.
3. Saar-tsechansky, M., and F. Provost. 2007. Handling missing values when applying classification models 8: 1625–1657.
4. Su, X., T.M. Khoshgoftaar, and R. Greiner. 2008. Using imputation techniques to help learn accurate classifiers. In *Proceedings of international conference on tools with artificial intelligence ICTAI* 1, vol. 1, 437–444.
5. Bashir, S., S. Razzaq, U. Maqbool, S. Tahir, and A.R. Baig. 2009. *Using association rules for better treatment of missing values 2 hybrid missing values imputation*. arXiv Prepr. arXiv 0904.3320, no. 4.
6. Mehala, B., P.R.J. Thangaiah, and K. Vivekanandan. 2009. Selecting scalable algorithms to deal with missing values 1 (2): 80–83.
7. Somasundaram, R.S. 2011. Evaluation of three simple imputation methods for enhancing preprocessing of data with missing values 21 (10): 14–19.
8. Malarvizhi, M.R., and A.S. Thanamani. 2012. K-NN classifier performs better than k-means clustering in missing value imputation 6 (5): 12–15.
9. Singhai, R. 2013. Comparative analysis of different imputation methods to treat missing values in data mining environment 82 (November): 34–42.

10. Ruiz, D.D.A., and L.C. Blomberg. 2013. Evaluating the influence of missing data on classification algorithm in data mining applications 0 (1): 734–743.
11. Rahman, G., and Z. Islam. 2013. Knowledge-based systems missing value imputation using decision trees and decision forests by splitting and merging records: Two novel techniques. *Knowledge-Based System* 53: 51–65.
12. Vohra, R. 2014. Estimation of missing values using decision tree approach 5 (4): 5216–5220.
13. Gimpy, M.D.R.V. 2014. Missing value imputation in multi attribute data set 5 (4): 5315–5321.
14. Rahman, G., and Z. Islam. 2015. Missing value imputation using a fuzzy clustering-based EM approach.
15. Purwar, A., and S.K. Singh. 2015. Hybrid prediction model with missing value imputation for medical data. *Expert System and Applications*.
16. Armitage, E.G. 2015. Missing value imputation strategies for metabolomics data 3050–3060.
17. Shohdy, S., Y. Su, and G. Agrawal. 2015. Accelerating data mining on incomplete datasets by bitmaps-based missing value imputation.
18. Yan, X., W. Xiong, L. Hu, F. Wang, and K. Zhao. 2015. Missing value imputation based on gaussian mixture model for the internet of things 2015.
19. Paul, A., J. Sil, and C. Das. 2017. Gene selection for designing optimal fuzzy rule base classifier by estimating missing value. *Applied Soft Computing Journal* 55: 276–288.
20. Mikkelsen, E.M., D. Cronin-fenton, N. R. Kristensen, and L. Pedersen. 2017. Missing data and multiple imputation in clinical epidemiological research 157–166.
21. Dalvand, B., F.H. Lotfi, and G.R. Jahanshahloo. 2017. Data envelopment analysis with missing values: An approach using neural network 17 (2): 2017.
22. Borges, H.B., and J.C. Nievola. 2007. Feature selection as a preprocessing step for classification in gene expression data 157–162.
23. Ming, W.U., and Y.A.N. Puliu. 2007. Feature selection based on difference and similitude in data mining 12 (3): 467–470.
24. Saeys, Y., I. Inza, and P. Larrañaga. 2007. A review of feature selection techniques in bioinformatics. *Bioinformatics* 23 (19): 2507–2517.
25. Yao, Y., and Y. Zhao. 2009. Discernibility matrix simplification for constructing attribute reducts 179 (5): 867–882.
26. Thangavel, K., and A. Pethalakshmi. 2009. Dimensionality reduction based on rough set theory: A review 9 (1–12).
27. Mahmud, W.M., H.N. Agiza, and E. Radwan. 2009. Intrusion detection using rough set parallel genetic programming based hybrid model 9 (10).
28. Lutu, P.E.N., and A.P. Engelbrecht. 2010. Expert systems with applications a decision rule-based method for feature selection in predictive data mining. *Expert Systems with Applications* 37 (1): 602–609.
29. Liu, H., H. Motoda, R. Setiono, and Z. Zhao. 2010. Feature selection: An ever evolving frontier in data mining. *Feature Selection Data Mining* 5: 4–13.
30. Hsu, H., C. Hsieh, and M. Lu. 2011. Expert systems with applications hybrid feature selection by combining filters and wrappers. *Expert Systems with Applications* 38 (7): 8144–8150.
31. Manimala, K., K. Selvi, and R. Ahila. 2011. Hybrid soft computing techniques for feature selection and parameter optimization in power quality data mining. *Applied Soft Computing Journal* 11 (8): 5485–5497.
32. Zhao, H., J. Chen, Y. Liu, Q. Shi, Y. Yang, and C. Zheng. 2011. The use of feature selection based data mining methods in biomarkers identification of disease. *Procedia Engineering*.
33. Ravisankar, P., V. Ravi, G.R. Rao, and I. Bose. 2011. Detection of fi nancial statement fraud and feature selection using data mining techniques. *Decision Support Systems* 50 (2): 491–500.
34. Novakovic, J., P. Strbac, and D. Bulatovic. 2011. Toward optimal feature selection using ranking methods and classification algorithms. *Yugoslav Journal of Operations Research* 21 (1): 119–135.

35. Suthaharan, S. 2012. Relevance feature selection with data cleaning for intrusion detection system.
36. Beniwal, S., and J. Arora. 2012. Classification and feature selection techniques in data mining 1 (6): 1–6.
37. Kannan, A., and G.Q. Maguire. 2012. Genetic algorithm based feature selection algorithm for effective intrusion detection in cloud networks.
38. Spolaˆ, N., E. Alvares, M. Carolina, and H. Diana. 2013. A comparison of multi-label feature selection methods using the problem transformation approach. *Electronic Notes in Theoretical Computer Science* 292: 135–151.
39. Azhagusundari, B., and A.S. Thanamani. 2013. Feature selection based on information gain 2: 18–21.
40. Sánchez-maroño, V.B.N. 2013. A review of feature selection methods on synthetic data 483–519.
41. Rabeti, H., and M. Burtscher. 2014. Feature selection by tree search of correlation-adjusted class distances. In *Proceedings of the international conference on data mining steering committee world congress in computer science computer engineering applied computer* no. Jan, 1.
42. Chen, J., S. Liu, W. Liu, X. Chen, Q. Gu, and D. Chen. 2014. A two-stage data preprocessing approach for software fault prediction.
43. Eesa, A.S., Z. Orman, A. Mohsin, and A. Brifcani. 2014. Expert Systems with applications a novel feature-selection approach based on the cuttlefish optimization algorithm for intrusion detection systems. *Expert Systems with Applications Journal* no. November, 1–10.
44. Tomar, D., and S. Agarwal. 2015. Twin support vector machine approach for diagnosing breast cancer, hepatitis, and diabetes.
45. Emary, E., H.M. Zawbaa, C. Grosan, and A.E. Hassenian. 2015. Feature subset selection approach by gray-wolf optimization 1–13.
46. Azar, A.T. 2015. A novel hybrid feature selection method based on rough set and improved harmony search.
47. Hu, Z., Y. Bao, T. Xiong, and R. Chiong. 2015. Hybrid filter—wrapper feature selection for short-term load forecasting. *Engineering Applications of Artificial Intelligence* 40: 17–27.
48. Elmagarmid, A.K., P.G. Ipeirotis, and V.S. Verykios. 2007. Duplicate record detection: A survey. *IEEE Transactions on Knowledge and Data Engineering* 19 (1): 1–16.
49. Huang, L., H. Jin, P. Yuan, and F. Chu. 2008. Duplicate records cleansing with length filtering and dynamic weighting. In *2008 fourth international conference on semantics knowledge and grid* 2003: 95–102.
50. Pei, Y., J. Xu, Z. Cen, and J. Sun. 2009. IKMC: An improved K-medoids clustering method for near-duplicated records detection 0–3.
51. Rehman, M., and V. Esichaikul. 2009. Duplicate record detection for database cleansing. In *2009 2nd international conference, machine vision, ICMV* 333–338.
52. Thinking, A.S., and P. Optimization. 2009. A property optimization method in support of approximately duplicated.
53. Yujiu, Y., W. Liu, and X. Liu. 2009. An integrated approach for detecting approximate 0–3.
54. Issa H. 2010. Application of duplicate records detection techniques to duplicate payments in a real business environment. *Rutgers Busch School Rutgers University.*
55. Zhe, L.I.U. 2010. An algorithm of detection duplicate information based on segment 156–159.
56. Su, W., J. Wang, F.H. Lochovsky, and I.C. Society. 2010. Record matching over query results from multiple web databases 22 (4): 578–589.
57. Beskales, G., M.A. Soliman, I.F. Ilyas, S. Ben-david, and Y. Kim. 2010. ProbClean: A probabilistic duplicate detection system 1193–1196.
58. Naumann, F., and M. Herschel. 2010. An introduction to duplicate detection.
59. Hua, Q., M. Xiang, and F. Sun. 2010. An optimal feature selection method for approximately duplicate records detecting.

60. Issa, H., and M. Vasarhelyi. 2011. *Duplicate records detection techniques: Issues and illustration*. Available SSRN 1910473, no. August, 2011.
61. Draisbach, U., and F. Naumann. 2011. A generalization of blocking and windowing algorithms for duplicate detection, 18–24.
62. Ektefa, M., H. Ibrahim, and S. Memar. 2011. A threshold-based similarity measure for duplicate detection, 37–41.
63. Subramaniyaswamy and C. Pandian. 2012. A complete survey of duplicate record detection using data mining techniques.
64. Draisbach, U., F. Naumann, S. Szott, and O. Wonneberg. 2012. Adaptive windows for duplicate detection, 1073–1083.
65. Papenbrock, T., A. Heise, and F. Naumann. 2014. Progressive duplicate detection 4347 (c): 1–14.
66. Skandar, A., M. Rehman, and M. Anjum. 2015. An efficient duplication record detection algorithm for data cleansing 127 (6): 28–37.
67. Dhivyabharathi, G.V., and S. Kumaresan. 2016. A survey on duplicate record detection in real world data. In *ICACCS 2016—3rd international conference on advanced computing communication system bringing to table, futuristic technologies from arround globe*, 1–5.

The Principle Internet of Things (IoT) Security Techniques Framework Based on Seven Levels IoT's Reference Model

Amira Hassan Abed, Mona Nasr and Basant Sayed

Abstract Basically, by increasing Internet of Things (IoT) paradigm involvement in our lives, a lot of threats and attacks regarding IoT security and privacy are realized and raised. If they are left without taking extensive considerations, such security and privacy issues and challenges can certainly threaten the IoT existence. Such security and privacy issues are derived from many reasons. One of these reasons is the unavailability of universal accepted appropriate IoT security techniques, methods, and guidelines. These methods and techniques will greatly guide IoT developers and engineering, draw the success road for developing, and implementing secure IoT systems. So, our contribution focusses on such objective in which, we propose a comprehensive IoT security and privacy framework based on the seven levels of IoT reference architecture introduced by Cisco, in which a set of proper security techniques and guidelines is specified for each level. Additionally, we identify several critical techniques which can be accomplished for blocking many possible attacks against the IoT seven levels.

Keywords Internet of things · IoT security techniques · IoT security framework · IoT reference model (RM) levels

A. H. Abed (✉)
Department of Information Systems Center, Egyptian Organization for Standardization and Quality, Cairo, Egypt
e-mail: mirahassan61286@gmail.com

M. Nasr
Faculty of Computers and Information Helwan University, Department of Information Systems, Helwan, Egypt
e-mail: m.nasr@helwan.edu.eg

B. Sayed
Department of Information Systems, Higher Institute of Qualitative Studies, Garden, Egypt
e-mail: basantaly.26@gmail.com

A. Z. Ghalwash et al. (eds.), *Internet of Things—Applications and Future*, Lecture Notes in Networks and Systems 114,
https://doi.org/10.1007/978-981-15-3075-3_15

1 Introduction

The Internet helped individuals to interface with static data accessible yet now its building association from individuals to individuals, individuals to physical articles, and physical items to other physical articles. IBM gauges that 90% of all information produced by gadgets like tablets and Smartphone and so on are never dissected. What's more, as much as 60% of this information begins losing an incentive inside milliseconds of being produced. According to gauges from late reports, there will be 30 billion Internet associated and sensor empowered gadgets by 2020. The fast development and joining of Internet information and procedures are making organized associations more applicable and important. Additionally, it makes energizing business open doors for ventures. IoT items and administrations will produce income past $300 billion and include $1.9 trillion in the worldwide economy by 2020.

1.1 What Is Internet of Things (IoT)

The Internet of Things (IoT) is a figuring idea that depicts a future where consistently physical items will be associated with the Internet and have the capacity to recognize themselves to different gadgets. IoT is a system of gadgets which imparts among itself utilizing IP availability without human obstruction. IoT biological community comprises keen articles, canny gadgets, smartphones, and tablets and so forth. It will utilize Radio-frequency identification (RFID), Quick Response (QR) codes, sensor or remote innovation to empower intercommunication between gadgets.

The IoT is a promising worldview, which coordinates a substantial number of heterogeneous and unavoidable items with different associating and figuring capacities, which are gone for giving a perspective of the physical world through the system [1]. IoT speaks to the advancement from a venture-driven Internet to a keen protest driven worldview, in which physical articles are being empowered with encompassing knowledge and system capacities permitting a consistent information trade. This data is totaled, handled and used to give inescapable and imaginative administrations, changing our encompassing condition into a keen and universal space [2] (Fig. 1).

It is normal that IoT innovation will make ready for momentous applications in a decent variety of territories, for example, medicinal services, security and observation, transportation, and industry, and that it will have the capacity to coordinate advances, for example, propelled machine-to-machine correspondence, autonomic systems administration, basic leadership, secrecy insurance and security, and distributed computing with cutting edge identification and activation advances.

Actually, IoT includes both static and dynamic objects of the physical world (physical things) and the data world (virtual world), which can be recognized and

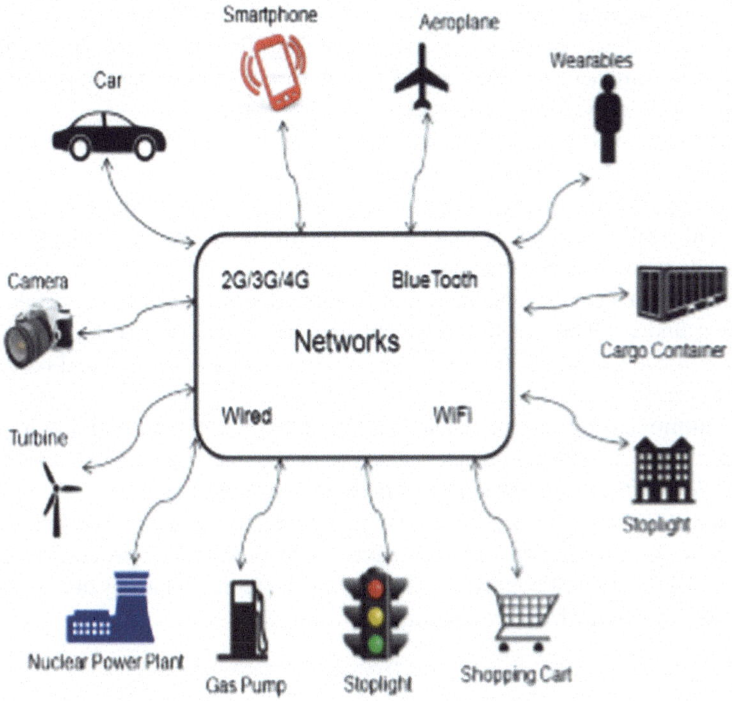

Fig. 1 Internet of Things (IoT)

coordinated into correspondence systems. The basic highlights of IoT include (I) interconnectivity, (ii) things-related administrations, for example, security assurance and semantic consistency, (iii) heterogeneity, (iv) support of dynamic changes in the state and the quantity of gadgets, and (v) tremendous scale [3]. Average cases can be Smartwatch conversing with your auto or ice chest conversing with market and so on. Internet of Things (IoT) incorporates everything from an implanted sensor in a petroleum pump, sensors in office building which can follow your area to show your records on the closest screen.

1.2 Characteristics of Internet of Things (IoT)

The disseminated groups can work over their areas to make tradeoffs in plan structures and plans of action utilizing Internet of Things (IoT). A few attributes of Internet of Things (IoT) as recorded beneath.

A. **Interconnected**: It encourages individuals to gadget and gadget to gadget interconnection.

B. **Smart Sensing**: The gadgets associated with IoT will have keen detecting abilities. For instance, utilization of movement sensors to turn lights on or off. The detecting innovation makes encounters that mirror a genuine familiarity with the physical world, individuals, and articles.

C. **Intelligence**: The IoT associated gadgets can have knowledge appended with them. For instance, Nest Learning Thermostats are Wi-Fi empowered, sensor-driven furnished with self-learning abilities. The Misfit Shine is a wellness tracker with rest observing capacities. The Misfit Shine can disseminate register assignments between a Smartphone and the cloud.

D. **Spare Energy**: IoT gadgets like Motion Sensor Light have in-fabricated movement finder which can turn the light on when it detects development. It can spare part of energy vitality from wastage and lift vitality collecting and productive usage of energy.

E. **Communicating**: IoT associated gadgets have one of a kind ability to advise the present state to other associated gadgets in the encompassing. It encourages better correspondence stream among human and machines.

F. **Safety**: IoT associated gadgets can help guarantee security of individual life. For instance, a moving auto tires can advise their present state to auto proprietor having brilliant auto dashboards, it will help anticipate miss chances due to blasting of auto tires because of overheating and so forth.

1.3 Why Internet of Things (IoT)

The point of IoT is to fill our hearts with joy to day life more secure and more productive. As indicated by examine from Business Insider, shrewd homes will turn into a $490 billion business by 2019. More IoT gadgets mean expanded productivity and more interconnected world. A portion of the advantages of IoT is examined underneath.

A. **Smarter Analytics**: More number of between associated gadgets implies more information and applying examination on all parts of the business will give a chance to enhance technique and the client encounter. For instance, Intel IoT Platforms arrangements produce significant data by running investigation programming on information for modern, retail, car businesses, and so forth.

B. **Improved Security**: The keen doorbells and reconnaissance frameworks will help distinguish and perceive the individual which will support security.

C. **Expanded Productivity**: IoT will encourage ideal usage of asset and time. For instance, if a printer is running low on ink, it will arrange more without anyone else and spare valuable time. What's more, it will send warning if printer machine is not working appropriately.

D. **Smart Inventory**: Businesses will have the capacity to follow merchandise in the production network with Internet-associated stock. It will give improved in-travel deceivability.

E. **More Secure Travel**: The Internet-associated autos can a superior feeling of constant movement conditions and vehicle diagnostics which will make travel more secure.

F. **Real-Time Demand Visibility**: The firmly coupled stockroom and request framework give better ongoing interest deceivability.

1.4 Internet of Things (IoT) Business Opportunities

According to Forbes report Internet of Things (IoT) in human services will be \$117 billion Market By 2020. The development of Internet of Things (IoT) makes open doors for organizations and individuals with information security arrange plan and information investigation abilities. The significant system suppliers like IBM, CISCO, GE, and Amazon have chosen to help IoT with option of Fog layer and plans to include Swarm layer which will disentangle and diminish cost for arrange availability. GE appraises that Industrial Internet can possibly add \$10 to \$15 trillion to worldwide Gross Domestic Product (GDP) throughout the following 20 years. While Cisco Systems report 99% of physical items will in the long run turn out to be a piece of an associated arrange. The retail and coordination's part is a key territory where Internet of Things (IoT) will have enormous effect. The SAP HANA Cloud Platform for the IoT gives the foundation to empower organizations to take advantage of a system of a great many associated gadgets.

The Internet of Things (IoT) in human services space is reclassifying tolerant care, healing center operations, medicine conveyance, secure record get to and so on. The carriers are putting resources into the Internet of Things (IoT) which can make another upset in the traveler encounter. The IoT speculations via aircrafts will be in the zones of registration, stuff drop, and claim. What's more, airplane terminal structures, types of gear, packs, and trolleys will be associated and imparting to each other. Aircrafts are utilizing Apple's iBeacons to upgrade client travel involvement. For instance, Virgin Atlantic is utilizing iBeacon innovation at Heathrow airplane terminal and American Airlines is sending iBeacons at Dallas/ Fort worth (DFW) International Airport. The systems administration goliath Cisco is growing IoT portfolio with 15 new items for organizing availability and security. It's pushing for idea of Internet of Everything since long time and its extended IoT methodology with the dispatch of the Cisco IoT framework which is a structure to streamline Internet of Things (IoT) organizations. The Internet of Things (IoT) showcase is anticipated > \$1 trillion for 2020. The oversaw administrations will contribute 58%, arrange administrations 39%, and equipment enablement 4% to general Internet of Things (IoT) advertise.

The Internet of Things (IoT) is at appropriation organize very like distributed computing amid 2008–10 and enormous information investigation around 2011–2013. As per Gartner report there will be a quarter billion associated vehicles out and about by 2020 with in-vehicle benefits and computerized driving abilities. The

Internet of Things (IoT) is assuming control Big Data as the most advertised innovation according to Gartner Hype Cycle for developing advancements 2014.

1.5 *Internet of Things (IoT) Challenges*

The security viewpoint is the greatest worry of Internet of Things (IoT) associated gadgets. The IoT application information can be close to home, venture, mechanical, or customer yet put away information ought to be secured against robbery, altering, and ensured in the travel and very still. For instance, an IoT application may stores streams and chronicled information of person's wellbeing, shopping conduct, area, funds and amount of stock, business orders, and so on. The IoT energizes another level of outsourcing, however, there are worries around benefit accessibility, adaptability, reaction time, value structure and protected innovation possession, and so forth.

In the meantime, various difficulties are impeding the IoT. Regarding versatility, IoT applications that require huge quantities of gadgets are frequently difficult to execute as a result of the confinements on time, memory, preparing, and vitality limitations. For instance, estimation of everyday temperature varieties around the majority of the nation may require a large number of gadgets and result in unmanageable measure of information. What's more, the sent equipment in IoT regularly have diverse working qualities, for example, examining rates and blunder dispersions, then sensors and actuators parts of IoT are constantly exceptionally mind boggling. These components add to the development of the heterogeneous system of IoT in which the information of IoT will be profoundly heterogeneous. Besides, it is costly to transmit immense volume of crude information in the unpredictable and heterogeneous system, so IoT requires information pressure and information combination to diminish the information volume. Thus, institutionalization of information handling mindfulness for future IoT is exceptionally wanted.

Security is a worry when information is transmitted over the Internet or even secured private systems and VPN burrows. The Government controls, for example, Health Insurance Portability and Accountability (HIPA) Act or confinements on transporting information crosswise over global fringes can be connected as wellbeing measures. The key IoT security undertakings ought to guarantee that appropriate application-level assurances like Distributed Denial of Service (DDoS) assault relief are set up. It should likewise fuse measures to affirm the character of elements asking for access to any information including multi-factor confirmation.

A. **Information Privacy**: Smart TVs are gathering information about survey propensities and in some cases, the shaft listened in discussions back to a maker.

B. **Information Security**: The Internet of Things (IoT) is enabling information to be exchanged consistently from observation gadgets to the Internet to empower live investigation. In any case, information security still remains a test here.

C. **Insurance Concerns**: The independent autos are including protection industry concerns. In any case, information will make it simpler to evaluate dangers and it gives a chance to new valuing models. For instance, protection premium tuning in light of wellbeing and driving information.

D. **Absence of Common Standards**: There is not kidding absence of brought together standard for IoT and accomplishing an industry wide acknowledgment of one bound together standard is an enormous test.

E. **Technical Concerns**: Each IoT gadget can create gigantic measure of information. It is a test to store, secure, and break down. The system ought to have the capacity to deal with high volume and thickness of the gadgets. Additionally, it ought to be fit to distinguish and separate amongst allowed and rebel gadgets.

F. **Social and Legal Concerns**: There is no instrument to address these social and lawful concerns. For instance, who claims the video gushing in from Google Glass and the social insurance-related date spilling from other wearable gadgets? What will happen when self-governing gadgets come up short on control? [4].

1.6 Internet of Things (IoT) Applications

There are numerous application areas of IoT, running from individual to big business conditions. The applications in individual and social space empower the IoT clients to communicate with their encompassing condition, and human clients to keep up and fabricate social connections. Another use of IoT is in transportation space, in which different shrewd autos, savvy streets, and brilliant movement signals fill the need of sheltered and helpful transportation offices. The undertakings and enterprises space include the applications utilized as a part of fund, keeping money, advertising, and so forth to empower distinctive between and interactivities in associations. The last application area is the administration and utility observing division which incorporates farming, rearing, vitality administration, reusing operations, and so on. The IoT applications have seen quick improvement as of late because of the advances in Radio-Frequency Identification (RFID) and Wireless Sensor Networks (WSN).

The RFID empowers the labeling or marking of each and every gadget, to fill in as the fundamental distinguishing proof component in IoT. Because of WSN, every "thing", i.e., individuals, gadgets, and so on turns into a remote identifiable protest and can convey among the physical, digital, and advanced world.

1.7 IoT Architecture

In IoT, each layer is characterized by its capacities and the gadgets that are utilized as a part of that layer. There are diverse suppositions with respect to the quantity of layers in IoT. In any case, as indicated by numerous analysts, the IoT mostly works on three layers named as Perception, Network, and Application layers. Each layer of IoT has inborn security issues related to it. Figure 2 demonstrates the

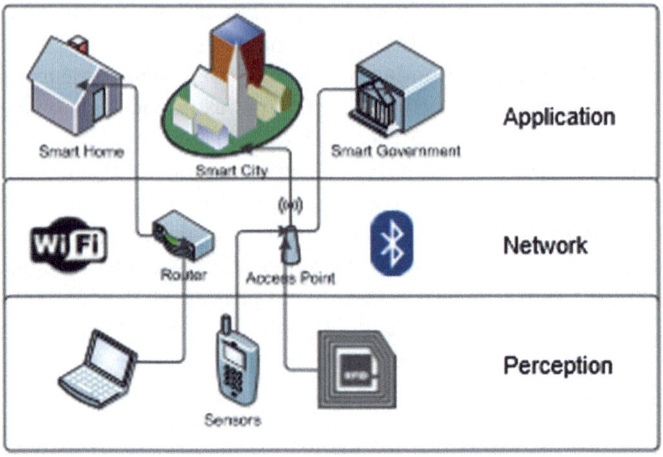

Fig. 2 IoT architecture

fundamental three-layer design system of IoT as for the gadgets and advances that envelop each layer.

A. **The "Perception Layer"**: Often known by the "Sensors" layer, it mainly subject to acquire and gather high volume of data from IoT devices-based extensively on several sensors and actuators. Thus, it has an interesting role on detecting, collecting, analyzing and processing data, and transmitting it to the next layer. Moreover, **Perception** layer executes the "IoT*'s nodes collaboration*" in locally and shortest range networks.

B. **The "Network Layer"**: It performs data routing and transmission functions to several IoT devices over the Internet. Basically, in this layer, several systems like cloud computing platforms, Internet gateways, switching, and routing devices are operating through depending on using many technologies such as Wi-Fi, LTE, 3G, Zigbee, etc.

C. **The "Application Layer"**: The application layer ensures the legitimacy, uprightness, and classification of the information. At this layer, the reason for IoT or the making of a keen domain is accomplished [5].

2 State-of-the-Art

In spite of the importance occupied by IoT security and privacy guidelines and their execution techniques and methods supported for securing many IoT systems, a few research contributions (almost are white papers) have been extensively proposed in the state-of-the-art, presented as the following:

By referring to reference [6], the **USDHS**, "*United States Department of Homeland Security*", it indicated several security significant practices and support an interesting principles list to improve awareness among IoT's stakeholders for enhancing IoT security. The "USDHS", however, not introduce any attacks that potentially affect IoT, nor support applicable security solutions to explain execution for the mentioned principles and practices.

In the work [7] introduced by the **OWASP**, "*Open Web Application Security Project*", it presented a group of guidance for enhancing IoT security based on various IoT dimensions including authorization and authentication, cloud interface schema, many network's services, and different physical objects. Furthermore, the OWASP determined interesting IoT stakeholders to be involving developers, customers, and manufacturers in which this guidance may help them to optimize IoT security, however, the OWASP not covers any countermeasures, attacks, and threats that possibly face IoT security.

In a research paper [8], the authors seriously proposed a comprehensive number of privacy guidelines for securing various IoT applications and middleware software that rely on it to evaluate two open IoT middleware paradigms: "Open IoT" and "Eclipse Smart Home". One limitation in their work, they do not talk about proper countermeasures to implement their presented guidelines and not identify feasible attacks blocking IoT privacy.

According to "*The* IoT *Security Foundation*" **IoTSF** [9] where they introduced a set of security and privacy guidelines organized for optimizing IoT applications, networks, operating systems, wireless and wired interfaces, cloud, object hardware, and mobile applications. Moreover, the work has identified the interesting parts who may extensively benefit for applying these guidelines. However, it not providing any suitable countermeasures to implement these guidelines or reported any challenges against IoT.

In addition to reference [10], in which the **BITAG** "*Broadband Internet Technical Advisory Group*" established an advanced-level category of IoT security and privacy guidance for various aspects including data, communication, object, and services. But actually, the **BITAG** did not report any comprehensive guidelines, nor describe any required countermeasures to establish these guidelines. In addition, attacks against IoT security remain also untouched.

Finally, in reference [11] The **ENISA** stands to "*European Union Agency for Network and Information Security*" it established a set of guidelines that can be used to mitigate many potential attacks and threats block IoT security. The base objective for this work is to support an extensive understanding for many requirements aspects needed to optimize IoT security and also to identify at an abstract level a number of IoT security challenges.

3 The Principle of IoT Security Techniques Framework

Based on the above state-of-the-art studies, for addressing the limitations in these previous works we must start by identifying IoT levels precisely which they are too complex, as IoT evolved from the combination of a large number of sophisticated applications, systems, tools, and technologies. Nevertheless, there are various IoT reference models (RM) which were introduced by many contributions such as a three-level model by [12]: WSN, cloud servers, and Application, a five-level model introduced by [13]: Edge nodes, Object abstraction, service management, service composition, and Application, and a seven-level model that proposed by Cisco: Edge nodes, Communication, Edge computing, Data accumulation, Data abstraction, Application and Users and Centers [14] have been proposed in the state-of-the-art.

In order to construct a security and privacy IoT framework supporting an extensive sets (blocks) of techniques and methods according to each IoT level, we effectively used the CISCO's IoT reference model RM that built in 2014 as a significant extension of the five levels and the three levels reference models in this work, as shown in Figs. 3, 4, 5. The reason why we using this model is, such the CISCO's IoT reference model is greatly simplified the complexity of IoT paradigm by divided it into several seven independent levels.

Fig. 3 3Layers model [12]

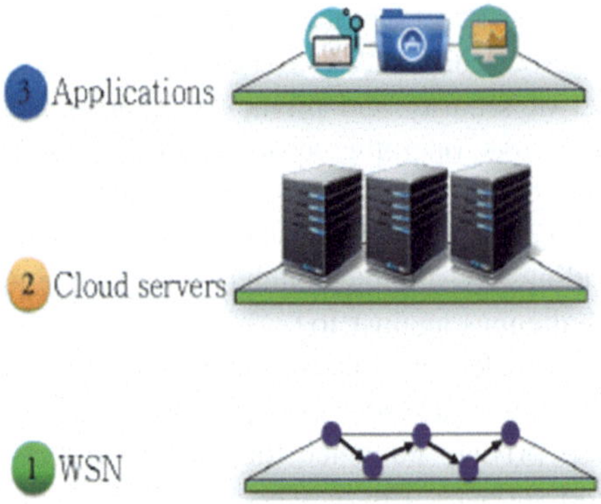

Fig. 4 5Layers model [13]

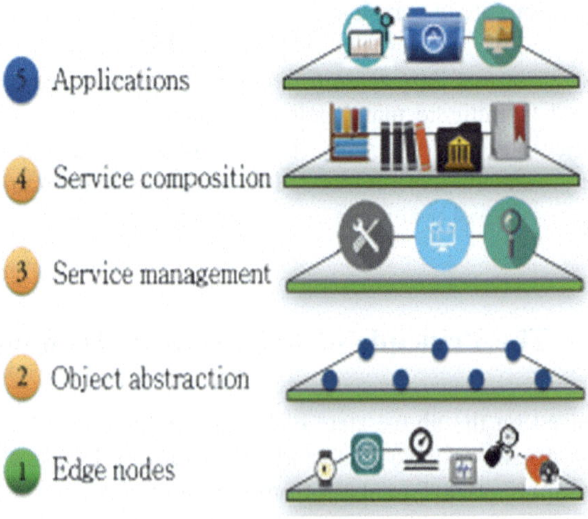

Fig. 5 CISCO's 7Layers model [14]

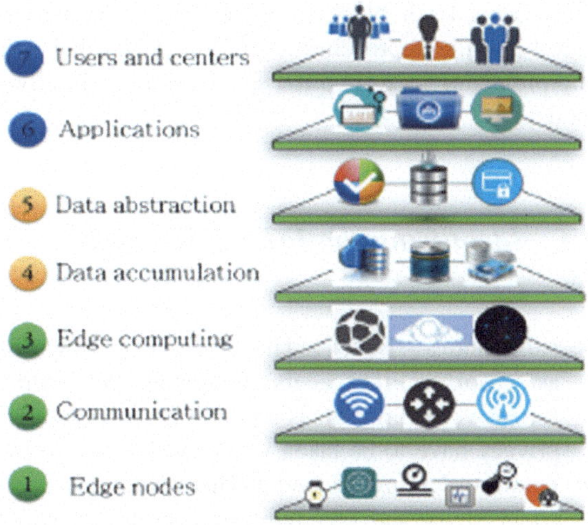

In the following, we proposed the IoT **security techniques framework** as presented in Fig. 6. The framework consists of two parts: the first side represents IoT's seven levels which would identify in the following section, and the second part reports a number of appropriate techniques and methods which organized on

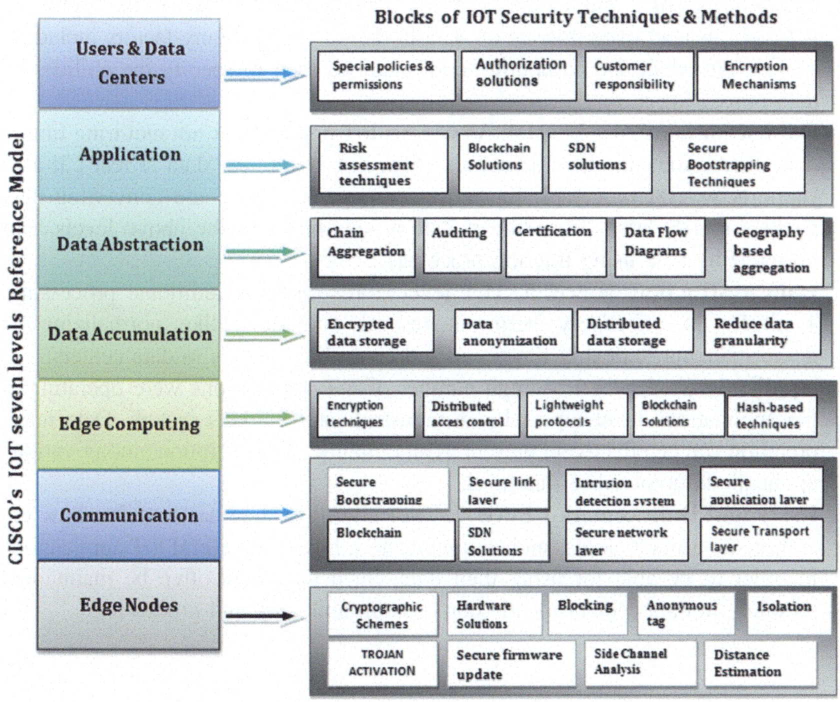

Fig. 6 The IoT security techniques framework

blocks/sets. Each block from the seven proposed blocks of security techniques specifying the approaches, tools, guidelines that will help effectively on securing and protecting its related IoT level. Now, we briefly identify each level of this IoT model.

3.1 Identifying the IoT's RM Levels

- **Edge nodes—Level 1**: This level is consisting of several computing nodes such as sensors, RFID readers, microcontrollers, and different types of RFID tags. Extensive security goals and objectives including integrity, confidentiality, and privacy have to be taken into consideration for this level seriously.
- **Communication—Level 2**: This level involves all enabler technologies like the connectivity and communication protocols which supporting commands and data transportation from objects in the Edge nodes level and objects located at the Edge computing level.
- **Edge computing—Level 3**: One of this level main objective is to execute simple data processing for reducing the computation workload in the above levels and supports a quick and fast response. It is powerful and suitable for the nature of real-time applications to easily operate on data closer to the fog of the network, instead of to operate on data in the far cloud. Many factors including service providers and computing nodes can be used to specify the abilities of data processing at this level.
- **Data accumulation—Level 4**: As various IoT applications not requiring timely data processing, this level transmits "data in motion" to "data at rest". It also supports several processes, the most common, of which are converting data packets to DB tables, deciding if data is significant to the above levels, and normalizing data using filtering processes.
- **Data abstraction—Level 5**: This level stores data for additional processing. Typically, it effectively performs several operations like normalizing or de-normalizing, indexing functions, and controlling access to data centers.
- **Applications—Level 6**: A high number of IoT applications were operating at the applications level especially in industries and markets sectors. Data interpretation can be effectively present as an outcome of cooperation among various applications involved in this level.
- **Users and Data centers—Level 7**: Typically, this level has to only allow the authorized users to access and communicate safely with several IoT applications in order to be able for using their data. Such data may often be maintained remotely in data centers for the purposes of analyzing and processing.

3.2 Security Techniques and Methods Proposed for IoT's RM Levels

Sid-channel analysis: It supports a significant techniques and tools which are able to detect potential attacks and threats like hardware Trojans and also malicious firmware on IoT objects.

1. Detection of Hardware Trojan: Detecting hardware Trojan has been indicated by many works through developing various side-channel signals analysis, such as Using Test vector selecting method and path delay architecture [15], providing off the shelf methods such as power analysis report [16] and comparing the power consumption between a Trojan-free IC and a Trojan-inserted IC as mentioned in work by [17]. The appearance of a Trojan within the IoT object usually has several effects on its components, the most well-known of these effects are on the "power and gates" also, it may change the distribution of heat on the IC. In addition, the authors [18] suggested "the lightweight dynamic permutation technique" for securing the integration of circuits from Trojan and side-channel threats and attacks through dynamic modifying the actual order of received data from sensors. The recent extensive surveys covering the hardware Trojan classification and serious detection were introduced by [19, 20].

2. Malicious Firmware Detection: The side-channel analysis capabilities in malicious firmware detection were introduced through many comprehensive researches such as works by [21, 22]. Malware detection method is able to analyze side-channel signals to find out any abnormal behaviors from any IoT objects.

Cryptographic schemes: The well-known and common cryptographic techniques, are encryption, hash-based functions, and lightweight protocols, which widely used in many researches to support many suggested security and privacy solutions that mitigate and protect possible attacks.

1. Encryption: Symmetric encryption and asymmetric encryption can effectively use as a powerful technique as the guidelines against various kinds of security and privacy attacks. Many indicated solutions for securing and protecting an instance's boot loader using "symmetric key methods" were reported and presented in [23]. On the other hand, a little number of solutions for protecting an instance boot loader using "asymmetric key methods" have been developed in reference [24]. One of recent published interested survey effort in this area was introduced by [25].

2. Hash-based techniques: These techniques are depending extensively on them for discovering RFID systems security issues, Different effective solutions have been supported in high number of literature. In work by [26], the authors indicated Interesting security methods and tools for enhancing RFID systems protection, one of them is mainly built on "hash function". Such a mention technique in their work providing two principles within every tag: (1) the locked state in which a tag can answer all received queries using own hashed value and (2) the unlocked state in which the tag conducts its normal process.

3. Lightweight protocols: Build RFID tags at lower cost considers one of the requirements in RFID technology, which hindering the traditional cryptographic techniques development. However, a lot of "lightweight cryptographic" protocols were executed by [26]. According to [27], the authors built an extensive "lightweight mutual authentication protocol" to secure RFID tags, in which the implementation required just "300 gates". The authors seriously reported that their protocol provides a security in an acceptable level for many applications.

Blocking: In research [28], the authors have developed a sophisticated method, which they referring to it as "blocking" to protect and secure the RFID tags privacy. In this approach, it requires a certain type of tags know called "a blocker tag". Also, according to [29], it developed "a soft blocking" technique which mainly based on a reader's configuration for compelling a list of policies supports with a system. These policies have a role of ensuring that readers are just reading public tags. Attacking such tag policies from a reader may detect by a device monitor.

Securing firmware update: There are two approaches can be provided in order to update firmware: (1) a remote updating and (2) a direct updating. In the cases of a remote firmware updating, a command for reporting the existence of a new firmware version is broadcast by a base server. Then, a server with a new updated version propagates this reporting to the closest nodes. Based on receiving such update, the nodes need to update their firmware compare their available firmware with new version and send accepts if they willing to do this update. Finally, the command starts transmitting data to the nodes. To ensure a secure technique for remote nodes updating their firmware, all the transmitting data, requests, and responses must be authenticated and encrypted.

Intrusion detection system: IDSs subject to making sure that the system policies are not attacked based on "a continuous monitoring process". Basically, IDSs support an effective technique against "battery draining "and "sleep deprivation" threats and attacks through successfully observing unusual requests to nodes. Several studies and research were conducted to focus on how to monitor the edge nodes and blocking any attacks within that level [30, 31].

Anonymous tag: The authors in [32],prepared a new method based on a lookup table hashing to secure the privacy of RFID tags. The significant contribution of this research is to maintain a mapping between both anonymous ID and genuine ID to not allow attacker to find the hashing structure to identify genuine ID from the anonymous user although emitting the anonymous IDs by tags, attackers may have abilities to track RFIDs if their IDs do not modify over time. Therefore, the anonymous ID must be modified continuously by blocking the tacking attack.

Distance estimation: Determining the distance from a tag to a reader is based mainly on signal to noise ratio as it explained in reference. The authors reported that it can determine a metric in which the distance of a reader attempting to reading a tag data is estimated. This supports the tag to just provide distance according to information.

Trojan activation methods: The key objective for a Trojan activation technique is to partially or fully enable the Trojan circuity thus it leads to detected hardware Trojan various Trojan activation methods were represented in many comprehensive researches [33, 34], another main objective for that techniques is the detection of the differences features between "Trojan-free circuit "and "Trojan-inserted circuit".

Isolation: One of the well-known approaches to protect and secure the RFID tags privacy is to isolate them from any EM waves. One suggested technique is the construction and utilization of the separation rooms. In this way, however, it is very expensive to do implement that in real world [35]. An additional recommended solution is established by an isolation container developed by metal in which it used to prevent EM waves. These containers were referred to them as the Faraday cages [36].

Customer responsibilities: Although the importance of the previous pointed security techniques to protect IoT against attacks and threats, the customers also have a critical role for preventing some IoT attacks. For instance, modifying the default passwords for IoT devices and systems lies in customer's to secure Dos attacks. In [37], the authors defining a number of IoT malwares such as "Mirai", "Carna", and "BASHLITE" which can lead to serious DDos attacks in IoT environment through using the default credentials, as almost IoT devices and systems are received to customers with default passwords in which they are not modified by customers.

Bootstrapping Security Techniques: In work [38], the authors indicated that the establishment of secure and successful bootstrapping techniques based extensively on their constructions fashion either by distributed manner or by centralized manner. For distributed construct, two IoT items can arrive for the agreement on a private using a "Die-Hellman algorithm."

But on a centralized architecture, the distribution of operational keys in every of the security domains relies on a single node holds certificates or pre-defined keys. The implementation of such centralized architectures in IoT environments was discovered in many studies such as [39] who recommend the utilizing of the Protocol for executing "Authentication for Network Access PANA", to convey the "Extensible Authentication EAP Protocol" communication between a PANA server and a PANA user.

Adding security at link layer: The IP-based communication between objects within IoT environment heavily relies on 6loWPAN protocol [40], which basically based on "*the IEEE 802.15.4*" link layer. It has the capability to supporting several security goals such as confidentiality and integrity [41]. As, the "IEEE 802.15.4 link-layer" provides hop to hop security where each part in the communication link must be trusted without existence of timely synchronized communications, authentication, and key management. To overcome the lack of communication time-synchronized, a modification of the "IEEE 802.15.4" was supported by IETF and known as "IEEE 802.15.4e".

Blockchain-based solutions: Blockchain is considered an emerging recent technology that has formed the shape of cryptocurrency such as bitcoin, objecting to building many kinds of transactions and communications between parts in a distributed schema without any requirements for supporting centralized trust bodies. Furthermore, a trust paradigm between objects and parts is not needed. In such technology, once the requested transaction is validated, it is impossible to reject it. Furthermore, its use in the cryptocurrency, high numbers of researchers and authors were focused on its roles as a powerful tool for different IoT security and privacy challenges. Several proposed works in that topic were tackled about the HTTPS protocol for IoT systems and devices [42], a multilayer security architecture designed for the smart cities [43], a decentralized IoT environment management [44], a suggested model IoT for industrial sector [45], and finally, the trust transactions chain [46].

Authorization solutions: In order to allowing system access to authorized requests, many authorization mechanisms must be applying during developing IoT systems [47]. Several authorization techniques and mechanisms were validated in which they should be suitably used if two parts participated in communication. The most well-known authentication methods are role-based access control and attribute-based access control. According to attribute-based access control changes privileges to a list of attributes specified to an item, whereas role-based access control changes privileges to a list of roles specified to an item. Another technique that can be effectively used to optimize authorization for IoT items is known as "Authentication and Authorization for Constrained Environments ACE" [48]. Several works are related to ACE can be found in references [48, 49].

Policies and permissions: Policies and permissions have to be stated and governed for accessing and controlling the IoT environments. Sending and receiving traffic and the access request of the system can be allowed or prevented depend on effective access control techniques.

Risk Assessment techniques: Risk assessment techniques have capabilities to detect and discover threats of the IoT system, so, the application layer should be secured by the risk assessment. For this purpose, updating the firmware of the system devices for successfully strengthen and optimize the security measures.

4 Conclusion

Basically, IoT subjects to integrate several advanced technologies of "communication, networking, cloud computing, sensing, and actuation," and pave the way for groundbreaking applications in a variety of areas, which will have a great effect on many areas in people's lives and provide them with a lot of conveniences means. Thus, the huge number of connected devices that are potentially vulnerable, highly significant attacks, threats and risks raise around the security, privacy, and governance of IoT.

This research paper focuses mainly on how to secure and protect IoT environment against many possible attacks, threats, and challenges, and presents a significant list of security techniques and guidance for this regard. Therefore, in this work, we develop an extensive IoT security framework based on the seven levels of CISCO'S reference model, namely, edge nodes and communication, edge computing to mitigate threats and attack associated with these levels. In this respect, we also illustrate the use of two new emerging technologies called blockchain and SDN to address some IoT security challenges.

References

1. Nitti, M., Pilloni, V., Colistra, G., and Atzori, L. 2017. The virtual object as a major element of the internet of things: a survey. *IEEE Communications Surveys & Tutorials* 18(2): 1228–1240.
2. Bernabe, J.B., Ramos, J.L.H., and Gomez, A.F.S. 2018. TACIoT: multidimensional trust-aware access control system for the internet of things. *Soft Computing* 20(5):1763–1779.
3. Atzori, L., Iera, A., and Morabito, G. 2017. The internet of things: a survey. *Computer Networks* 54(15):2787–2805.
4. Singh, S., and Singh, N. 2018. Internet of Things (IoT): security challenges, business opportunities & reference architecture for E-commerce. In: *2018 International conference on green computing and internet of things (ICGC*IoT*)*. IEEE 1577–1581.
5. Mahmoud, R., Yousuf, T., Aloul, F., and Zualkernan, I. 2017. Internet of Things (IoT) security: current status, challenges and prospective measures. In: *2015 10th International conference for internet technology and secured transactions (ICITST)*. IEEE, 336–341.
6. U.S. Department of Homeland Security. 2016. Strategic principles for securing the Internet of Things (IoT) Introduction and overview; U.S. Department of Homeland Security: Washington, DC, USA, 1–17.
7. OWASP. 2019. IoT security guidance OWASP. https://www.owasp.org/index.php/Main_Page Accessed 8 April 2019.
8. Perera, C., McCormick, C., Bandara, A.K., Price, B.A., Nuseibeh, B. 2016. Privacy-by-Design framework for assessing internet of things applications and platforms. In: *Proceedings of the 6th international conference on the internet of things*—IoT*'16*, Stuttgart, Germany, 7–9 Nov 2016, 83–92.
9. IoT Security Foundation. 2016. *IoT security compliance framework*. IoT Security Foundation: West Lothian, Scotland.
10. BITAG. 2016. Internet of Things (IoT) Security and privacy recommendations. BITAG: Denver, CO, USA.
11. Ross, M., A.J. Jara, and A. Cosenza. 2017. *Baseline security recommendations for IoT*. ENISA: Heraklion, Greece.
12. Gubbi, J., R. Buyya, S. Marusic, and M. Palaniswami. 2013. Internet of Things (IoT): a vision, architectural elements, and future directions. *Future Generation Computer Systems* 29 (7): 1645–1660.
13. Atzori, L., A. Iera, and G. Morabito. 2010. The internet of things: a survey. *Computer Networks* 54 (15): 2787–2805.
14. Cisco. 2014. The internet of things reference model. *Internet of Things World Forum*, 1–12.
15. Nejat, A., Mohammd Hossein Shekarian, S., Saheb Zamani, M. 2014 *A study on the efficiency of hardware Trojan detection based on path-delay fingerprinting*. Microprocessors microsystems, vol. 38, 246–252.
16. Rooney, C., Seeam, A., Bellekens, X. Creation and detection of hardware trojans using non-invasive off-the-shelf technologies. *Electronics*, 7, 124.
17. Iwase, T., Nozaki, Y., Yoshikawa, M., Kumaki, T. 2015. Detection technique for hardware Trojans using machine learning in frequency domain. In: *Proceedings of the 2015 IEEE 4th global conference on consumer electronics (GCCE)*, Osaka, Japan, 27–30 Oct 2015, 185–186.
18. Dofe, J., Frey, J., Yu, Q. 2016. Hardware security assurance in emerging IoT applications. In: *Proceedings of the IEEE international symposium on circuits and systems*, Montreal, QC, Canada, 22–25.
19. Li, H., Q. Liu, and J. Zhang. 2016. A survey of hardware Trojan threat and defense. *Integration* 55: 426–437.
20. Venugopalan, V., Patterson, C. 2018. Surveying the hardware trojan threat landscape for the internet-of-things. *Journal of Hardware Systems Security* 2, 131–141.

21. Msgna, M., Markantonakis, K., Naccache, D., Mayes, K. 2014. *Verifying Software Integrity in Embedded Systems: A Side Channel Approach.* Springer: Cham, Switzerland, 261–280.
22. Stergiou, P., Maniatakos, M., Konstantinou, C., Robison, P., Lee, S., Kim, S., Wang, X., Karri, R. 2016. Malicious firmware detection with hardware performance counters. *IEEE Transactions on Multi-Scale Computing Systems* 2, 160–173.
23. Lau, D. 2012. *Secure bootloader implementation.* https://www.nxp.com/docs/en/application-note/AN4605.pdf. Accessed 8 April 2019.
24. Das, A., Rolt, J.D., Ghosh, S. 2013. To cite this version: secure JTAG implementation using Schnorr Protocol. *Journal of Electronic Testing* 29, 193–209.
25. Vishwakarma, G., and W. Lee. 2018. Exploiting JTAG and its mitigation in IoT: a survey. *Future Internet* 10: 121.
26. Weis, S.A., Sarma, S.E., Rivest, R.L., Engels, D.W. 2003. Security and privacy aspects of low-cost radio frequency identification systems. *Security* in *Pervasive Computing*, 2802, 201–212.
27. Peris-Lopez, P., Hernandez-Castro, J.C., Estevez-Tapiador, J.M., Ribagorda, A. 2006. AAP: a minimalist 1286 mutual-authentication protocol for low-cost RFID tags. In: *Proceedings of the International Conference on Ubiquitous Intelligence and Computing*, Orange County, CA, USA, 17–21 Sept 2006.
28. Juels, A., Rivest, R.L., Szydlo, M. 2003. The blocker tag: selective blocking of RFID tags for the blocker tag: selective blocking of RFID tags for consumer privacy. In: *Proceedings of the 10th acm conference on computer and communication security—CCS '03*, Washington, DC, USA, 27–30 Oct 2003, 103.
29. Juels, A., Brainard, J. 2004. Soft blocking: flexible blocker tags on the cheap. In: *Proceedings of the 2004 ACM workshop on privacy in the electronic society*, Washington, DC, USA, 28 Oct 2004.
30. Saiful Islam, M., Sultanul, A., Sakhawat, M., Hayat, M. 2012. *Policy based intrusion detection and response system in hierarchical WSN architecture.* arXiv:1209.1678.
31. Zhijie, H., Ruchuang, W. 2012. Intrusion detection for wireless sensor network based on traffic prediction model. In: *2012 International conference on solid state devices and materials science, physics*. Procedia, 25, 2072–2080.
32. Chen, Y.Y., Lu, J.C., Chen, S.I., Jan, J.K. 2009. A low-cost RFID authentication protocol with location privacy protection. In: *The 5th International conference on information assurance and security*, China, 109–113.
33. Ye, X., Feng, J., Gong, H., He, C., Feng, W. 2015. An anti-trojans design approach based on activation probability analysis. In: *Proceedings of the 2015 IEEE international conference on electron devices and solid-state circuits, EDSSC 2015*, Singapore, 1–4 June 2015, 443–446.
34. Lesperance, N., Kulkarni, S., Cheng, K.T. 2015. Hardware trojan detection using exhaustive testing of k–bit subspaces. In: *Proceedings of the 20th Asia and South Pacific design automation conference*. Chiba, Japan, 19–22 Jan 2015, 755–760.
35. Syamsuddin, I., Dillon, T., Chang, E., Han, S.A 2008. Survey of RFID authentication protocols based on hash-chain method. In: *Proceedings of the 3rd international conference on convergence and hybrid information technology, ICCIT 2008*. Busan, Korea, 11–13 Nov 2008, 559–564.
36. Samyde, J.J.Q. 2001. Electro Magnetic Analysis (EMA): measures and counter-measures for smart card. In: *The International conference on research in smart cards (E-smart 2001)*. Cannes, France, 19–21.
37. Angrishi, K. 2017. Turning Internet of Things (IoT) into Internet of Vulnerabilities (IoV): IoT Botnets. arXiv:1702.03681.
38. Heer, T., Garcia-Morchon, O., Hummen, R., Keoh, S.L., Kumar, S.S., Wehrle, K. 2011. Security challenges in the IP-based Internet of Things. *Wireless Personal Communications* 61, 527–542.
39. Sarikaya, B., Ohba, Y., Moskowitz, R., Cao, Z., Cragie, R. 2012. *Security bootstrapping solution for resource- constrained devices*; Technical Report for the Internet Engineering Task Force; IETF: Fremont, CA, USA, 22 June 2012.

40. Montenegro, G., Kushalnagar, N., Hui, J., Culler, D. 2007. *Transmission of IPv6 Packets over IEEE 802.15.4 Networks*, Technical Report for Internet Engineering Task Force; IETF: Fremont, CA, USA.
41. Granjal, J., Monteiro, E., Sa Silva, J. 2015. Security for the Internet of Things: a survey of existing protocols and open research issues. *IEEE Communications Surveys and Tutorials* 17, 1294–1312.
42. Gaurav, K., Goya, P.V.A. 2015. IoT transaction security. In: *Proceedings of the 5th international conference on the Internet of Things (*IoT*)*. Seoul, Korea, 26–28.
43. Biswas, K., Muthukkumarasamy, V. 2016. Securing smart cities using blockchain technology. In: *Proceedings of the 18th ieee international conference on high performance computing and communications, 14th ieee international conference on smart city and 2nd ieee international conference on data science and systems*, HPCC/SmartCity/DSS, Sydney, Australia, 12–14 Dec 2016, 1392–1393.
44. Kokoris-Kogias, L., Gasser, L., Kho, I., Jovanovic, P., Gailly, N., Ford, B. 2016. Managing Identities using blockchains and CoSi. In: *Proceedings of the 9th workshop on hot topics in privacy enhancing technologies (HotPETs)*. Darmstadt, Germany, 19–22.
45. Bahga, A., Madisetti, V.K. 2016. Blockchain platform for industrial internet of things. *Journal of Software Engineering Applications* 9, 533–546.
46. Otte, P., de Vos, M.; Pouwelse, J. 2017 Trust chain: a sybil-resistant scalable blockchain. *Future Generation Computing Systems.*
47. Cirani, S., Ferrari, G., Veltri, L. 2013. Enforcing security mechanisms in the IP-based internet of things: an algorithmic overview. *Algorithms*, 6, 197–226.
48. Aragon, S., Tiloca, M., Maass, M.. Hollick, M., Raza, S. 2018. ACE of spades in the IoT security game: a flexible IPSEC security profile for access control. In: *Proceedings of the 2018 IEEE conference on communications and network security*, CNS, Beijing, China, 30 May–1 June 2018.
49. Ab, E., Seitz, L. 2019. *Object Security for Constrained RESTful Environments (OSCORE).* https://tools.ietf.org/html/draft-ietf-core-object-security-16. Accessed 20 June 2019.

Using Artificial Intelligence Approaches for Image Steganography: A Review

Jeanne Georges and Dalia A. Magdi

Abstract Security of confidential information has been always an important issue. It has always been an interesting topic for researchers to develop secure techniques to send and receive data without revealing it to anyone rather than the receiver. Researchers are working every day to achieve a high level of security and consistency for their data during its transmission over the network. Cryptography is a way used to encrypt the data during its transmission so that when there is any attack, the message seen is not the real message, as it is encrypted. On the other hand, Steganography is a method of hiding secret messages in a cover object (image, audio, video, text) while communication takes place between sender and receiver. Steganography is often used with cryptography so that the information is doubly protected; First, it is encrypted and then hidden so that the attacker has to first detect the presence of information that is already hidden (a difficult task) and then decrypt it. In this paper, we will discuss different artificial intelligence techniques used to achieve maximum information hidden and high quality of image as the low quality is a sign of distortion which means that there is something hidden.

Keywords Cryptography · Steganography · Security · Encryption · Artificial intelligence

J. Georges (✉) · D. A. Magdi
French University in Egypt, Elshorouk, Egypt
e-mail: jane.georges252@gmail.com

D. A. Magdi
e-mail: daliamagdi@gmail.com

D. A. Magdi
Computer and Information System Department, Sadat Academy for Management Sciences, Cairo, Egypt

A. Z. Ghalwash et al. (eds.), *Internet of Things—Applications and Future*,
Lecture Notes in Networks and Systems 114,
https://doi.org/10.1007/978-981-15-3075-3_16

1 Introduction

1.1 Cryptography

Image encryption has great importance in the field of information security. Most of the image encryption techniques have some security and performance issues [1]. Over the last two decades, the fast development of internet requires confidential information to be protected from unauthorized users; this is done by data hiding which is a method of hiding secret messages into a cover medium so that an unauthorized observer will not be aware of the existence of a hidden message. This is done by steganography. The similarity between steganography and cryptography is that both are used to conceal information. But the difference is that the steganography does not reveal any secret information about what is hidden from the attacker. Therefore, the attackers will not try to decrypt information because it seems as an empty message.

1.2 Steganography

The term steganography is retrieved from the Greek words steganos means the cover and grafia meaning writing defining it as covered writing [2].

The architecture of the steganography model as shown in Fig. 1 represents the embedding process of a text inside an image using secret key and the secret message (which will be hidden) and the cover media (image), the resultant image after the embedding process is sent by the sender as a normal image, on the other side (the receiver), the message (text hidden) is extracted to obtain the secret key to decrypt the text and obtain the hidden secret message).

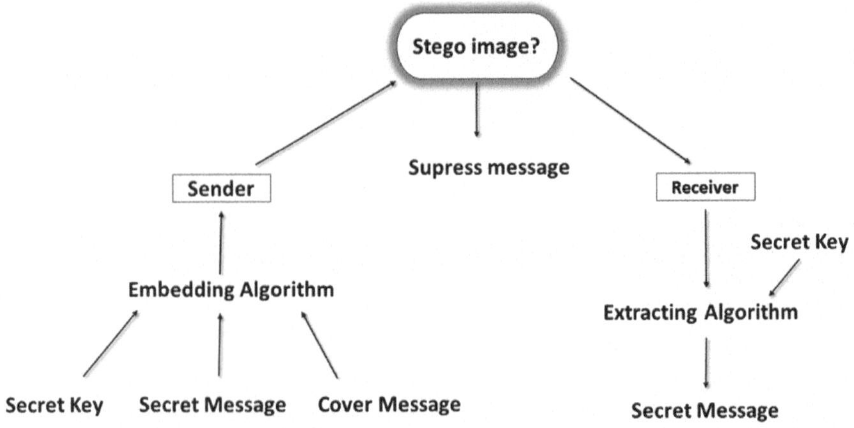

Fig. 1 Architecture of the steganography model

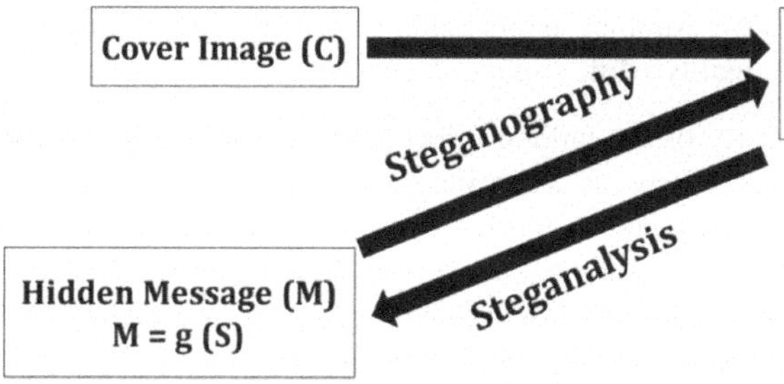

Fig. 2 The embedding of a message in a cover image

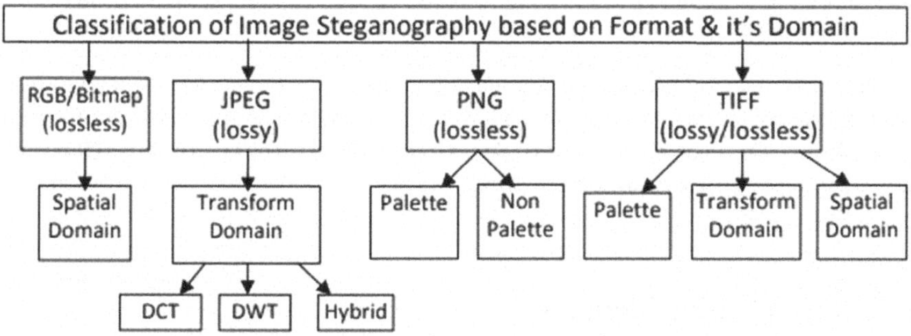

Fig. 3 Classification of steganography methods based on image formats and embedding domain [4]

Many researchers use whether spatial domain or transform domain, the major challenge is to increase the hiding capacity while maintaining the good visual quality of the cover image. Most of the steganographic techniques either use spatial domain or transform domain to embed the secret message [3] The Fig. 2 illustrates the classification of image steganography based on format and domain of the image as Spatial and Transform (Frequency) Domain (Fig. 3).

1.3 Challenges Using Steganography

As discussed, Security of confidential information has always been a major issue for researchers to develop secure techniques to send data without revealing it to anyone rather than the receiver. Unlike text messages, the multimedia information including image data has some special characteristics like High capacity, Redundancy, and High correlation among pixels. Our main Challenge is sharing sensitive or confidential information through a common communication channel without being secured (e.g., Credit Card). Also, the capacity of the hidden information is somehow low. Another main challenge is when hiding images, some data will be lost and the quality of the image decreases.

Data Capacity: The capacity of the hidden information inside a cover object (image) has always been a great issue as when increasing the capacity, the quality of the image decreases then it may be a sign for the attacker that there is something hidden.
Visibility: The inability for computers or humans to detect any distortion in the stego-image, so that it appears normal during transmission.
Detectability: The inability for a computer to differentiate between the cover image (if it is already saved) and the stego-image or if the attacker has access to the original image, he can easily detect that there is something hidden.
Robustness: The ability of a message to resist despite of compression or any other modifications [5]

There are different techniques to apply steganography to embed text data into an image; one of them is artificial intelligence techniques that include Genetic Algorithms, Artificial Bee Colony, Ant Colony Optimization, Firefly Algorithm Some techniques like Neural Networks.

2 Materials and Methods

Over the years, many algorithms for hiding data in images have been proposed and developing newer algorithms are a topic of current researchers [6] There are different techniques to apply steganography to embed text data into an image; one of them is artificial intelligence techniques that include Genetic Algorithms, Artificial Bee Colony, Ant Colony Optimization, Firefly Algorithm, some other techniques like Neural Networks.

2.1 Artificial Intelligence

Genetic Algorithm
This algorithm's main idea is trying to find best place for embedding modified secret data in the host image to achieve a high level of security. The process of embedding is ac-complished in two main steps:

1. To modify secret bits and second to embed it into host image.
2. Different places in the host image are defined by the order of scanning host pixels and the starting point of scanning and best LSBs of each pixel.

- The genetic algorithm is utilized to find the best starting point, scanning order, and other options such that the PSNR of the stego-image maximized [7].

After embedding the secret message in LSB (least significant bit) of the cover image, the pixel values of the stego-image are modified by the genetic algorithm to keep their statistic characters. Thus, the existence of the secret message is very hard to be detected by the Reactive Search (RS) analysis [8].

Swarm Intelligence
PSO (Particle Swarm Optimization) algorithm is used to find pixel optimum position. The cost function is proposed to evaluate the fitness of the pixel position. Cost function considers factors such as entropy, edge, and individual seed point edge. Initialization is performed by selecting the particles randomly and searching for an optimal solution with the help of the updated position of all particles.

Best identified position undergoes embedding. The following are the steps followed in PSO.

Step 1: Initialization In dimensional size area X*Y, let's consider n no of populations which are given as

$$\mathrm{P} = \{P1, P2, \ldots, Pn\}$$

Each population has some particles associated with them, initialize the position, as well as velocities associated with the particle by random selection.

Step 2: Fitness evaluation: the fitness function is used to find a personal best position of particles. The fitness of all particles is determined. The fitness function is based on cost matrix.

Step 3: Determine the optimal solution In population, the fitness value of all particles is evaluated by the fitness function. This search is done to find best position of particles and embedding is done

Step 4: Update the position Entire search is done to find the optimal position and is updated. We consider a series of the previous iteration to find the optimal position

Step 5: Iteration Repeat the above steps 2–4 times until you get an optimal position for embedding… [9].

BCO (Bee Colony Optimization)

Bee colony optimization is one of the artificial intelligence techniques applied to both combinatorial optimization and continuous optimization. It is based on evolutionary computing and can be used in solving steganography problems. In this system, artificial bees fly around in a multidimensional search space and some (employed and onlooker bees) choose food sources and adjust their positions depending on the experience of themselves and their nest mates. Some (scouts) fly and choose the food sources randomly without using experience. If the amount of nectar of a new source is higher than that of the previous one in their memory, they memorize the new position and forget the previous one. Thus, BCO system combines local search methods, carried out by employed and onlooker bees, with global search methods, managed by onlookers and scouts, attempting to balance exploration and exploitation process.

BCO can be used to find optimal pixel position in the cover image for hiding the secret image. In other words they used BCO to find the similarity between the cover image and the secret image in order to make less distortion not to be a sign for hidden information [10].

Firefly Algorithm

The firefly module finds the optimum location using an objective function. These locations are then used for the secret data embedding. The firefly algorithm is an iterative algorithm. In each iteration, the stego-image block for each firefly is obtained by embedding 'r'bits of secret data in that firefly position. Once this process is completed then the Structure Similarity Index Measure (SSIM) for the cover and stegoimage is calculated. Then the BER is calculated by extracting the secret bits from stego-image block. The best location is found when the following conditions occur

Number of iteration exceeds the maximum number of iterations.

No improvement is obtained in the successive iterations.

An acceptable result has been found.

Firefly allows good localization both in time and spatial frequency domain and it has higher compression ratio to avoid blocking artifacts [11].

3 Findings

Firefly Algorithm simulates the behavior of fireflies searching for brightness [11] The ABC (Artificial Bee Colony) algorithm is based on the food foraging behavior of honeybees [12]. Generally, an optimization problem can be considered as a minimization or maximization problem. In the optimization process, the absolute best decision of a suitable objective function must be taken. The decision should be taken in the presence of the collection of feasible constraints. An objective function also known as fitness function expresses the main aim of the model which is either to be minimized or maximized.

Firefly and PSO (Particle Swarm Optimization) algorithms are good search techniques from steganography point of view, firefly algorithm gives more accurate results than PSO algorithm for the following: Firefly algorithm selects the best hiding positions as shown in Fig. 4 These figure show that firefly hiding positions are deployed all over the cover images, while PSO hiding positions are concentrated in specific locations [12].

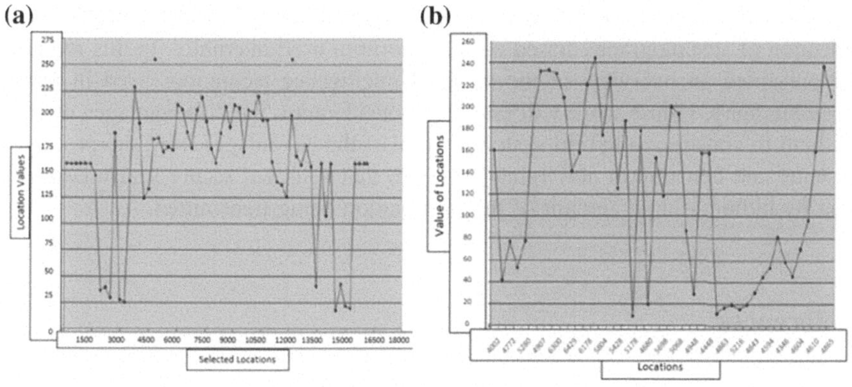

Fig. 4 Relation between locations and values of **a** Firefly algorithm **b** PSO algorithm

4 Discussion

The main challenges in image steganography are to hide secret data inside a cover image without being detected, highly secured, great consistency against intruder attacks that may attack high payload capacity. Many researchers are still working, the fulfillment of these above said requirements is not yet completely achieved. Since the steganographic properties are mutually related, enhancing some properties may decrease the efficiency in other aspects. Some of the steganographic concepts are given here for improving the efficiency of current image steganographic techniques [13] as the security is our main target, doing the cryptography beside the steganography so the information is doubly protected will help the confidentiality of the data being transmitted.

Many artificial intelligence techniques have been discussed, new techniques as deep learning and machine learning agents are suggested and being implemented nowadays.

5 Conclusion and Future Work

Image steganography is a considerably new dimension in the field of information hiding. Though there have been many active researchers in the field many research issues are yet to be explored. The tremendous improvement in information hiding techniques has a lot of attention nowadays, and becomes a dynamic topic in both private and government sectors. In order to protect the system integrity and prevent the exploitation of digital media from the criminal activities that have malicious intent. Steganography and cryptography techniques are the two known subdisciplines of information hiding in current years [14, 15]. Steganography, as discussed, is used to hide data into images or any cover medium without letting hackers and other attackers detect the information being transmitted. There is a different classification of steganography based on the algorithm used internally. In this research we presented an overview of the artificial intelligence techniques used in image steganography. Future work will take into consideration the disadvantages of each of them to construct a hybrid system that takes the advantages from the presented systems and construct a new reliable system that achieves secure transmission of data by hiding a large amount of the information being transmitted.

References

1. Shrija, Somaraj., and Mohammed Ali Hussain, F. 2015. Performance and security analysis for image encryption using key image. *Indian Journal of Science and Technology.*
2. Kurukshetra University, Kurukshetra, Haryana. 2016. Image steganography using enhanced LSB technique. *International Journal of Scientific & Engineering Research.*
3. Ansari, Arshiya Sajid, Mohammad Sajid Mohammadi, Mohammad Tanvir Parvez. 2019. A Comparative Study of Recent Steganography Techniques for Multiple Image Formats. *International Journal of Computer Network and Information Security.*
4. Ram MR Seshan. 2016. Assistant professor, Department of computer science, Challenges in steganography. *International Journal for Research in Applied Science & Engineering Technology (IJRASET).*
5. Roy, Ratnakirti, Suvamoy Changder, Anirban Sarkar, Narayan C Debnat. 2013. Department of computer applications, National institute of technology, Durgapur, India, Department of computer science, Winona state university, MN, USA. *Evaluating Image Steganography Techniques: Future Research Challenges*, IEEE.
6. Kanan, Hamidreza Rashidy, Bahram Nazeri. 2014. A novel image steganography scheme with high embedding capacity and tunable visual image quality based on a genetic algorithm. *Expert Systems with Applications.*
7. Wang Shen, Bian Yang and Xiamu Niu. 2010. A secure steganography method based on genetic algorithm. *Journal of Information Hiding and Multimedia Signal Processing.*
8. Sanjutha, MK. 2018. M.Tech, Computer science engineering, An image steganography using particle swarm optimization and transform domain. *International Journal of Engineering & Technology.*
9. Ibrahim Saleh University of Mosul. 2016. Alauddin Al-Omary University of Bahrain, Apply Bee Colony Optimization for Image Steganography with Optical Pixel Adjustment. *IJCDS Journal.*
10. Ziyad Tariq Tikrit University. 2017. Ziyad Tariq Mustafa Al-Ta'i University of Diyala-College of Science, Image steganography between firefly and PSO algorithms. *International Journal of Computer Science and Information Security.*
11. Yang, X.S. 2010. Firefly Algorithm, Levy Flights and Global Optimization, Research and Development in Intelligent Systems XXVI: Incorporating Applications and Innovations in Intelligent Systems XVII, Springer, London, 209–218.
12. Vaishali Kulkarni M. S. 2018. Ramaiah University of Applied Sciences, Bengaluru, India, Veena V Desai Gogte Institute of Technology, Raghavendra V. Kulkarni M. S. Ramaiah University of Applied Sciences, Comparison of Firefly, Cultural and the Artificial Bee Colony Algorithms for Optimization, *RUAS-SASTech Journal.*
13. Kadhim, Inas Jawad, Prashan Premaratne, Peter James Vial, Brendan Halloran. 2018. Comprehensive survey of image steganography: Techniques, Evaluations and trends in future research, Elsevier.

14. Ms. Ankita Kadu, Prof. Arun Kulkarni, Prof. Deepali Patil. 2016. Secure data hiding using robust firefly algorithm. *International Journal of Computer Engineering In Research Trends.*
15. Kale K.V., Najran N.H. Aldawla, and M.M. Kazi. 2012. Steganography enhancement by combining text and image through wavelet technique. *International Journal of computer & Applications (IJCA)*, 51(21):0975–8887.

Application of Hyperspectral Image Unmixing for Internet of Things

Menna M. Elkholy, Marwa Mostafa, Hala M. Ebeid and Mohamed F. Tolba

Abstract Hyperspectral Unmixing (HU), also known as spectral mixture analysis, is a challenging problem that decomposes a mixed spectrum into a collection of endmembers and their abundance fractions. In this paper, we extended the autoencoder network (Palsson et al. in IEEE Access 6:25646–25656, [1]) for blind hyperspectral nonlinear unmixing. The proposed autoencoder architecture consists of two networks encoder and decoder. The encoder network has the same as the original architecture. The architecture of the decoder network was altered to handle nonlinear unmixing. Our proposed encoder network has four fully connected layers, each with the number of neurons equal to the dimension of the end members. Experiments were conducted using nonlinear synthetic dataset sampled from the USGS library, and the performance was evaluated using both SAD and SID metrics. Four categories of experiments were carried out; accuracy assessment, weight initialization techniques, learning rate, and robustness to the noise. Experimental results show that the proposed autoencoder outperforms traditional endmember extraction algorithms in nonlinear cases. Finally, we introduced the application of hyperspectral image unmixing algorithm in the Internet of things (IoT) environment.

Keywords Hyperspectral unmixing · Spectral mixture analysis · Endmember extraction · Synthetic dataset

---

M. M. Elkholy (✉) · H. M. Ebeid · M. F. Tolba
Faculty of Computer and Information Sciences, Ain Shams University, Cairo, Egypt
e-mail: menna.elkholy@cis.asu.edu.eg

H. M. Ebeid
e-mail: halam@cis.asu.edu.eg

M. F. Tolba
e-mail: fahmytolba@cis.asu.edu.eg

M. Mostafa
Data Reception, Analysis and Receiving Station Affairs, National Authority for Remote Sensing and Space Science, Cairo, Egypt
e-mail: marwa@narss.sci.eg

A. Z. Ghalwash et al. (eds.), *Internet of Things—Applications and Future*,
Lecture Notes in Networks and Systems 114,
https://doi.org/10.1007/978-981-15-3075-3_17

1 Introduction

Hyperspectral images have been widely used in various applications due to their ability to cover hundreds of contiguous bands across the electromagnetic spectrum with abundant ground information [2, 3]. Recently, hyperspectral remote sensing field has attracted much attention. Compared with its rich spectral resolution, the spatial resolution of hyperspectral sensors is relatively low. Each pixel spectrum represents a region on the earth's surface that inevitably mixed with various microscopic material mixing. Hyperspectral unmixing (HU) is an essential processing step for various hyperspectral remote sensing applications, such as Plant Phenotyping, nature recognition and monitoring, classification, and compression, etc.

Hyperspectral unmixing [2] is a technique aiming at separating the pixels into several constituent components, called endmembers. Each of these endmembers has an associated areal proportion called abundances. Unmixing problem considers both linear and nonlinear models. The linear mixing model (LMM) [4, 5] is the most popular to describe the spectral mixing in hyperspectral imagery. The LMM can be described as

$$X = Ma + n, \tag{1}$$

$$\text{s.t. } \alpha_{ij} \geq 0, \sum_{j=1}^{M} \alpha_{ij} = 1 \tag{2}$$

where $M = \{m_j : j = 1, \ldots, M\}$ is the endmember set, and α_{nj} is the abundance fraction for the jth endmember at the nth pixel, i is the number of pixels, M is the number of endmembers, $n \in R^B$ is additional noise, and is typically assumed to be Gaussian noise with corresponding abundances fractions that are positive, and their sum is equal to one.

Different approaches had been developed to estimate both the endmember signatures and abundance maps using the linear unmixing model, such as DEAN [6] and nonnegative matrix factorization (NMF) [7]. Some approaches assume that pure pixels (pixels that contain a single endmember or material) exist for each endmember such as Pixel purity index (PPI) algorithm [8], Vertex Component Analysis (VCA) algorithm [9], N-FINDR [10], Minimum Volume Simplex Analysis (MVSA) algorithm [11], Simplex Identification Via Variable Splitting and Augmented Lagrangian (SISAL) [12] and simplex growing algorithm (SGA). To estimate the abundance map, approaches such as non-negativity constrained least squares linear unmixing (NCLS) [13], fully constrained least squares linear unmixing (FCLS) [14], and weighted least squares [15] were introduced based on linear unmixing.

Although the linear mixing model is commonly used, nonlinear mixing [4, 5] may occur in real scenarios, such as vegetation scenery and intricate mineral mixtures. Many techniques have been developed to cope with these nonlinearities: Fan model [16], which derived from the Taylor series expansions by assuming nonlinear function and then simplified by only reserving the first-order item.

Generalized Bilinear Model (GBM) [17] is an extension of the Fan model which implies that the interaction between two endmembers is regulated by introducing an additional coefficient. Another extension is MGBM [18], which adds endmembers' self-product and also considers the effects of materials from nearby pixels. Polynomial Post Nonlinear Model (PPNMM) [19] is inspired by the Weierstrass approximation theorem, which indicates that polynomials can uniformly approximate any bounded continuous function with any precision.

Deep Learning has been extensively utilized in remote sensing applications such as hyperspectral classification, material identification, and image registration. Recent works on unmixing problem estimate both endmembers and abundances using deep learning. However, almost all researches addressed the linear unmixing problem. Deep Autoencoder Network (DEAN) [6] consists of two parts, stacked autoencoder (SAE) for initialization, and variational autoencoder for nonnegative matrix factorization. The proposed DEAN successfully handles the blind unmixing problem and real-data outlier problem. In [20], a 3D convolution neural network (CNN) was presented, then a multilayer perceptron (MLP) was added in the last layer to obtain the abundances of each pixel. Deep learning approaches achieved better results compared with the classical approach in unmixing.

In this paper, we utilized a deep autoencoder network for blind nonlinear hyperspectral unmixing. This work is an extension of [1], which investigated the performance of autoencoder architecture in linear unmixing. We replaced the shallow decoder network architecture in [1] with a deep nonlinear autoencoder to capture the nonlinearity of hyperspectral data. The performance of the proposed model was evaluated on nonlinear synthetic datasets using two different objective metrics: spectral information divergence (SID), and spectral angler distance (SAD). Also, we investigate the effect of different network parameters such as weight initialization techniques and learning rate. The experiment results show that the proposed deep autoencoder exhibit superior performance compared to classical endmember extraction algorithms.

The remainder of this paper is organized as follows: In Sect. 2, the proposed deep autoencoder network is described in detail. Section 3 presents the results of the experiments. Section 4 presents the hyperspectral image unmixing algorithm in IoT and finally, in Sect. 5, conclusions are drawn.

2 Proposed Method

2.1 Problem Definition

In order to consider multiple scattering effects caused by multi-factors such as multiple scattering and shadowing. We consider the general nonlinear unmixing problem as

$$X = \Psi(Ma) + n \quad (3)$$

where Ψ is a nonlinear function that defines nonlinear interactions between the endmembers in matrix M parameterized by a. In this paper, the nonlinear unmixing problem is scrutinized using a deep autoencoder network. We seek an estimation of the endmember signature matrix (m_j), and their abundance maps gives a hyperspectral data (y_i) and the number of endmembers as a priori knowledge.

2.2 *Deep Autoencoder Network*

The proposed deep autoencoder network for nonlinear hyperspectral unmixing is shown in Fig. 1. In this work we extend the work in [1], Autoencoder consists of two networks encoder and decoder. We modify the architecture to consider nonlinear hyperspectral unmixing. After the entire network got trained, the proposed autoencoder effectively reconstructs the endmember signatures from the hyperspectral input image as the decoder's weights.

2.2.1 Encoder

The encoder network consists of six layers similar to the network described in [1]. Hidden layers from 1 to 4 have the same activation function, which can be Rectified with Linear unit (ReLU), Leaky ReLU (LReLU), and Sigmoid to introduce nonlinearity. We replaced the batch normalize layer in layer 5 with the dropout layer. The dropout layer prevents over-fitting which happens when neurons develop co-dependency among each other during training which curbs the individual power of each neuron leading to over-fitting. Hidden layer 6 is the thresholding layer,

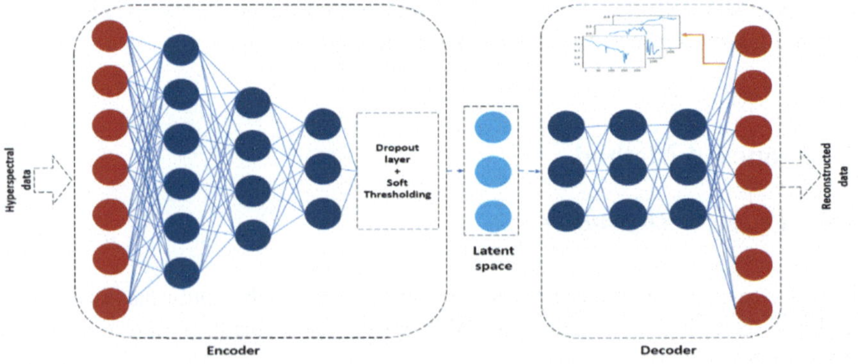

Fig. 1 An illustration of the proposed deep encoder-decoder architecture for nonlinear hyperspectral unmixing

that uses ReLU activation function with dynamic learnable parameter α for each unit in the layer, where α is a very small number.

2.2.2 Decoder

We extended the decoder network in [1]. Four fully connected layers were added. The number of nodes in the three layers is equal to the dimension of the endmembers. We utilized a Leaky ReLU (LReLU), rectified linear unit (ReLU) or Sigmoid activation function in the three-layer and linear activation in the fourth layer. Considering this, we train the proposed model using Adam optimizer.

3 Experimental Results

In this section, we introduced a nonlinear synthetic hyperspectral data in Sect. 3.1. Section 3.2 introduced the objective function. Finally, Sect. 3.3 evaluates the effectiveness of our proposed nonlinear autoencoder in terms of accuracy assessment, weight initialization techniques, learning rate, and robustness to the noise. The proposed autoencoder was compared with traditional endmember extraction algorithms namely: Vertex Component Analysis (VCA) [9], Pixel Purity Index (PPI) [8], N-FINDR [10], and Minimum Volume Simplex Analysis (MVSA) [12].

3.1 Nonlinear Synthetic Hyperspectral Images

This section presents the methodology to generate a complex nonlinear synthetic hyperspectral data. First, different numbers of random spectral signatures from the United States Geological Survey (USGS) spectral library [21] were selected. Second, we used the Dirichlet distribution method [22] to generate the abundance maps. Finally, hyperspectral imagery was created using nonlinear FAN model defined in [16] as

$$y = \sum_{i=1}^{m} a_i e_i + \sum_{i=1}^{m-1} \sum_{j=i+1}^{m} a_i a_j e_j \cdot e_j + n \tag{4}$$

To simulate real hyperspectral image, we add the zero-mean Gaussian noise to the generated image data using the Signal-to-Noise Ratio (SNR).

$$\mathrm{SNR} = 10 \log_{10} \left\{ \frac{\mathbb{E}[y^t y]}{\mathbb{E}[n^t n]} \right\}. \tag{5}$$

3.2 Objective Function

In this paper, two metrics were used to measure the accuracy of our proposed autoencoder. Spectral Angler Distance (SAD) is defined as

$$\Delta x_{\mathrm{SAD}} = \frac{1}{p}\sum_{p=1}^{p} \cos^{-1}\left(\frac{x_p \cdot \hat{x}_p}{\|x_p\| \cdot \|\hat{x}_p\|}\right) \tag{6}$$

where Δx is the spectral angle between a true spectrum and a reference spectrum, p is the number of bands, x_p: true spectrum and $\hat{x}_p$: reference spectrum.

The Spectral Information Divergence (SID) objective function is given by [23]

$$\Delta x_{\mathrm{SID}} = \frac{1}{p}\sum_{P=1}^{P}\sum_{n=1}^{B} p_n \log\left(\frac{p_n}{q_n}\right) + \sum_{n=1}^{B} q_n \log\left(\frac{q_n}{p_n}\right). \tag{7}$$

where

$$p_n = \frac{x_{i,n}}{\sum_{k=1}^{M} x_{i,k}}, q_n = \frac{x_{i,n}}{\sum_{k=1}^{M} \hat{x}_{i,n}} \tag{8}$$

where $p_n \, and \, q_n$ are the probability vectors for the spectral signatures $x_{i,n} \, and \, \hat{x}_{i,n}$, respectively.

3.3 Results

In the first experiment set, we used a nonlinear hyperspectral synthetic scene with size 100 × 100 pixels (N = 100), and a different number of endmembers P ranging from 4 to 30 to evaluate the performance of the proposed deep autoencoder network. Figure 2 shows the computed SAD and SID of different unmixing algorithms with a different value of endmembers in terms of SAD and SID. The proposed autoencoder and N-FINDR get hardly affected by the change in the number of endmembers. With $P \leq 10$, PPI displays the worst performance, but when $P > 10$, MVSA becomes the worst one. Autoencoder has the lowest SAD and SID values in all cases. N-FINDER has SAD, and SID values better than both VCA, MVSA, and PPI. Overall, the proposed deep autoencoder outperforms other classical unmixing algorithms in all cases. Figure 3 shows that the estimated endmembers using the proposed autoencoder network hardly differ compared with the ground truth in the USGS spectral library.

In the second set of experiment, we assessed the impact of the weight initialization technique on the proposed model performance. We used the nonlinear hyperspectral synthetic scene with size 100 × 100 pixels ($N = 100$), and four

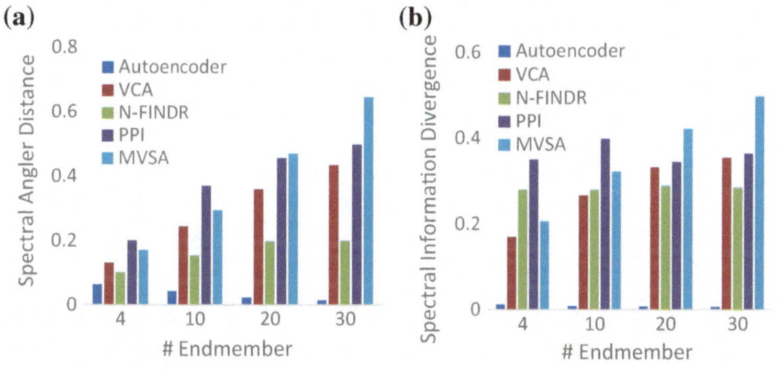

Fig. 2 The results obtained by varying the number of endmembers in terms of **a** spectral angle mapper and **b** spectral information divergence

endmembers ($P = 4$). The weight initialization step can be critical to the model's ultimate performance, and it requires the right method. A very large initialization leads to exploding gradients, and a very small initialization leads to vanishing gradients. Different weight initialization techniques were investigated namely: Random, He Normal, He Uniform, Xavier, Lecun Uniform, Orthogonal, Variance Scaling, Truncated Normal, and Random Uniform. Results obtained using the aforementioned initialization techniques are shown in Fig. 4. In terms of SAD and SID, the Orthogonal initialization technique attains the worst result. He Normal, He Uniform, Xavier, and Random Uniform initialization techniques have almost identical results.

The third experiment set was performed to study the impact of the learning rate on the proposed model. Learning rate is the most crucial hyperparameter which controls the rate or speed at which the model learns to control the amount of change

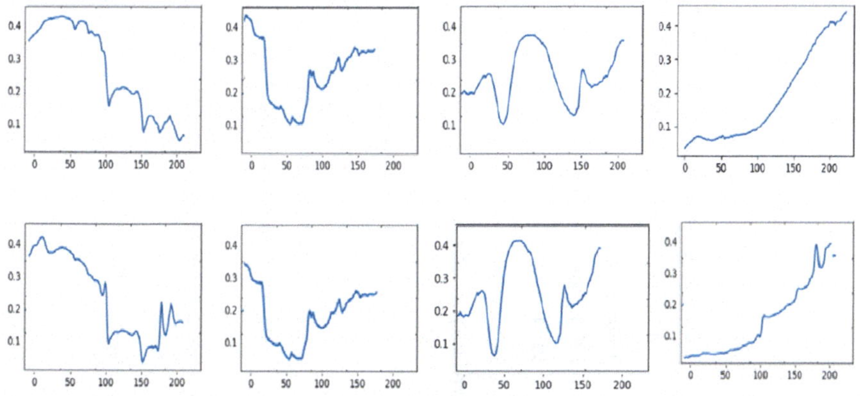

Fig. 3 Estimated endmember signatures for the nonlinear synthetic data set. The images in the first row are the ground truth signatures and the images in the second row were obtained by the proposed method

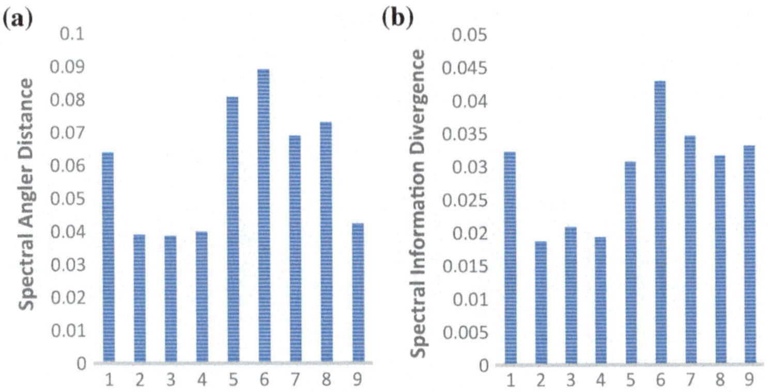

Fig. 4 The results obtained by nine weight initialization techniques in terms of **a** spectral angle distance and **b** spectral information divergence charts

to the model during each step, or the step size, and it provides them to tune for autoencoder to achieve the best performance. We used the nonlinear hyperspectral synthetic scene with size 100 × 100 pixels ($N = 100$), and four endmembers ($P = 4$). In this experiment, we use different values of learning rate in addition to step decay learning rate, which drops the learning rate by a factor every few epochs. Figure 5 shows the results obtained using different values of learning rates in terms of SAD and SID. Learning rate 10^{-3} achieved the best performance in terms of both SAD and SID. Large learning rate (0.5 and 0.1) leads to the worst results. Low learning rate values 0.001, 10^{-4} and 10^{-5} have almost identical results, and their results are very close to 10^{-3}.

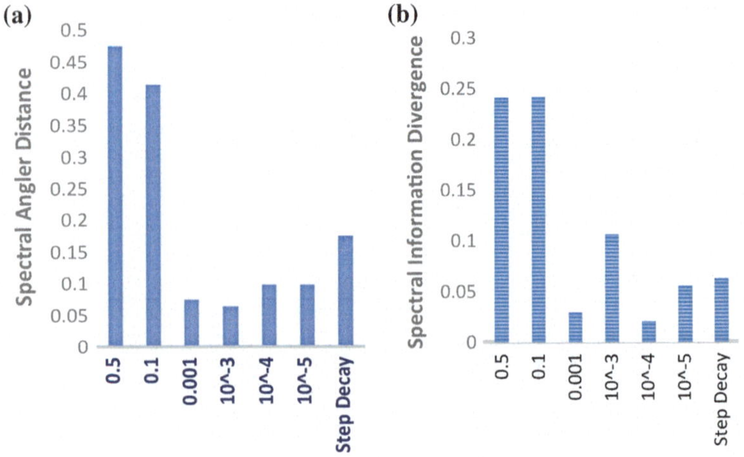

Fig. 5 The results obtained by varying the learning rates in terms of **a** spectral angle distance and **b** spectral information divergence charts

Table 1 SAD and SID obtained by proposed autoencoder for the nonlinear synthetic data with different SNR values

	SNR				
	90	80	70	60	50
SAD	0.0638	0.0647	0.0670	0.0698	0.0738
SID	0.0123	0.0125	0.0142	0.0192	0.0214

Finally, the last set of experiments was performed to quantitatively assess the proposed model robustness to noise. A nonlinear hyperspectral synthetic scene with size 100 × 100 pixels ($N = 100$), and four endmembers ($P = 4$) were used. Additive Gaussian noise with varying value was added to the synthetic scene. Table 1 depicted the obtained results in terms of SAD and SID using different SNR. As demonstrated in Table 1, the performance of the proposed autoencoder slightly decreased with the higher ratio of additive noise.

4 Application

In this paper, we proposed a novel hyperspectral unmixing algorithm that can be applied in the Internet of Things (IoT) environment. ZigBee technology [24] was selected as a well-designed wireless technology approach that is efficient in both power consumption and deployment scalability. The network-based ZigBee consists of three node types, namely: coordinator node, router node, and end node. Zigbee technology supports several topologies [25] such as mesh, bus, star, tree, and ring. In the proposed application, the ZigBee network is composed of a coordinator node, a routing, and several terminal nodes. The selected topology structure is the hybrid mesh and star topology structure which is easy to build, has the function of self-healing with the network structure and the data transmission reliability is high. An illustration of the proposed system is shown in Fig. 6.

When the drones carrying a hyperspectral imager is flying over the area under investigation, the data center will acquire these images and process them. The hyperspectral unmixing algorithm proposed in this paper will be utilized to handle these imageries and relevant hyperspectral profiles, and their associate abundance maps will be obtained. The target profiles and their associated geometric information are transmitted to the Zigbee network, the personnel in the vicinity of the investigation area can process the whole area in time through the portable terminal and send feedback information to the data center through the Zigbee network.

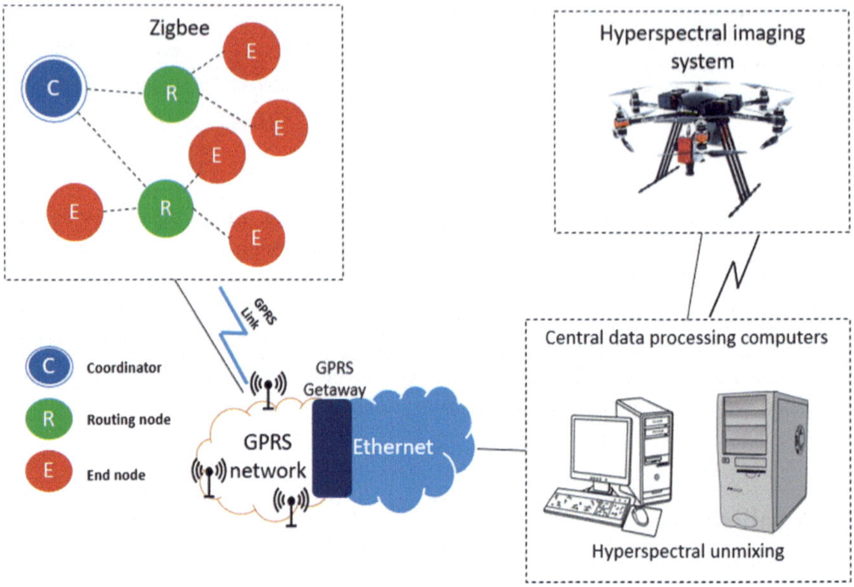

Fig. 6 The proposed system of HSI unmixing based on Zigbee network

5 Conclusion

Hyperspectral unmixing separates mixed pixel into a set of endmembers, with their associated areal proportion called abundances. In this paper, we proposed a deep autoencoder network. The proposed architecture extended the work in [3] for nonlinear unmixing. Two different objective functions namely: SAD and SID were utilized in our experiments. We conducted various experiments to compare the proposed architecture with benchmark endmember extraction algorithms. The proposed autoencoder network has significantly achieved higher accuracy. Several experiments were conducted to evaluate weight initialization techniques and learning rate to the performance of the proposed architecture. Experimental results show that He Normal, He Uniform, Xavier, and Random Uniform weight initialization techniques have almost identical results while the Orthogonal weight initialization technique had the worst performance. The best value for the learning rate was 10^{-3}. The obtained results show robustness to noise, using both SAD and SID objective functions. Finally, we utilized the proposed unmixing algorithm in the Internet of Things environment in a wide range of applications such as precision farming, mineral exploration activities, and crop monitoring.

References

1. Palsson, B., J. Sigurdsson, J.R. Sveinsson, and M.O. Ulfarsson. 2018. Hyperspectral unmixing using a neural network autoencoder. *IEEE Access* 6: 25646–25656.
2. Bioucas-Dias, J.M., A. Plaza, N. Dobigeon, M. Parente, Q. Du, P. Gader, and J. Chanussot. 2012. Hyperspectral unmixing overview: Geometrical, statistical, and sparse regression-based approaches. *IEEE Journal of Selected Topics in Applied Earth Observations and Remote Sensing* 5 (2): 354–379.
3. Moustafa, M., H.M. Ebeid, A. Helmy, T.M. Nazmy, and M.F. Tolba. 2016. Rapid real-time generation of super-resolution hyperspectral images through compressive sensing and GPU. *International Journal of Remote Sensing* 37 (18): 4201–4224.
4. Dobigeon, N., Y. Altmann, N. Brun, and S. Moussaoui. 2016. Linear and nonlinear unmixing in hyperspectral imaging. *Data handling in science and technology*. Elsevier, 185–224.
5. Elkholy, M.M., M. Mostafa, H.M. Ebeid, and M.F. Tolba. 2019. Comparative analysis of unmixing algorithms using synthetic hyperspectral data. In *International conference on advanced machine learning technologies and applications*. Springer.
6. Su, Y., J. Li, A. Plaza, A. Marinoni, P. Gamba, and S. Chakravortty. 2019. DAEN: Deep autoencoder networks for hyperspectral unmixing. *IEEE Transactions on Geoscience and Remote Sensing*.
7. Bao, W., Q. Li, L. Xin, and K. Qu. 2016. Hyperspectral unmixing algorithm based on nonnegative matrix factorization. In *2016 IEEE international geoscience and remote sensing symposium (IGARSS)*. IEEE.
8. Chaudhry, F., C.-C. Wu, W. Liu, C.-I. Chang, and A. Plaza. 2006. Pixel purity index-based algorithms for endmember extraction from hyperspectral imagery. *Recent advances in hyperspectral signal and image processing*, vol. 37, no. 2, 29–62.
9. Nascimento, J.M., and J.M. Dias. 2005. Vertex component analysis: A fast algorithm to unmix hyperspectral data. *IEEE Transactions on Geoscience and Remote Sensing* 43 (4): 898–910.
10. Plaza, A., and C.-I. Chang. 2005. An improved N-FINDR algorithm in implementation. In *Algorithms and technologies for multispectral, hyperspectral, and ultraspectral imagery XI*. International Society for Optics and Photonics.
11. Li, J., A. Agathos, D. Zaharie, J.M. Bioucas-Dias, A. Plaza, and X. Li. 2015. Minimum volume simplex analysis: A fast algorithm for linear hyperspectral unmixing. *IEEE Transactions on Geoscience and Remote Sensing* 53 (9): 5067–5082.
12. Bioucas-Dias, J.M. 2009. A variable splitting augmented lagrangian approach to linear spectral unmixing. In *2009 first workshop on hyperspectral image and signal processing: Evolution in remote sensing*. IEEE.
13. Bro, R., and S. De Jong. 1997. A fast non-negativity-constrained least squares algorithm. *Journal of Chemometrics: A Journal of the Chemometrics Society* 11 (5): 393–401.
14. Heinz, D.C. 2001. Fully constrained least squares linear spectral mixture analysis method for material quantification in hyperspectral imagery. *IEEE Transactions on Geoscience and Remote Sensing* 39 (3): 529–545.
15. Chang, C.-I., and B. Ji. 2006. Weighted abundance-constrained linear spectral mixture analysis. *IEEE Transactions on Geoscience and Remote Sensing* 44 (2): 378–388.
16. Fan, W., B. Hu, J. Miller, and M. Li. 2009. Comparative study between a new nonlinear model and common linear model for analysing laboratory simulated-forest hyperspectral data. *International Journal of Remote Sensing* 30 (11): 2951–2962.
17. Halimi, A., Y. Altmann, N. Dobigeon, and J.-Y. Tourneret. 2011. Nonlinear unmixing of hyperspectral images using a generalized bilinear model. *IEEE Transactions on Geoscience and Remote Sensing* 49 (11): 4153–4162.
18. Qu, Q., N.M. Nasrabadi, and T.D. Tran. 2013. Abundance estimation for bilinear mixture models via joint sparse and low-rank representation. *IEEE Transactions on Geoscience and Remote Sensing* 52 (7): 4404–4423.

19. Altmann, Y., A. Halimi, N. Dobigeon, and J.-Y. Tourneret. 2012. Supervised nonlinear spectral unmixing using a postnonlinear mixing model for hyperspectral imagery. *IEEE Transactions on Image Processing* 21 (6): 3017–3025.
20. Zhang, X., Y. Sun, J. Zhang, P. Wu, and L. Jiao. 2018. Hyperspectral unmixing via deep convolutional neural networks. *IEEE Geoscience and Remote Sensing Letters* 99: 1–5.
21. Kokaly, R.F., R.N. Clark, G.A. Swayze, K.E. Livo, T.M. Hoefen, N.C. Pearson, R.A. Wise, W.M. Benzel, H.A. Lowers, and R.L. Driscoll. 2017. *USGS spectral library version 7*. US Geological Survey.
22. Minka, T. 2000. *Estimating a Dirichlet distribution*. MIT: Technical report.
23. Chang, C.-I. 1999. Spectral information divergence for hyperspectral image analysis. In *IEEE 1999 international geoscience and remote sensing symposium. IGARSS'99* (*Cat. No. 99CH36293*). IEEE.
24. Chang, H.-Y. 2019. A connectivity-increasing mechanism of ZigBee-based IoT devices for wireless multimedia sensor networks. *Multimedia Tools and Applications* 78 (5): 5137–5154.
25. Agarwal, S.M.M. 2019. Network topologies in wireless sensor networks in real-time system. *Journal Current Science* 20 (2).

A New Vision for Contributing to Prevent Farmlands in Egypt to Become Uncultivable Through Monitoring the Situation Through Satellite Images

Hesham Mahmoud

Abstract The problem of turning plantation lands in Egypt to uncultivable lands is one of the complicated and aggravated problems faced by the Egyptian Ministry of Agriculture, and the problem will get more serious in the few coming years with the continuation of the increase of the population and the tendency of most Egyptians to reside around Nile banks. This research presents a contemporary view to contribute to limit the problem by suggesting a new minimum cost implementation mechanism that achieves the best results and within a short period of plan and completely executed by Egyptians.

Keywords Major problems of agricultural lands in Egypt · Lands information systems · New vision

1 Introduction

The contemporary business environment comprises information, computers, and digital establishments' era environment, or more definite; the information technology systems are based on computers and the Internet. There ensues a new recognition that knowledge relevant to information systems is considered a basic necessity for any organization aiming at survival and continuation.

First/The Research Problem: Making agricultural lands uncultivable is one of the most important problems faced by the Egyptian ministry of agriculture and the gravity of the problem will be increased in the following years with the rise in population rates [1].

Second/The Research Importance: The ministry of agriculture is one of the most important ministries that perform a very important role on diverse economic

H. Mahmoud (✉)
Management Information System, Modern Academy for Computer Science and Management Technology in Maadi, Cairo, Egypt
e-mail: dr.heshammahmoud122@gmail.com

A. Z. Ghalwash et al. (eds.), *Internet of Things—Applications and Future*,
Lecture Notes in Networks and Systems 114,
https://doi.org/10.1007/978-981-15-3075-3_18

levels, and the scientific importance of this research can be summed up with the following points:

- Contributing to finding new and modern scientific solutions for the problem of making farmlands uncultivable.
- Improving Egyptian agricultural economics.
- Contributing to the achievement of Egypt Strategy 2020/2030.
- Achieving integration with World economic system.

Third/ Research Specialized Goals: This study aims at achieving the following: Design an applicable suggestion to solve the problem of rendering farmlands uncultivable through the use of satellite images and lands information systems.

Fourth/ Egyptian Ministry of Agriculture: The Ministry of Agriculture is considered one of the most important entities in the contemporary era of Egypt, especially since agriculture still represents the most important source of Egyptian Gross National Product (GNP).

4/1 Goals of the Ministry of Agriculture:

- Improvement of agricultural policies and new lands cultivation policies that guarantee the achievement of coordination and integration between all stakeholders in a way that implements national development plans and creates ties between diverse governmental entities and improves them according to the most advanced scientific and technological methods to achieve the best economic outcome.
- Improvement of agricultural assets and increasing the area of cultivable lands and bettering developing countryside economics.

4/2 The Most Important Functions Performed by the Ministry of Agriculture (MOA):

- Development of the general policy for cooperation in solving agricultural issues and dealing with uncultivable lands. MOA works to increase the area of newly cultivated lands and deal with uncultivable and desert lands according to the provisions of the Law and supervision and coordination between entities working in the fields of agriculture and land cultivation in a way that achieves a precise and speedy performance of the above.

4/3 Area of Lands rendered uncultivable in Egypt

Ancient agricultural lands in the Nile Valley and Delta, known as the "green belt" are under the danger of becoming completely uncultivable. That area has always been considered as a "lung" for Egypt since it helps in reducing pollution. That area is now becoming incrementally occupied with residential buildings, and Egypt is annually losing about 20,000.00–50,000.00 acres of this very valuable land because of building thereon for residential purposes since farmers and Nile valley and delta residents are increasing in numbers and desert lands are mostly uninhabited for many reasons relevant to the uneven distribution of people inside Egypt.

Some scientists predict that by 2050, Egypt will have lost 17% of the Nile Delta area as a result of random construction on planted lands.

Fifth/Geographic Information System (GIS)

GIS is a system that relies on computers that collect, store, analyze, output, and distribute spatial data and information.

There are other systems that work to collect, input, process, analyze, display, and output spatial and descriptive information for definite purposes. Those systems aid in planning and decision taking concerning agriculture, city planning, and residential areas expansion. They also help in making the infrastructure of any city readable and easily comprehensible through the creations of layers through three dimensional visual diagrams.

This system helps us in the display of geographical information such as maps, aerial images, satellite visuals, and descriptive information illustrated in the form of names and tables. The system also helps in the aforesaid information processing such as their error correction, storing, retrieval, interpretation, and provision of spatially and statistically analyzed results thereof, and 3 display of the results on computer screens or printing them on papers as maps, reports, diagrams or their display on websites [2].

An example of a GIS defining map as in Fig. 1.

Sixth/A New Vision for Solving the Problem

The Egyptian Ministry of Higher Education and Scientific Research (MOHSER), through the National Authority of Remote Sensing and Space Sciences (NARSS) affiliated to aforesaid ministry, performs the task of providing the Ministry of Agriculture with digital photos taken by the Egyptian Satellite and this occurs once every three months, and such photos show transgressions against cultivable lands,

Fig. 1 An example of a GIS defining map

and solving the problem of the increase of transgressions happening through the following steps:

First Step: NARSS downloads the required photos through satellites as data to be processed in Earth Stations.
Second Step: Entry of the aforesaid processed data into the relevant Geographic Information System (GIS); i.e., Land Information Systems.
Third Step: Development of modern digital maps showing the current cadastral situation of lands transgressed against in great precision.
Fourth Step: Providing the Ministry of Agriculture with digital maps.
Fifth Step: The Ministry of Agriculture is to take the necessary disciplinary procedures against transgressors on the lands.

Anticipated Results of the Execution of the Aforesaid Suggestion

The Egyptian Ministry of Agriculture (MOA), through maps provided by GISs, shall be able to perform comparisons on quarterly basis between the situations of cultivable lands before and after interval period, and therefore, MOA will be able to identify transgressions against on cultivable lands and that would result in the preservation of the cultivable land area in Egypt.

Recapitulation and Final Results

Upon the foregoing, it is obvious that this research sheds light on a number of basic points concerning and rendering farmlands in Egypt uncultivable and suggestion of up to date solutions, and this can be summed up in the following points:

1. The problem of turning agricultural lands into uncultivable lands is one of the major problems facing the Egyptian MOA.
2. Shedding light on Land Information Systems.
3. Suggestion of a solution for the problem with the least cost possible through connecting Egyptian authorities and coordination between them.

Future Work

Satellite Imaging Systems and Geographic Information Systems are some of the important information systems used in Egypt, and they are used widely in the field of Agriculture. Therefore, those systems should be utilized in making two research papers or more in other different fields of the environment, mining, petroleum exploration, studying population distribution across the land and other fields where the aforesaid systems can be used to achieve advancement for Egyptians and achievement of the State goals.

References

1. Report of the central agency for public mobilization and statistics (CAPMAS) to be found at: http://www.arc.sci.eg/Default.aspx?lang=ar.
2. Report of the Agricultural Research Center (ARC) to be found at: http://www.arc.sci.eg/ViewPubsOfType.aspx?TabId=1&NavId=6&TYPE_ID=1&OrgID=&lang=ar.
3. The World Bank (WB) report to be found at: http://data.albankaldawli.org/topic/agriculture-and-rural-development.
4. Report of the United Nations Food and Agriculture Organization to be found at: http://www.fao.org/home/ar/.
6. Report of the Egyptian government to be found at: www.Egypt.gov.eg.
7. Huisman, Otto, and Rolf A.de. 2009. Principal of geographic information systems. In *The international institute for geo-information science and earth observation ITC—Northland, 4th edn.*
8. Campbell, Jonathan, and Michael Shin. 2011. *Essentials of geographic information systems.* Flat World Knowledge.
9. Alam, Bhuiyan Monwar. 2012. *Application of geographic information systems*. In Tech.
10. Mahmoud, Hesham, and Ahmed Attia. 2014. *New quality function deployment integrated methodology for design of big data e-government system in Egypt, big data conference.* Harvard University, USA, 14–16 December 2014 to be found at: https://www.iq.harvard.edu/files/iqss-harvard/files/conferenceprogram.pdf.
11. Mahmoud, Hesham. 2012. *Electronic government.* Cairo: Administrative Professional Expertise Center—Bambak.
12. Mahmoud, Hesham. 2015. Transform the city of Sharm El Sheikh into an electronic city. *Application Model, Egyptian Computer Science "ECS"* 2: 36–48 to be found at: http://ecsjournal.org/JournalArticle.aspx?articleID=471.
13. Mahmoud, Hesham. 2016. Proposal for Introducing NFC technology into the electronic government system in Egypt. *Egyptian Computer Science "ECS"* 1: 45–52 to be found at: http://ecsjournal.org/JournalArticle.aspx?articleID=485.
14. Mahmoud, Hesham. 2016. A suggested novel application for the development of electronic government portal in Egypt. In *sixth international conference on ICT in our lives information systems serving the community*, December, 2016. Alexandria University, Egypt. ISSN 2314-8942.

Applying Deep Learning Techniques for Heart Big Data Diagnosis

Kamel H. Rahouma, Rabab Hamed M. Aly and Hesham F. A. Hamed

Abstract The medical deep learning is considered one of the important analysis techniques and many evolutionary techniques and algorithms that have been improved to extract the features of images or classify and diagnose the different types of medical data. On another hand, deep learning is more practical especially for the classification process and it is noticed in speed and accuracy. In this paper, we apply different algorithms of deep learning such as Alex net and google net. Furthermore, we will test the algorithms on big data and try to achieve the result at a minimum time as possible. The second target of this paper is how to apply some of machine learning methods for ECG big data. Finally, we compare the results of the methods of deep learning and machine learning to obtain the best accuracy in classification. The accuracy of deep learning classification achieved 96.5% and more accurate in big data on CPU and machine learning approved good accuracy but we divide the data into parts. Deep learning takes big data without separating it. Based on the previous, deep learning achieved suitable results in ECG big data.

Keywords ECG · Google net · Deep learning · Alex net · Big data

1 Introduction

Artificial intelligence and deep learning techniques are considered one of the most practical methods especially in heart diagnosis. Furthermore, machine learning plays an important role in the detection and diagnosis of ECG. In this paper, we apply a different method of deep learning such as Google net and Alex net. On another hand, researchers have used computer vision with neural networks to extract the features and classify the types of MRI images [1]. A lot of researchers

K. H. Rahouma (✉) · H. F. A. Hamed
Electrical Engineering Department, Faculty of Engineering, Minia University, Minia, Egypt
e-mail: kamel_rahouma@yahoo.com

R. H. M. Aly
The Higher Institute for Management Technology and Information, Minia, Egypt

A. Z. Ghalwash et al. (eds.), *Internet of Things—Applications and Future*,
Lecture Notes in Networks and Systems 114,
https://doi.org/10.1007/978-981-15-3075-3_19

have suggested adaptive neural networks to analyze dangerous diseases and to diagnose the degree of tumors such as brain cancer, lung cancer, breast cancer, and so on. These techniques have utilized different tools such as the Polynomial Neural Networks (PNN), the Fuzzy Logic (FL), the Optimization Techniques (OT), Recurrent Neural Network (RNN), Google Net-deep learning, KERAS deep learning and Alex Net …., etc. In this paper, we have used Google net and Alex net to analyze the ECG signal to classify the degree of heart disease. MATLAB software is used to simulate the results with MIT-BIH Arrhythmia Database ECG datasets [2].

The paper consists of six sections. Section 1 gives an introduction and Sect. 2 depicts a literature review. Section 3 describes the methodology and Sect. 4 explains the results and compares them with a set of algorithms from the literature review. Section 5 depicts some conclusions.

2 Literature Review

Recently, deep learning and machine learning are considered two of the most important techniques in analysis and diagnosing research. They play an important role in detection, classifications, and prediction of the behaviors of dangerous diseases. One of the popular methods of machine learning is Adaptive Neural Networks (ANN) based on fuzzy inference system and sometimes called (ANFIS). There is another method of machine learning called Neural Gas Network (NGN) algorithm. On the other hand, there is another method of machine learning called Neural Gas Network (NGN). The learning of the NGN is either supervised or unsupervised [3].

In the following, we will introduce some of the important previous researches which have focused on machine learning and deep learning in medical classification.

In [4], authors have introduced the deep neural classification for CT images of the sick heart. There are a lot of difficulties faced by the researchers to get a correct diagnosis so the authors introduced this method based on CT for heart fetal to help in diagnosis. This residual network mechanism achieved great performance than other techniques.

Furthermore, the fuzzy cluster is playing an important role in classification and detection [5]. In [5], authors described the Fuzzy c-means to extract the features from Magnetic Resonance Images (MRI) of brain tumors. Then, they used Support vector machine (SVM) to classify the degree of tumors. The method accuracy is found to be from 97.9% to 98.8.

In [6], authors applied CNN to classify thyroid nodules based on infrared images. They tested the method based on Google Net architecture, Alex net architecture, and VGG architecture. The result achieved high accuracy for Google Net architecture compared to the other architectures by 86.2%. In this paper, we introduced the Google Net as architecture for classification data with wavelet analysis.

In [7], authors introduced how to apply artificial recurrent neural network (RNN) architecture based on Long short-term memory (LSTM) to classify ECG data. The proposed algorithm achieved high accuracy with lightweight.

In [8], authors introduced the neural network to help dentists in medical clinical images. They introduced two methods based on the classification to extract features from 3D images and the second one is giving information in range images. The accuracy of correct classification achieved 93.34%.

In [9], authors applied CNN to extract features of the breast cancer images. They used SVM for classification. The sensitivity of the cancer cases achieved 95.65%.

Some authors introduced a deep neural classification method for environment images [10]. In [10], authors applied deep learning for hyper spectral remote sensing images. They applied classification methods to achieve multi features learning and the accuracy is 99.67%.

In this paper, we use, combine, and modify some of the mentioned methods to diagnose ECG big data, classify, and predict future complications based on deep learning algorithm and compare with another machine learning introduced in [11].

3 Methodology

The method of this part consists of three parts as shown in Fig. 1:

1. Preparing ECG big data sets from online MIT datasets [2].
2. The analysis of ECG data sets based on Continuous wavelet transform (CWT).
3. The classification process and we apply in this part two methods of deep learning are Google Net and Alex Net and then we compared them and with ANFIS method as one of machine learning methods.

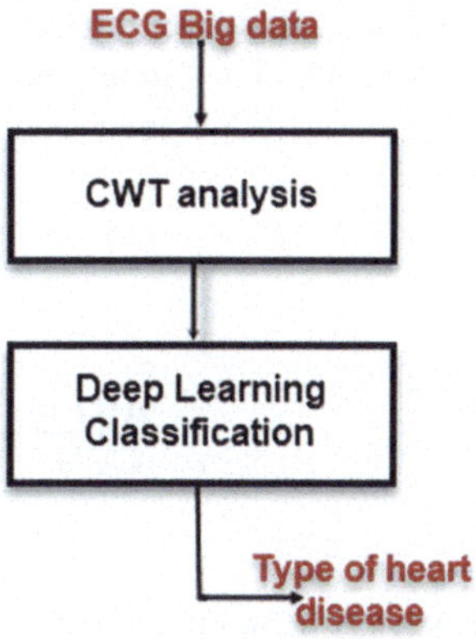

Fig. 1 Block diagram of the deep learning system

3.1 Preparing Data: ECG Big Data

Actually, in this part, we collect a group of ECG disease data from MIT [2]. The file contains 200 cases of different patient and heart diseases. The types of diseases are Three as three diagnosis categories

1. Arrhythmia.
2. Normal Sinus Rhythm.
3. Congestive Heart Failure.

The Data field is a 200-by-65536 matrix where each row is an ECG recording sampled at Fs = 128 Hz.

3.2 Analysis of Data Using Continuous Wavelet (CW)

The main concept of *continuous wavelet transform* is the inner products as shown in Eq. 1 (see Fig. 2).

$$C(a, b; f(t), \omega(t)) = \int_{-\infty}^{\infty} f(t)\frac{1}{a}\omega^{*}\left(\frac{t-b}{a}\right)\mathrm{d}t \tag{1}$$

where ω is analyzing function is a wavelet, scale parameter, $a > 0$, and position b.

3.3 Deep Learning Classification Method

3.3.1 Deep Learning Based on Google Net

From previous work, some authors introduced the convolution neural network in general architecture for Google net [12] as shown in Fig. 3.

Actually, The Google Net model is significantly more complex and deeper than all previous CNN architectures. It is more important and it also introduces a new

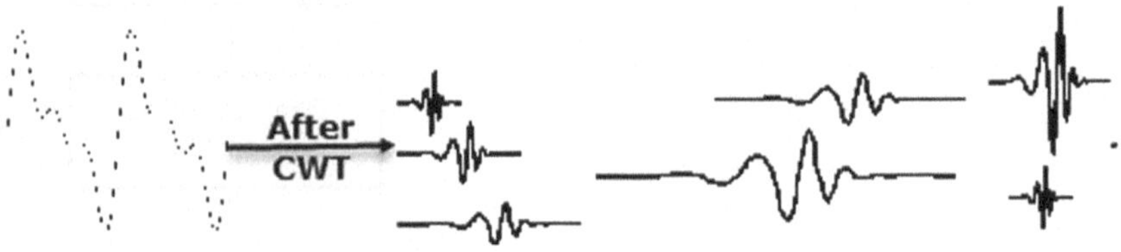

Fig. 2 CWT for ECG signal

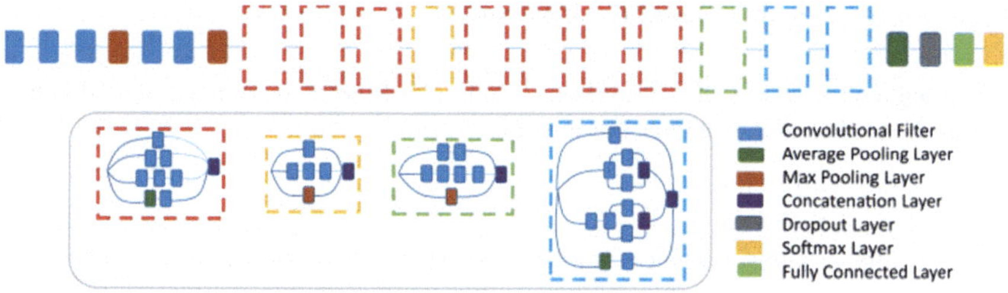

Fig. 3 CNN general architecture based on [12]

module called "Inception", which concatenates filters of different sizes and dimensions into a new one arrangement. Google Net consists of:

- Two convolution layers.
- Two pooling layers.
- Nine ''Inception'' layers and each "Inception" layer consists of six convolution layers and one pooling layer.

The image net for google net is 256 × 256 images categorized under 1000 object class categories.

There are more than 1000 training images per class. The database is organized according to the WordNet hierarchy, which currently contains only nouns in 1000 object categories (see Fig. 4). We applied this architecture for ECG big data and it approved a suitable accuracy near to 97%.

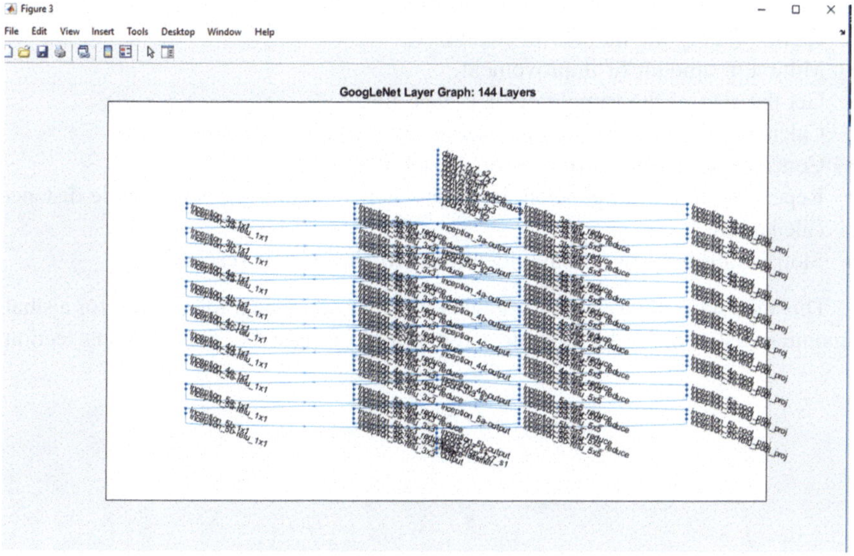

Fig. 4 The structure of Google Net after

3.3.2 Deep Learning Based on Alex Net

The ImageNet consists of 1.2 million 256 × 256 images belonging to 1000 categories. Note that, the AlexNet consists of

- Five convolution layers.
- Three pooling layers.
- Two fully connected layers with approximately 60 million free parameters.

We applied also this architecture and it achieved accuracy of 94% [12].

3.3.3 Adaptive Neuro Fuzzy Inference System Classification-FCM (ANFIS)

We applied this method as a machine learning technique to evaluate it and compared it with the deep learning method.

Adaptive neuro fuzzy based on Fuzzy c-means (FCM) is one of the most popular clustering techniques. Some authors introduced it as one of the best methods in clustering such as [11, 13]. In medical datasets Fuzzy C-Mean, we will use the cluster processing to classify and predict data. Fuzzy C-means iterates based on the number of clusters it comes across on the datasets. The FCM classification algorithm is near to the algorithm of segmentation based on a system of clusters shown as the following:

- Prepare data sets to Fuzzy system by dividing it into parts because it is big data sets.
- Identify Maximum number of iterations = 100 (the number clusters are approximately the number of iterations).
- Minimum amount of improvement.
- Get the size of the sample of ECG data file.
- Calculate the distance possible size using repeating structure.
- Concatenate the given dimension for the data size.
- Repeat the matrix to generate large data items in carrying out possible distance calculations.
- Stop Iteration when possible identification classifier is reached.

This method is approved for segmented file accuracy near to 93% but for a small amount of data not for the total file and it will be discussed in the following section.

4 Results and Discussions

The main concept of this paper is the diagnosis of the ECG based on Deep Learning techniques (DL) and one of the methods of Machine Learning technique (ML) (ANFIS-FCM) classification method. The ECG datasets contain 3 categories of different types of heart diseases. First, we applied CW to filter the ECG data and prepare it for the classification process. We applied our system on a laptop with (core i3 2.5 GH and 6 GB Ram) and using MATLAB 2019a.

We applied Google Net as a classification process and we used the default structure and category as shown in Fig. 4 we applied automatic iteration optimal classification result. The Google net achieved accuracy from 95 to 96% we got a total time of 30 min to finish all processes of Google Net.

On the other hand, we applied Alex Net as a classification process and we used the default structure and category a. we applied an automatic iteration optimal classification result. The Alex Net achieved an accuracy of 94%. We got a total time of 35 min to finish all processes of Alex Net (see Table 1 of results).

In the following, comparision between our method in classification and the accuracy with the other pervious works have been discussed(Table 2).

Table 1 Compare between results of DL methods and ML method

Method for classification	Maximum iteration	Support big data	Limited size of layers	Accuracy
Google Net (DL)	Automatic iteration for optimal classification result	Yes	More accurate for big data	Up to 96.6%
Alex Net (DL)	Automatic iteration for optimal classification result	Yes	More accurate for big data	Up to 94%
ANFIS-FCM (ML)	100	Limited layers	More accurate in limited size	Near to 99% for limited size of data

Table 2 Compare with pervious work

Paper	Number of data sets images	Method	Epochs	Processor	Accuracy
Manit et al. [14]	900 categories (4096 image)	DNN	200	NVIDIA GTX 980 GPU and an Intel® CoreTM i5 CPU (3.40 GHz), running Ubuntu 14.04	90.4%

(continued)

Table 2 (continued)

Paper	Number of data sets images	Method	Epochs	Processor	Accuracy
Mukherjee et al. [15]	10000 images	CNN	500	Intel i7 processor with 32 GB RAM and with two linked GPU of Titan	90.58%
Ayan and Ünver [16]	10000 images	CNN-Keras	25	Two intel processors E5-2640 Xenon with 64 GB RAM and NVIDIGPU card	82%
Maia et al. [17]	200 images	VGG19 + LR	Up to 100	Non limited	92.5%
This work	200 different datasets	Classification using ANFIS -FCM	100	CPU (core i3 2.5GH and 6 GB Ram)	Up to 98%
This work	200 different datasets with 1000 categories of Google Net	Google Net	Automatic for optimal	CPU (core i3 2.5GH and 6 GB Ram)	Reached to 96%
This work	200 different datasets with 1000 categories of Alex Net	Alex Net	Automatic for optimal	CPU (core i3 2.5GH and 6 GB Ram)	Reached to 94%

5 Conclusion

This paper introduced two methods of deep learning and one method of machine learning for classification techniques. The Google Net achieved a good accuracy for a big data in a suitable time and also in Alex Net. Furthermore, we obtained that machine learning (ANFIS-FCM) can achieve big data but after separating it into small data files and then applying the method for each file separately.

References

1. Zahedinasab, Roxana, and Hadis Mohseni. 2018. Enhancement of CT brain images classification based on deep learning network with adaptive activation functions. In *2018 8th international conference on computer and knowledge engineering (ICCKE)*. IEEE.
2. https://www.physionet.org/physiobank/database/mitdb/.
3. Azizi, Navid, Mashallah Rezakazemi, and Mohammad Mehdi Zarei. 2019. An intelligent approach to predict gas compressibility factor using neural network model. *Neural Computing and Applications* 31 (1): 55–64.
4. Lei, Li, et al. 2018. A deep residual networks classification algorithm of fetal heart CT images. In *2018 IEEE international conference on imaging systems and techniques (IST)*. IEEE.
5. Srinivas, B, and G. Sasibhushana Rao, et al. 2019. Performance evaluation of fuzzy C means segmentation and support vector machine classification for MRI brain tumor. *Soft Computing for Problem Solving* 355–367. Singapore: Springer.
6. Özyurt, Fatih, et al. 2018. A novel liver image classification method using perceptual hash-based convolutional neural network. *Arabian Journal for Science and Engineering* 1–10.
7. Saadatnejad, Saeed, Mohammadhosein Oveisi, and Matin Hashemi. 2019. LSTM-based ECG classification for continuous monitoring on personal wearable devices. *IEEE Journal of Biomedical and Health Informatics*.
8. Raith, Stefan, et al. 2017. Artificial neural networks as a powerful numerical tool to classify specific features of a tooth based on 3D scan data. *Computers in Biology and medicine* 80: 65–76.
9. Araújo, Teresa, et al. 2017. Classification of breast cancer histology images using convolutional neural networks. *PloS One* 12 (6): e0177544.
10. WANG, Lizhe, et al. 2017. Spectral–spatial multi-feature-based deep learning for hyper-spectral remote sensing image classification. *Soft Computing* 21 (1): 213–221.
11. Rahouma, Kamel H., et al. 2017. Analysis of electrocardiogram for heart performance diagnosis based on wavelet transform and prediction of future complications. *Computer Science Journal* 41. Egypt.
12. Shin, H.C., H.R. Roth, M. Gao, L. Lu, Z. Xu, I. Nogues, J. Yao, D. Mollura, and R.M. Summers. 2016. Deep convolutional neural networks for computer-aided detection: CNN architectures, dataset characteristics and transfer learning. *IEEE Transactions on Medical Imaging* 35 (5): 1285–1298.
13. Ajala Funmilola, A., et al. 2012. Fuzzy kc-means clustering algorithm for medical image segmentation. *Journal of Information Engineering and Applications* 2225–0506, ISSN 22245782.
14. Manit, Jirapong, Achim Schweikard, and Floris Ernst. 2017. Deep convolutional neural network approach for forehead tissue thickness estimation. *Current Directions in Biomedical Engineering* 3 (2): 103–107.
15. Mukherjee, S, A. Adhikari, and M. Roy. 2019. Malignant melanoma classification using cross-platform dataset with deep learning CNN architecture. In *Recent trends in signal and image processing* 31–41. Singapore: Springer.
16. Ayan, E, and H.M. Ünver. 2018. Data augmentation importance for classification of skin lesions via deep learning. In *2018 Electric Electronics, Computer Science, Biomedical Engineerings' Meeting (EBBT)* 1–4. IEEE.
17. Maia, L.B., A. Lima, R.M.P. Pereira, G.B. Junior, J.D.S. de Almeida, and A.C. de Paiva. 2018. Evaluation of melanoma diagnosis using deep features. In *2018 25th international conference on systems, signals and image processing (IWSSIP)*, 1–4. IEEE.

Bone X-Rays Classification and Abnormality Detection

Manal Tantawi, Rezq Thabet, Ahmad M. Sayed, Omer El-emam and Gaber Abd El bake

Abstract Computer-aided diagnosis (CAD) has become an urgent demand in the last decades. One of the frequently occurred diseases among individuals is bone fractures. This paper proposes a method for classifying and detecting abnormalities (fractures) of extremity upper bones through two-stage classification step. Two convolution neural network (CNN) models, namely, ResNet-50 and Inception-v3 are investigated for both classification stages. After needed enhancement, each bone X-ray image is classified into one of seven bones: shoulder, humerus, forearm, elbow, wrist, hand, and finger in the first stage. Thereafter, the bone image is directed to a specific classifier in the second stage to check if it is normal or not, according to the bone type defined from the first stage. The results reveal the superiority of Inception model for all classifiers of both stages. The proposed method can deal with any extremity upper bones, instead of only one bone as the majority of existing studies. Moreover, the average accuracy achieved is better than that achieved by existing studies. All experiments were carried out using Mura dataset.

Keywords Computer-aided diagnosis (CAD) · Image enhancement · Classification · Convolution neural network (CNN)

1 Introduction

Computer-aided diagnosis (CAD) can give clinicians great assistance. Besides reducing time and effort, it reduces human mistakes, since the performance of human experts is usually affected by their physical and psychological status. In addition, it can help researchers to search for certain cases; they need in their research [1]. Input for CAD can be biological signals (ex: ECG or EEG) [2–4] or medical imaging such as Magnetic Resonance Imaging (MRI) or X-rays [5–14].

M. Tantawi (✉) · R. Thabet · A. M. Sayed · O. El-emam · G. A. El bake
Faculty of Computer and Information Sciences, Ain Shams University, Cairo, Egypt
e-mail: manalmt@cis.asu.edu.eg

A. Z. Ghalwash et al. (eds.), *Internet of Things—Applications and Future*,
Lecture Notes in Networks and Systems 114,
https://doi.org/10.1007/978-981-15-3075-3_20

X-rays (Radiographs) images have been used by clinicians as a powerful tool for detecting different bone fractures for decades. Recently, several attempts for developing an automatic diagnosis tool for abnormality (fracture) detection from X-ray images have been carried out resulting in many publications with promising results. However, majority of these studies consider only one type of bones (ex: hand, finger, long bone of leg…, etc.) [7–11] due to lack of data or the strong variations among bones which may confuse the abnormality detection classifier and significantly limits its performance. De facto, developing a tool for each bone is not convenient, how many tools will be needed to cover the bones of all the body! Thus, developing a tool that can cover a wide range of bones as much as possible is recommended.

In this paper, instead of considering one bone as done in existing studies, seven bones of extremity upper bones (shoulder, humerus, forearm, elbow, wrist, hand, and finger) are considered. Moreover, the suggested method provides information first about bone type, and then it detects abnormality (fractures) if exits through two-stage classification step. Hence, stage 1 includes one classifier that predicts the type of bone in the X-ray image as one of the seven considered extremity upper bones. On the other hand, stage 2 includes seven binary classifiers; each of them is trained to detect abnormality of one of the seven bones. Thus, for a bone X-ray image, its type is defined by stage 1 and then the corresponding classifier in stage 2 provides yes/no decision about fracture existence. The introduced two-stage classification step is accomplished by convolution neural networks (CNN). Two efficient CNN models were examined namely ResNet-50 and Inception-v3. Mura dataset [11] was utilized for training and testing purposes. Better average accuracy has been achieved by the proposed method compared to the existing study that considered the same dataset.

The remainder of this paper is organized as follows: Sect. 2 presents a review on the existing studies. Section 3 gives a detailed discussion of the proposed method. Section 4 provides the achieved results and finally Sect. 5 presents the conclusion and future work.

2 Related Work

Many Computer-aided diagnosis (CAD) studies have been published in the last decades with very promising results. Regarding automatic detection of bone fractures from X-ray images, a brief survey for the key studies is provided in this section as follows:

Al-Ayyoub and Al-zghool [7] proposed a method for classifying abnormalities of long bones. X-ray images are smoothed using a set of filters to reduce darkness, brightness, blurring, Poisson and Gaussian noise. Thereafter, features are extracted using edge detection, corner detection, peak detection, texture features and parallel and fracture lines. Support Vector Machine (SVM), Decision Tree (DT), Nave Bayes (NB), and Neural Network (NN) were examined for classification. Two classification problems are addressed in this study: binary classification (fracture exists or not) and fracture type (five classes). 300 X-ray images were gathered from hospitals and websites for training and testing purposes. SVM provides the best results, above 85% accuracy for both considered problems.

Chung et al. [8] trained a CNN model for classifying normal and four proximal humerus fracture types, namely, greater tuberosity, surgical neck, 3-part and 4-part. 1891 images were gathered for training and testing purposes. Images are down sampled to 256 × 256 before being fed to the CNN. The best results achieved are 96%, 99%, and 97% for accuracy, sensitivity, and specificity, respectively.

Al-Ayyoub et al. [1] developed a method for hand bone fractures detection. First, the images are filtered to reduce noise and then edge detection using Sobel operator is applied. Wavelet transform and Curvelet transform are applied on images after edge detection to extract features. In addition, textural features are extracted using Gray Level Co-occurrence Matrix (GLCM) method from the filtered images. Only 84 Features are selected using Weka's supervised attribute filter. Finally, Bayesian Networks (BN), Naive Bayes (NB), Neural Network (NN), and Decision Tree (DT) were examined for abnormality detection. Data were gathered from hospitals and websites, and the best accuracy is 91.8% achieved by BN after applying boosting then bagging scheme.

Mahendran and Baboo [9] considered detecting fractures in long bones of legs called Tibia. X-ray images are enhanced by Simultaneous Automatic Contrast adjustment, Edge Enhancement, and Noise Removal (SACEN) algorithm. Subsequently, two segmentation algorithms are used to first segment the bone image from the X-ray image and then identify the diaphysis region of the bone image. Textural features are extracted using Gray Level Co-occurrence Matrix (GLCM). Thereafter, Feed Forward Back Propagation Neural Networks (BPNN), Support Vector Machine (SVM), and Naïve Bayes (NB) classifiers were examined. Moreover, a fusion approach is proposed by using the three classifiers together. Each classifier gives its decision (there is fracture or not) and the final decision is drawn by majority voting. 1000 images were gathered for training and testing purposes and accuracy above 90% is achieved.

Akter and Badrul Alam Miah [10] considered fractures of hand fingers. Images are denoised, resized, and then turned into binary images. GLCM features, moments, entropy, major axis length, minor axis length, orientation, eccentricity, area, convex area, filled area, equivalent diameter, solidity, extent, mean, standard deviation, perimeter, correlation coefficient, median, variance, width, height, pixel count, and Euclidian distance are extracted as 32 features that represent hand fingers X-ray image. Artificial Neural Network (ANN) and Support Vector Machine (SVM) were examined for classification. The best accuracy is 92.24% for fracture detection and it is achieved using ANN. In future, the authors are looking forward to succeed in classifying fracture type, not only detecting its existence.

A 169-layer convolutional neural network with Dense Convolutional Network architecture is used by Rajpurkar et al. [11] to detect abnormality in upper extremity bones which include shoulder, forearm, humerus, elbow, wrist, hand, and finger. Images are resized to 320 × 320 before being fed into the CNN model. Mura dataset [11] was used for training and testing purposes. Mura dataset includes more than 40,000 X-ray images for upper extremity bones. Moreover, data is augmented by using random lateral inversions and rotations of up to 30°. The best average accuracy achieved for the seven bones is 70.5%.

To sum up, we can have the following observations: (1) most of the existing studies consider detecting abnormality for only one bone, this can be justified by the strong variations of bones shapes which by its role affect the accuracy of fracture detection; (2) few studies not only detect fracture existence, but also consider classifying fracture type; (3) lack of public available datasets and difficulty of gathering data from hospitals may be the reason why not all body bones are covered by existing studies. In this work, seven bones are considered together by a two-stage classification step which detects both bone type and fracture existence.

3 Proposed Method

In this work, a two-stage upper extremity bone's classification method is introduced as shown in Fig. 1. Regarding the first stage, bone type is detected by classifying bone X-ray image into one of seven bone classes, namely: shoulder, forearm, humerus, elbow, wrist, hand, and finger. Thereafter, the abnormality of bone is checked by the second stage through a specific classifier for each bone type. Thus, seven classifiers exist in the second stage corresponding to the seven considered bones by this study. Each classifier is trained to decide whether a specific bone is normal or abnormal (fracture exists or not). The needed preprocessing steps and the examined convolution neural network (CNN) models for the proposed method are discussed in the next subsections.

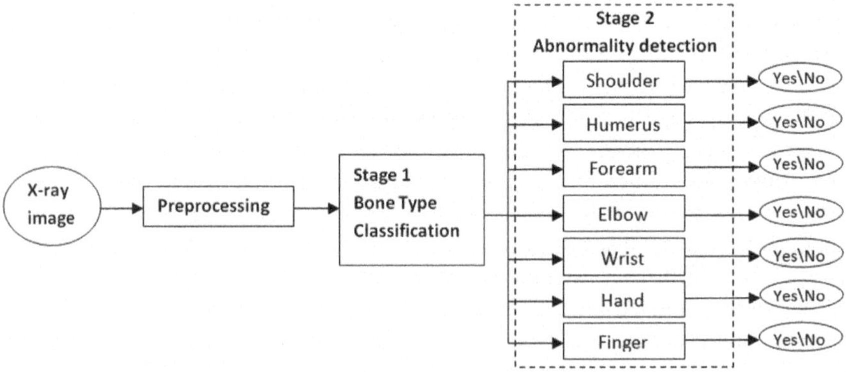

Fig. 1 Flow of the proposed method

3.1 Preprocessing

For more prominence of bones in images, the difference between the bone and its surrounding tissues or background in the X-ray should be maximized as much as possible. Hence, adaptive histogram equalization is applied to all considered images to improve contrast. Histogram equalization changes the intensity values of an image with the aim to adjust the intensity distribution of that image to be uniform (flat or near flat histogram) [15]. On the other hand, adaptive histogram equalization divides the image into small regions and then histogram equalization is applied locally to each region resulting in improving the local contrast and sharpening the edges of each region. Thus, it provides better results [15].

In this study, adaptive histogram equalization is applied and each image is divided into 8×8 regions. The threshold for contrast limiting is set empirically to be 2. Fig. 2a–c shows original bone X-ray image, after histogram equalization and finally after adaptive histogram equalization. It is obvious that the elbow in Fig. 2c is more prominent than in Fig. 2a, b. Moreover, each image is normalized, so the intensity values are ranged from 0 to 1 instead of 0 to 255. Finally, data augmentation is performed by flipping the images along the y-axis. Thus, the size of the considered dataset is doubled.

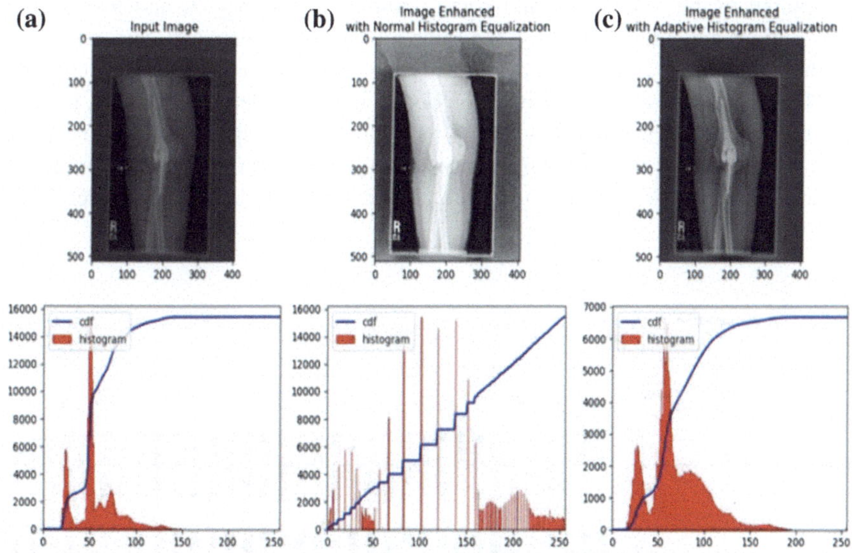

Fig. 2 Image enhancement: **a** original image, **b** after histogram equalization and **c** after adaptive histogram equalization

3.2 Feature Extraction and Classification

As discussed before, a two-stage classification step is proposed. CNN is utilized for feature extraction and classification for the eight considered classifiers (one classifier for the first stage and seven classifiers for the second stage). Two different pretrained CNN models were examined in this study for the task in hand, namely, ResNet-50 and Inception-v3. For each of the two models, the initial values for weights are equal to those of ImageNet [16]. A brief description of the considered CNN models is as follows:

(A) ResNet-50 model
Residual Neural Network (ResNet) is a CNN model provided by Kaiming, He et al. [17]. ResNet is a solution for training very deep networks (ex: more than 150 layers) efficiently. Instead of fitting a desired mapping directly, ResNet layers fit a residual mapping. Each layer learns from subtracting the output of the previous layer from the final loss [17]. This model can be optimized easier than traditional deep networks. Moreover, the power of ResNet is its ability to overcome vanishing gradients problem, which occurs usually in case of very deep neural networks. In this study, ResNet-50 consisted of 50 layers is used for the task in hand. Fig. 3 shows the architecture of ResNet-50 model.

(B) Inception V.3
Inception v3 model [18] distinguishes itself by different convolution filters applied to the same input and some pooling. 1 × 1, 3 × 3, and 5 × 5 convolutions are computed within the same module of the network. The outputs of these filters are concatenated before going to the next layer of the network. Hence, multi-level feature extraction occurs in this model. For example, general and local features are extracted by 5 × 5 and 1 × 1, respectively. The network is 48 layers and the number of nodes in the outer layer equals the number of considered classes. Fig. 4 shows the architecture of Inception v3 model.

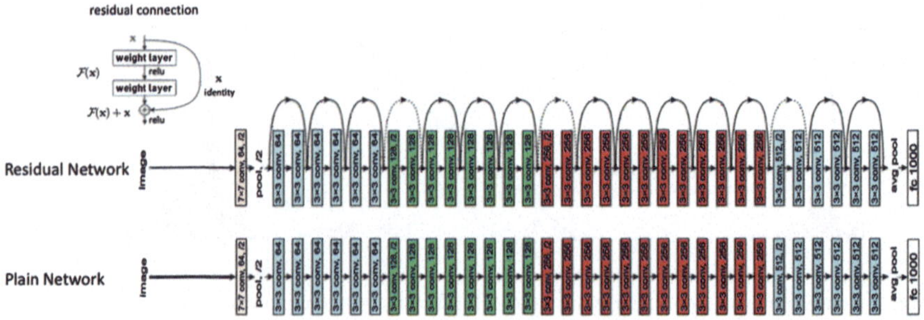

Fig. 3 The ResNet architecture

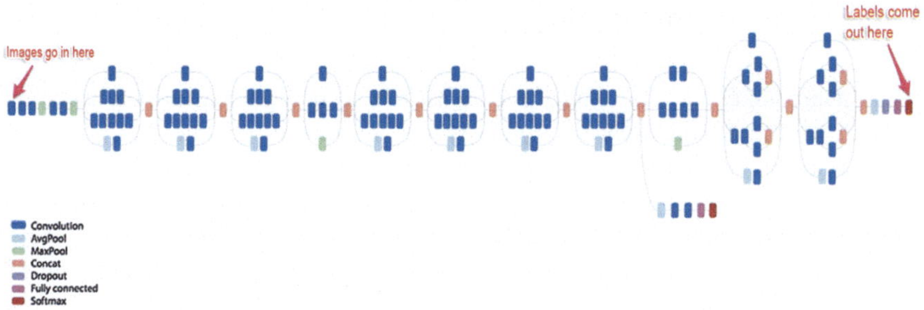

Fig. 4 The Inception architecture

4 Experimental Results

Mura dataset [11] has been used for training and testing purposes. It is a large public dataset for musculoskeletal radiographic studies. It encompasses 14,863 studies of upper extremity from 12,173 patients. Upper extremity bones include shoulder, humerus, elbow, forearm, wrist, hand, and finger. The number of normal and abnormal studies is 9,045 and 5,818, respectively. Each study includes one or more views which is labeled by radiologists as normal or abnormal resulting in 40,561 multi-view X-ray images. For reliable accuracies, fivefold cross-validation strategy is considered. RGB Images are preprocessed by applying adaptive histogram equalization and normalization as discussed in Sect. 3.1. Thereafter, two-stage classification is applied using two different CNN models: ResNet-50 and Inception-v3. The next subsections provide the results achieved by each stage separately and then the results of second stage depending on the first one.

4.1 Results of First Stage

Regarding the first stage, the considered CNN model is trained to assign the input bone X-ray image to one of seven classes representing the seven extremity upper bones, namely, shoulder, humerus, elbow, forearm, wrist, hand, and finger. Hence, the output layer of CNN includes seven nodes. Table 1 summarizes the average accuracy achieved for the seven bones by the two considered CNN models.

The results reveal that Inception-v3 is superior and provides the best average results, in addition to its stability over trails.

Table 1 Average accuracy achieved for the first stage by the two CNN models

CNN model	Accuracy (%)
ResNet-50	78.2
Inception-v3	95.6

4.2 Results of Second Stage

In this stage, the bone X-ray image is classified as normal (no fracture) or abnormal (fracture exists). Seven CNN binary classifiers are trained separately in this stage. Each binary classifier can detect the abnormality of a specific bone. Hence, each binary classifier gives yes/no decision about the fracture existence. For each classifier, the two considered CNN models in this study were examined. Table 2 shows the accuracy for abnormality detection achieved by each bone classifier using the two CNN models.

The results prove the superiority of Inception-v3. It provides the best abnormality detection accuracy for all bones except forearm. However, although ResNet-50 provides slightly higher accuracy than Inception-v3 for forearm bone, Inception-v3 shows more stability in accuracy over the accomplished trails. Thus, Inception-v3 is the winner model of this study.

4.3 Results of Second Stage Depending on First One

In the previous subsection, the results of stage 2 are provided independently, in other words, without considering the results of stage 1. In this subsection, the final results of stage 2 are provided dependant on stage 1. Each testing image is classified by the first stage classifier to know the bone type, and then it propagates to a specific classifier in the second stage for abnormality detection according to bone type defined from the first stage. To the best knowledge of authors, the study provided by Rajpurkar et al. [11] is the only one that considers all extremity upper bones using the same dataset. Table 3 shows the final accuracies for abnormality detection of the seven bones in comparison with the accuracies achieved by Rajpurkar et al. [11].

As shown in Table 3, the proposed method has achieved higher average accuracy. Furthermore, all accuracies are above 60%. Finally, the proposed method not only provides information about the bone abnormality, but also it provides information about its type with average accuracy more than 95%.

Table 2 Accuracy achieved for each bone by the two CNN models for the second stage

Bone type	ResNet-50 (%)	Inception-v3 (%)
Shoulder	74.9	**75.6**
Humerus	79.9	**82.3**
Elbow	80.2	**81.7**
Forearm	**77.7**	76.8
Wrist	78.4	**80.1**
Hand	66.6	**71.8**
Finger	71.6	**75.7**

Table 3 Results of stage 2 dependant on stage 1 compared to the results of Rajpurkar et al. [11]

Bone type	Accuracy of stage one for each bone (bone type) (%)	Accuracy of stage two for each bone (normal or not) (%)	Accuracies achieved by Rajpurkar et al. [11] (%)
Shoulder	99.29	**74.60**	72.90
Humerus	92.01	**77.43**	60
Forearm	85.38	64.45	**73.70**
Elbow	98.28	**80.00**	71.60
Wrist	96.51	77.69	**93.10**
Hand	95.22	68.48	**85.10**
Finger	96.75	**73.32**	38.90
Average accuracy	**95.6**	**73.71**	70.50

5 Conclusion

To sum up, this paper proposes a method for abnormality detection of extremity upper bones including shoulder, humerus, forearm, elbow, wrist, hand, and finger. Mura dataset has been used for training and testing purposes. Images are preprocessed to improve contrast and be normalized. Two CNN models, namely, ResNet-50 and Inception-v3 were examined for feature extraction and classification. The proposed method introduces two-stage classification step. In the first stage, the bone in the X-ray is classified into one of the seven extremity upper bones. Subsequently, the second stage checks the bone abnormality by a classifier (one of seven classifiers) trained to do so for this type of bones (type defined from previous stage). Inception-v3 model gives the best results for all classifiers of both stages. The advantages of the proposed method can be summarized as follows: (1) the average accuracy has increased by 3% compared to related work; (2) two detection problems (bone type and abnormality) are considered, instead of one as done in literature; (3) the proposed method can deal with any type of extremity upper bones, not only one type of bone as done by most of the existing studies that consider bone fractures; (4) scalability, it is not needed to retrain all classifiers if new bone is added. In future, we are looking forward to improve the results of abnormality detection stage and add other bones (ex: chest and leg) to the proposed diagnostic tool.

References

1. Al-ayyoub, M., I. Hmeidi, and H. Rababah. 2013. Detecting hand bone fractures in X-ray images. *Journal of Multimedia Processing and Technologies* 4 (3): 155–168.
2. El-Saadawy, H., M. Tantawi, H. Shedeed, and M.F. Tolba. 2018. Hybrid hierarchical method for electrocardiogram heartbeat classification. *IET Signal Processing* 12 (4): 506–513.

3. Ye, C., B.V.K.V. Kumar, and M.T. Coimbra. 2012. Heartbeat classification using morphological and dynamic features of ECG signals. *IEEE Transactions on Biomedical Engineering* 59 (10): 2930–2941.
4. Nasr, A., M. Tantawi, H. Shedeed, and M.F. Tolba. 2019. Detecting epileptic seizures using Abe entropy, line length and SVM classifiers. In *5th international conference on advanced intelligent systems and informatics*. Egypt: Springer. doi:https://doi.org/10.1007/978-3-030-14118-9_17.
5. Gong, T., R. Liu, C.L. Tan, N. Farzad, C.K. Lee, B.C. Pang, Q. Tian, S. Tang, and Z. Zhang. 2007. Classification of ct brain images of head trauma. In *Pattern Recognition in Bioinformatics, PRIB 2007. Lecture Notes in Computer Science*, vol 4774, 401–408. Berlin, Heidelberg: Springer.
6. Shree, N.V., and T.N.R. Kumar. 2018. Identification and classification of brain tumor MRI images with feature extraction using DWT and probabilistic neural network. *Brain Informatics* 5 (1): 23–30.
7. Al-ayyoub, M., and D. Al-zghool. 2013. Determining the type of long bone fractures in X-ray images. *WSEAS Transactions on Information Science and Application* 10: 261–270.
8. Chung, S.W., S.S. Han, J.W. Lee, K. Oh, N. Kim, J. Yoon, J.Y. Kim, S.H. Moon, J. Kwon, H. Lee, Y. Noh, and Y. Kim. 2018. Automated detection and classification of the proximal humerus fracture by using deep learning algorithm. *Acta Orthopaedica* 89 (4): 468–473.
9. Mahendran, S., and S. Baboo. 2011. An enhanced tibia fracture detection tool using image processing and classification fusion techniques in X-ray images. *Global Journal of Computer Science and Technology* 11 (14): 23–28.
10. Akter, A., M.D. Badrul Alam Miah. 2018. Detecting and determining the types of hand bone fracture using K-means clustering. *Journal of Computer Science Applications and Information Technology* 3(3): 1–10.
11. Rajpurkar, P., J. Irvin, A. Bagul, D. Ding, T. Duan, H. Mehta, B. Yang, K. Zhu, D. Laird, R. L. Ball, C. Langlotz, K. Shpanskaya, M.P. Lungren, and A.Y. Ng. MURA: large dataset for abnormality detection in musculoskeletal radiographs. In *1st conference on medical imaging with deep learning*. arXiv:1712.06957 [physics.med-ph].
12. Fesharaki, N.J., and H. Pourghassem. 2012. Medical X-ray images classification based on shape features and Bayesian rule. In *4th international conference on computational intelligence and communication networks (CICN)*, 369–373. IEEE, India.
13. Dimililera, K. 2017. IBFDS: Intelligent bone fracture detection system. In *9th international conference on theory and application of soft computing, computing with words and perception*, 260–267. Budapest, Hungary: Elsevier.
14. Yang, A.Y., and L. Cheng. 2019. *Long-bone fracture detection using artificial neural networks based on contour features of X-ray images*. ArXiv: abs/1902.07897.
15. Gonzalez, R.C., and R.E. Woods. 2017. *Digital image processing*, 4th ed. Pearson.
16. Russakovsky, O., J. Deng, H. Su, J. Krause, S. Satheesh, S. Ma, Z. Huang, A. Karpathy, A. Khosla, and M. Bernstein, et al., 2014. *Image net large scale visual recognition challenge*. ArXiv:abs/1409.0575.
17. Kaiming, H., X. Zhang, S. Ren, J. Sun. 2015. *Deep residual learning for image recognition*. CoRR, abs/1512.03385.
18. Szegedy, C., V. Vanhoucke, S. Ioffe, J. Shlens, Z. Wojna. 2015. Rethinking the inception architecture for computer vision. CoRR, abs/1512.00567.

TCP/IP Network Layers and Their Protocols (A Survey)

Kamel H. Rahouma, Mona Sayed Abdul-Karim and Khalid Salih Nasr

Abstract Computer and communication networks are arranged according to certain models. The network models are composed of layers. The number of layers depends on the used model. For the general Open System Interconnected (OSI) model, there are seven layers. In other models, the number of layers is reduced because more than one layer may be compressed or included in one layer. Thus, in other models, four or five layers may be used. In the following review, we consider a model of five layers which is called TCP/IP model. The five layers of the TCP/IP network are generally introduced and then each layer is studied in a separate section. The layer is defined, then its function is explained and the important protocols and processes are explained. An example of the protocols and processes is given when it is needed.

Keywords Network layer · TCP/IP · Protocols · Application · Transport · Network · Data link · Physical

1 Introduction

The Transmission Control Protocol/Internet Protocol (TCP/IP) protocol suite is a staple of nowadays' global communications. In other words, the TCP/IP suite is indeed the industry-standard method of interconnecting hosts, networks, and the Internet. Its standards have been maturely evolved to meet the changing needs of our communities. This provides a wide and flexible foundation on which, the infrastructure of so many applications is built such as entertainment, business, financial transactions, delivering services, and much more. However, it might be hard to keep track of new TCP/IP functionality and it might be hard to identify new

K. H. Rahouma (✉) · M. S. Abdul-Karim · K. S. Nasr
Electrical Engineering Department, Faculty of Engineering, Minia University, Minia, Egypt
e-mail: kamel_rahouma@yahoo.com

M. S. Abdul-Karim
e-mail: mona.sayed.abdelkarim@mu.edu.eg

A. Z. Ghalwash et al. (eds.), *Internet of Things—Applications and Future*,
Lecture Notes in Networks and Systems 114,
https://doi.org/10.1007/978-981-15-3075-3_21

possibilities. Thus, the TCP/IP technical survey may serve as a reference for those who seek to keep their TCP/IP skills aligned with current standards.

The TCP/IP supports a host of applications. Both standard and nonstandard applications could not exist without the foundation of a set of core protocols. Also, to understand the capability of TCP/IP applications, these core protocols must be realized and understood.

This paper aims to discuss the networks which are based on the TCP/IP and the core protocol suite that must be implemented in any stack to interface with the physical network media including the protocols belonging to the IP and transport layers. Also, we may consider some protocols which can be very useful for certain operational needs.

The rest of this paper explains a general overview of TCP/IP network layers in Sect. 2. Each layer is presented in detail in a separate section. Section 3 introduces the application layer and Sect. 4 explains the transport. Section 5 focuses on the internetworking layer and Sect. 6 explores the interface layer. Section 7 depicts the physical layer. Each section starts by defining the layer and its function. Then the important protocols of the layer are presented. Some examples are explained and when it is needed, the algorithm of the protocol is also given. A conclusion is highlighted in Sect. 8, and a list of the used references is given at the end of the paper.

2 An Overview of the TCP/IP Network Layers

The TCP/IP protocol suite is named by its most two important protocols. These are the Transmission Control Protocol (TCP) and the Internet Protocol (IP). In the official Internet standards documents, it is called the Internet Protocol Suite. It is commonly nowadays known between users and researchers by the TCP/IP, to refer to the entire protocol suite. The design objective of TCP/IP was mainly to build an interconnection of networks that is called internetwork, or internet, which provided communication services universally over the heterogeneous physical networks. This enabled communication between hosts on different networks which perhaps are separated by large geographical areas [1].

The word internet is simply composed of the highlighted underlined initials of the phrase <u>inter</u>connected <u>net</u>work. It always starts with the capital I such as Internet. The Internet consists of the following groups of networks:

(1) The Backbones which are large and they connect other networks. They may be also known as network access points (NAPs) or Internet Exchange Points (IXPs). Nowadays, the backbones consist of commercial entities.
(2) The Regional Networks which are used to connect, for example, universities and colleges.
(3) The Commercial Networks which are used to connect the subscribers and the commercial organizations to the backbones for use the Internet.

(4) The Local Networks which may be used for campus-wide universities, banks, …, etc.

Networks may be limited in size by the number of users of the network, or by the geographical distance which the network can span, or by the applications and services which the network can provide. The ability to connect different networks in a certain hierarchical and organized fashion enables any two hosts in this connection to communicate. To have this facility, a standardized abstraction of the communication mechanisms is created to provide the form of connection between the different networks. This means that each network has its own communication interface (a programming interface) which provides the basic communication functions. The TCP/IP enables the communication services which run between the programming interface of a network and the user applications.

The network architecture is hidden from both the user and the developer of the application. A code is written to enable the standardized communication abstraction to function under any type of network and operating platform. Thus, a computer is attached to both networks and can forward data packets from one network to the other. This machine is called a router. The term "IP router" is used because the routing function is part of the Internet Protocol portion of the TCP/IP protocol suite. Each host is identified by an address, called the IP address. When a host has multiple network adapters or interfaces, each interface has a unique IP address. The IP address consists of two parts:

IP address = < network number > <host number>

Where the network number identifies the network within the Internet and it is assigned by a central authority which resides with the organization controlling the network. This number must be unique throughout the internet.

In a similar fashion of most networking software, TCP/IP is programmed (modeled) in layers which may be called the protocol stack, referring to the stack of layers in the protocol suite. Figure 1 shows the TCP/IP network protocol stack. It can be used for positioning the TCP/IP protocol components against others.

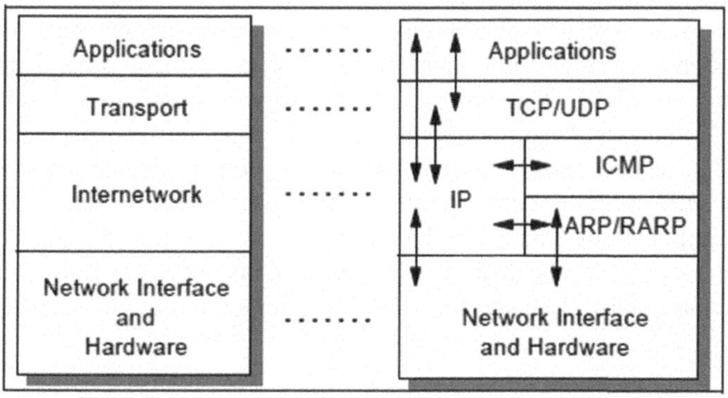

Fig. 1 The TCP/IP protocol stack: Each layer represents a package of functions [1]

Examples of layered framework are the Systems Network Architecture (SNA) and the Open System Interconnection (OSI) model.

Software layering allows ease of writing the codes and their implementation as well as developing alternative layer implementations. Adjacent layers communicate together via concise interfaces. The upper layers make use of the services of the layers directly below them while the lower layers provide services to the layers directly above them. For instance, IP layer is directly under the transport layer. Thus, the IP layer provides the service of transferring data from one host to another regardless its reliability and duplication. The Transport protocols such as TCP make use of this service to provide applications with reliable, in order, data stream delivery.

Figure 1 illustrates the detailed architecture model of the TCP/IP network. These layers of the TCP/IP network include [1].

2.1 The Application Layer

This layer is used for communication. An application is a user program or process that cooperates with another program or process on the same host or on a different host. Examples of applications are Telnet and the File Transfer Protocol (FTP). The application and transport layers are interfaced by port numbers and sockets. A socket is defined as one endpoint of a two-way communication link between two programs or processes running on the network. A socket is bound to a port number so that the TCP layer can identify the application that data is destined to be sent to [2].

2.2 The Transport Layer

This layer provides the end-to-end data transfer where it delivers data from an application to its remote peer. Multiple applications can simultaneously be supported. The most-used and important transport layer protocol is the Transmission Control Protocol (TCP). This protocol allows (1) connection-oriented reliable data delivery, (2) data duplication suppression, (3) congestion control, and (4) flow control. The second important transport layer protocol is the User Datagram Protocol (UDP). It allows connectionless, unreliable, best-effort service. Thus, if it is desired, applications which use the UDP as a fast transport mechanism or protocol have to provide (1) their own end-to-end integrity, (2) flow control, and (3) congestion control. Usually, the UDP can tolerate the loss of some data.

2.3 The Internetwork (Internet) Layer

The Internetwork (Internet) layer provides an image of the "virtual network" or "Internet". This layer shields the higher levels from the physical network architecture below it. The most important protocol in this layer is the Internet Protocol (IP). It is a connectionless protocol. It does not assume reliability from lower layers. It also does not provide (1) reliability, (2) flow control, or (3) error recovery. These functions are to be provided at a higher level.

The IP simply provides a routing function which attempts to deliver transmitted messages to their destination. The message units in an IP network are called IP datagrams. The datagram is the basic unit of information which is transmitted across the TCP/IP networks. Other Internet-layer protocols are IP, ICMP, IGMP, ARP, and RARP.

2.4 The Network Interface Layer

The network interface (link or data link) layer is the interface to the actual network hardware. This interface may (or may not) provide reliable delivery. It may be packet or stream oriented. The TCP/IP does not specify any protocol here. It can use almost any available network interface. This illustrates the flexibility of the IP layer. Examples of the link-layer protocols are the IEEE 802.2, X.25 which is reliable in itself, the ATM, FDDI, and SNA. The TCP/IP specifications only standardize ways of accessing those protocols from the internetwork layer (Fig. 2).

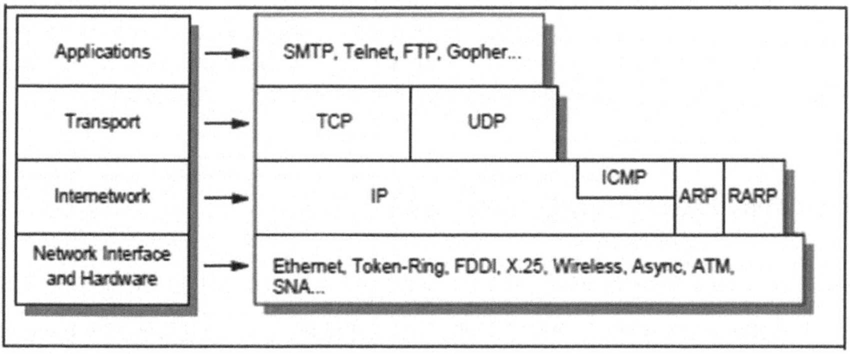

Fig. 2 Detailed architectural model of the TCP/IP network [1]

2.5 The Physical Layer

The physical layer is the lowest layer in the network architecture and it allows the transport of data across the network media. The physical layers aim to

a. Create the electrical, optical, or microwave signal that represents the bits in each frame which are then sent on the media one at a time to either an end device or an intermediate device.
b. Retrieve these individual signals from the media, from either an end device or an intermediate device, restore them to their bit representations, and pass the bits up to the Data Link layer as a complete frame.

The data delivery across the local media requires the availability of the following elements:

a. The physical media and its associated connectors
b. A representation of bits on the media
c. Data encoding and information control
d. Transceiver circuitry to transmit and receive on the network devices.

3 The Application Layer

3.1 Definition of Application Layer

Application layer is the most important and most visible layer in computer networks. Applications reside in this layer and human users interact via those applications through the network [3, 4]. Application layer contains a variety of protocols that are commonly needed by the users such as Hypertext Transfer Protocol (HTTP), File Transfer Protocol (FTP), Simple Mail Transfer Protocol (SMTP), Post Office Protocol (POP), Domain Name System (DNS), Telnet and etc. [3, 5, 6].

3.2 Principles of Application Layer

There are two important models used to organize a networked application [3, 7]:

(1) The client-server model which is the first and oldest one. In this model, a server provides services to clients that exchange information with it. This model is highly asymmetrical, clients send requests, and servers perform actions and return responses.
(2) The peer-to-peer model: In this model, all hosts act as both servers and clients and they play both roles. The peer-to-peer model has been used to develop various networked applications, ranging from Internet telephony to file sharing or Internet-wide file systems.

3.3 Application Layer Protocols

In the following subsections, we will define the most common protocols of the application layer.

HTTP protocol. The HTTP protocol defines how web clients request web pages from web server and how server transfers web pages to clients. The HTTP is implemented in two programs; client and server. Client and server talk to each other by exchanging HTTP messages. The HTTP protocol defines the structure of these messages and how client and server exchange it [3]. When a user (client) requests a web page or an object in a web page, user browser sends an HTTP request message for the object in that page to the server. Server receives the request and responds with HTTP response message that contains the object. HTTP uses TCP as its underline transport protocol so there are two behaviors for HTTP; non-persistent and persistent connections [3].

HTTP with non-persistent connections. To show the steps of transferring web pages from server to clients in case of non-persistent connections, suppose a page consists of a base HTML file and 10 JPEG files. All 11 files reside on the same server. Suppose the URL of the HTML file is http://www.someschool.edu/somedepartment/home.index [3]:

(1) The HTTP client process starts a TCP connection to the server www.someschool.edu on port number 80. Associated with TCP connection, there are two sockets one at the client and one at the server.
(2) The HTTP client sends an HTTP request message via its socket. The request message includes the path name/somedepartment/home.index.
(3) The HTTP server process receives the request message through its socket, retrieves the object/somedepartment/home.index from its storage (RAM or disk), encapsulates the object in an HTTP response message and sends it to the client through its socket.
(4) The HTTP server process tells the TCP to close its connection. The TCP terminates the connection only when it is sure that the client received the response message intact.
(5) The HTTP client receives the response message. The TCP connection terminates. The message indicates that the encapsulated object is an HTML file. The client extracts the file from the response message, examines the HTML file, and finds references to the 10 JPEG object.
(6) The first four steps are then repeated for each of the referenced JPEG objects.

HTTP with Persistent Connections. The HTTP with persistent connections, also called "the HTTP keep-alive", or "the HTTP connection reuse". This means that it uses the same TCP connection to send and receive multiple HTTP requests/responses. This is different from the non-persistent connection type in which a new TCP connection is opened for every single request/response pair. Using persistent connections improves the HTTP performance [8].

HTTP message format. RFCs [1945] and [2616] give the HTTP specifications including the definitions of the message formats. As explained in the following subsections.

Format of the HTTP request message

The HTTP messages are written in an ASCII text format. The message consists of a request line, header lines, a blank line, and entity-body. Figure 3 shows the format of the request message.

Here we present an example of the HTTP request message [3]:

GET/somedir/page.html HTTP/1.1
Host: www.someschool.edu
Connection: close
User-agent: Mozilla/5.0
Accept-language: fr

This example is explained as follows:

(a) The request line:

The object/***somedir/page.html HTTP/1.1*** is requested by the browser using the GET method. In this example, the browser uses HTTP/1.1 version.

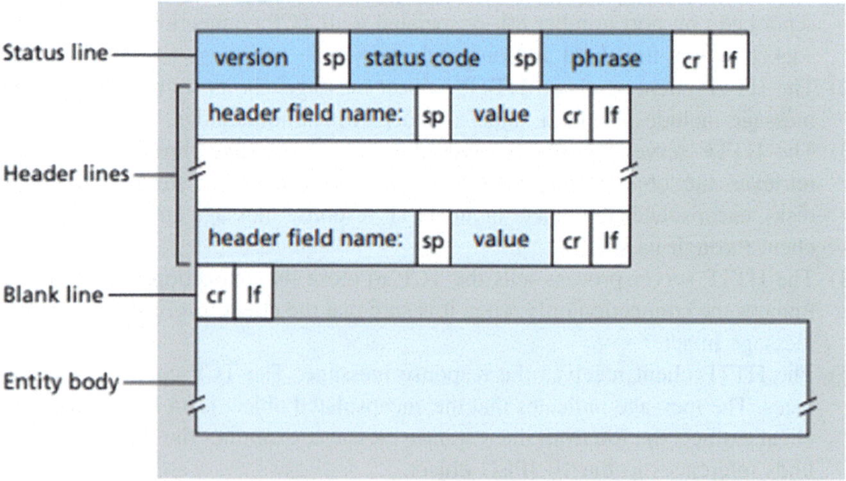

Fig. 3 General format of an HTTP request message [3]

(b) The header lines:

(1) *Host:* www.someschool.edu→ It indicates the host on which the requested object resides.
(2) *Connection: close* → It indicates that the browser asks the server to close the connection after sending the requested object.
(3) *User-agent: Mozilla/5.0* → It specifies the browser sending the request. Here, it's the Mozilla/5.0 Firefox browser.
(4) *Accept-language: fr* → It indicates that the user prefers to receive a French version of the object if it exists on the server. Otherwise, the server sends the default version.

HTTP Response Message
In the following figure, the general format for the response HTTP message is shown (Fig. 4).

Let's explain the message format for a given example that is a typical HTTP response message corresponding to the previous example request message [3].

HTTP/1.1 200 OK
Connection: close
Date: Tue, 09 Aug 2011 15:44:04 GMT
Server: Apache/2.2.3 (CentOS)
Last-Modified: Tue, 09 Aug 2011 15:11:03 GMT
Content-Length: 6821
Content-Type: text/html
(data data data data data...)

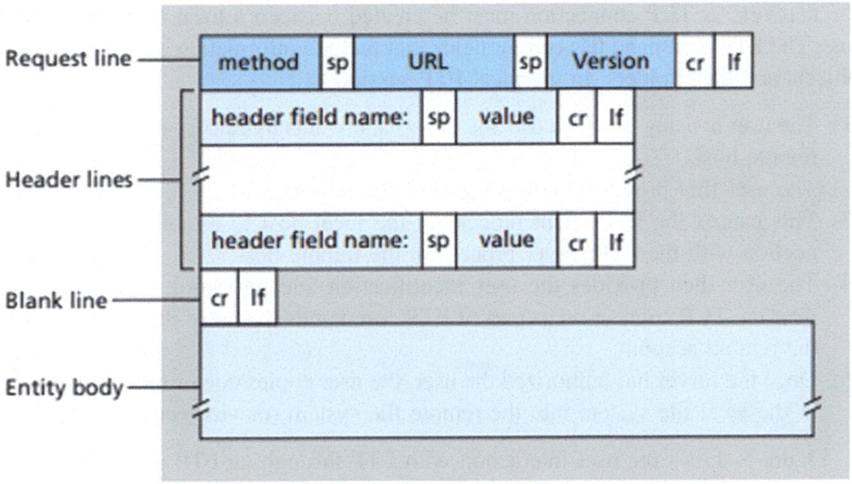

Fig. 4 General format of an HTTP response message [3]

The message includes three sections: an initial status line, some header lines, and then the entity-body. In the following, we describe these sections in the above message [3].

(a) The status line:

The line *HTTP/1.1 200 OK* →It indicates that the server uses *HTTP/1.1* version and that everything is running OK.

(b) The header lines:

  (1) The first header line *Connection close:* → It tells the client that the server is going to close the TCP connection after receiving the message.
  (2) The header line *Date: Tue, 09 Aug 2011 15:44:04 GMT* → *It* gives the time and date of sending the HTTP response.
  (3) The header line *Server: Apache/2.2.3 (CentOS)* → It shows that the message was generated by an Apache Web server.
  (4) The header line *Last-Modified*: *Tue, 09 Aug 2011 15:11:03 GMT* → It indicates the time and date when the object was modified (or created if no modification has been done).
  (5) The header line *Content-Length:* → It indicates the number of bytes in the requested object.
  (6) The header line *Content-Type:* → It indicates that the requested object in the entity body is HTML text. Notice that the object type is indicated by the *Content-Type*: and not by the file extension.

(c) The entity lines. The entity body is the main part of the message → It contains the requested object itself which is represented by: *data data data data.*

File Transfer Protocol (FTP)

The FTP is used by the application layer for transferring files between computers in the network. A TCP connection must be created between a local host and remote one. The FTP command has certain fields that include information about the moved objects and their folders. In a typical FTP session [3]:

(1) The user is using one host (the localhost) and wants to transfer files to or from a remote host.
(2) The user first provides the hostname of the remote host.
(3) This causes the FTP client process in the local host to establish a TCP connection with the FTP server process in the remote host.
(4) The user then provides the user identification and password, which are sent over the TCP connection as part of FTP commands to enable the user to access the remote account.
(5) Once the server has authorized the user, the user copies one or more files stored in the local file system into the remote file system (or vice versa).

Figure 5 shows the user interaction with FTP through an FTP user-agent.

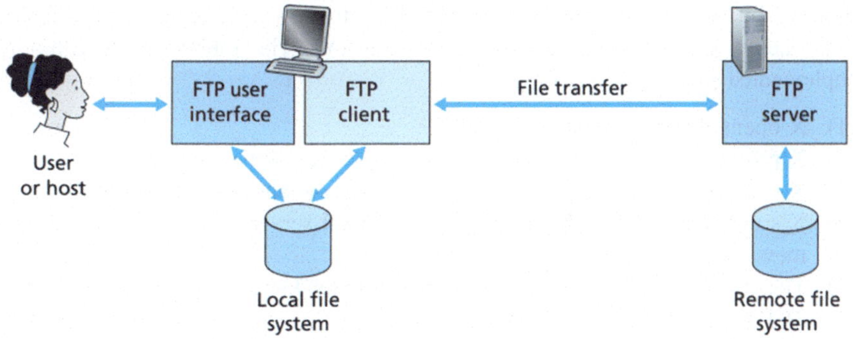

Fig. 5 FTP protocol moves files between local and remote file systems [3]

It is important to notice that the HTTP and FTP are both file transfer protocols. Thus, they may have many common characteristics. For example, they both run on top of TCP. However, the two protocols have some important differences. The most striking difference is that FTP uses two parallel TCP connections to transfer a file. The first TCP connection is a control connection. It is used for sending control information between the two hosts. This information includes the username, password, commands to change remote directory, and commands to "put" and "get" files. The second TCP connection is a data connection. In the following, examples of the FTP protocol commands are given. The protocol has two parts: one is at the local host and the other is at the remote host [3].

FTP protocol commands
The following commands are used by the ftp protocol [3]:

(1) USER username: is used to send the user identification to the server.
(2) PASS password: is used to send the user password to the server.
(3) LIST: is used to ask the server to send back a list of all the files in the current remote directory.
(4) RETR filename: is used to get a file from the current directory of the remote host.
(5) STOR filename: is used to store a file into the current directory of the remote host.

FTP protocol reply examples:

(1) 331 Username OK, password required.
(2) 125 Data connection already open; transfer starting.
(3) 425 Can't open data connection.
(4) 452 Error writing file.

The Simple Mail Transfer Protocol (SMTP). The SMTP protocol transfers messages from a sender's mail server to a recipient's mail server. It uses the TCP as its underline transport protocol. SMTP restricts the body of all mail messages to simple 7-bit ASCII. To show how the SMTP protocol works, assume that a sender (X) sends a simple ASCII mail to a receiver (Y). The following algorithm is implemented [3]:

(1) X opens the user-agent for e-mail.
(2) X provides Y's e-mail address (for example, X@someschool.edu)
3) X composes a message, and instructs the user-agent to send the message.
(4) X's user-agent sends the message to X's mail server, where it is placed in a message queue.
(5) The client-side of SMTP sees the message in the message queue and it opens a TCP connection to an SMTP server.
(6) A handshaking process is performed between X's and Y's SMTPs.
(7) SMTP client sends X's message into the TCP connection.
(8) At Y's mail server, the server-side of SMTP receives the message.
(9) The TCP connection is closed.
(10) Y's mail server then places the message in Y's mailbox.
(11) Y opens its user-agent to read the message at its convenience.

The following commands show how to exchange the message between the client (C) and the server (S) [3]:

S: 220 hamburger.edu
C: HELO crepes.fr
S: 250 Hello crepes.fr, pleased to meet you
C: MAIL FROM: <alice@crepes.fr>
S: 250 alice@crepes.fr... Sender ok
C: RCPT TO: <bob@hamburger.edu>
S: 250 bob@hamburger.edu... Recipient ok
C: DATA
S: 354 Enter mail, end with "." on a line by itself
C: Do you like ketchup?
C: How about pickles?
C:.
S: 250 Message accepted for delivery
C: QUIT
S: 221 hamburger.edu closing connection

The domain name service (DNS) protocol. The DNS protocol allows hosts to query the distributed database. It runs over UDP and uses port 53. The DNS is commonly employed by other application layer protocols including HTTP, FTP, and SMTP to translate user-supplied hostnames to IP addresses. The DNS protocol uses a large number of servers which are organized hierarchically and distributed around the world. No single DNS server has all the mappings of all hosts in the Internet. Instead, these mappings are distributed across the DNS servers. For that purpose, there are three classes of DNS servers:

(1) The root DNS servers T
(2) Top-level domain (TLD) DNS servers.
(3) Authoritative DNS servers.

Figure 6 shows a portion of the hierarchy of DNS servers [3].

To understand the DNS levels interaction, assume a DNS client wants to determine the IP address of the hostname www.amazon.com. Then, the following events take place [3]:

(1) The client first contacts one of the root servers, which returns IP addresses for TLD servers including the top-level domain com.
(2) The client then contacts one of these TLD servers, which returns the IP address of an authoritative server for amazon.com.
(3) Finally, the client contacts one of the authoritative servers for amazon.com, which returns the IP addresses for the hostname www.amazon.com.

Let's suppose a simple example to show how DNS works. Assume that the host cis.poly.edu desires the IP address of gaia.cs.umass.edu. Also, assume that the Polytechnic's local DNS server is called dns.poly.edu and that an authoritative DNS server for gaia.cs.umass.edu is called dns.umass.edu. Figure 7 shows the steps of the protocol which are given as follows [3]:

(1) The host cis.poly.edu sends a DNS query message to its local DNS server, dns. poly.edu including the hostname gaia.cs.umass.edu.
(2) The local DNS server forwards the query message to a root DNS server.
(3) The root DNS server takes note of the ***edu*** suffix and returns to the local DNS server a list of IP addresses for TLD servers responsible for ***edu***.
(4) The local DNS server then resends the query message to one of these TLD servers.
(5) The TLD server takes note of the umass.edu suffix and responds with the IP address of the authoritative DNS server of the University of Massachusetts, namely, dns.umass.edu.
(6) Finally, the local DNS server resends the query message directly to dns.umass. edu, which responds with the IP address of gaia.cs.umass.edu.

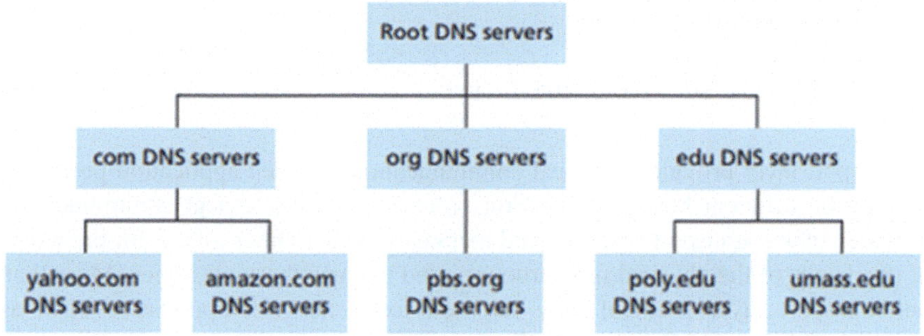

Fig. 6 A Portion of the hierarchy of DNS servers [3]

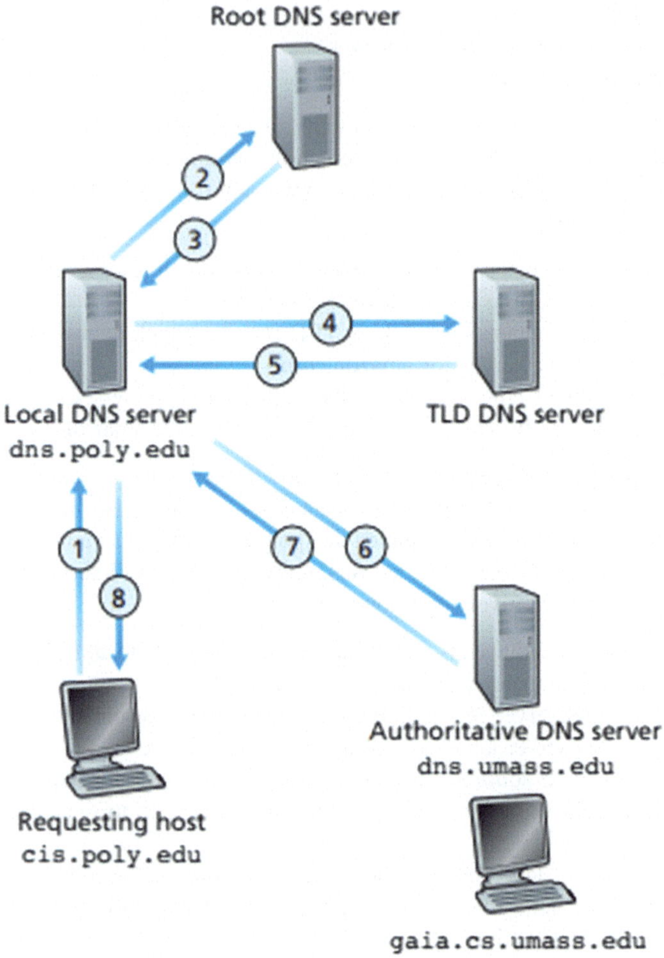

Fig. 7 Interaction between the various DNS servers [3]

4 The Transport Layer

4.1 Definition of Transport Layer

A transport layer provides a logical communication between application processes running on different hosts. Application processes use the logical communication provided by the transport layer to send messages to each other, free from the worry of the details of the physical infrastructure used to carry these messages. More than one transport layer protocol may be used for a network application such as Transmission Control Protocol (TCP), User Datagram Protocol (UDP), Datagram Congestion Control Protocol (DCCP) and Stream Control Transmission Protocol

(SCTP). Each of those protocols provides a set of services to the invoking application [3, 9].

4.2 Principles of Transport Layer

The following steps show the mechanism of the transport layer in brief [3]:

(1) At the sending side, the transport layer converts the application layer messages firstly into small chunks.
(2) Adding transport layer header to each chunk to create transport layer segments.
(3) The transport layer then passes the segments to the network layer at the sending end system, where the segment is encapsulated within a network layer datagram and then is sent to the destination.
(4) At the receiving side, the network layer extracts the transport-layer segment from the datagram.
(5) Network layer passes the segment to the transport layer.
(6) The transport layer processes the received segment, making the data in the segment available to the receiving application.

4.3 Services Provided by Transport Layer Protocols

Connection-oriented services. The transport layer provides a connection-oriented service to the upper layers. Before the transfer of data, a virtual connection is established between the source and the destination. After all, data is transferred, the connection is terminated. Transport layer provides a stream-oriented mechanism [10–12].

Connectionless services. The transport layer also provides a connectionless service to the upper layers. There is no need to establish a virtual connection between the source and the destination to transfer data. The connectionless service is used for faster data transfer than connection oriented but when there is no congestion [10–12].

Multiplexing and demultiplexing. If there are many processes at the sender side and they need to send data but there is only one transport layer protocol. This requires multiplexing and the receive side requires demultiplexing to get these many processes. The single protocol accepts data from different processes differentiating them by port numbers [10–12].

Slow start and congestion control. Once a connection is made, the sender starts sending packets, slowly at first, so it does not overwhelm the network. If congestion is not bad, it picks up the pace. This is called "slow start." Later, congestion controls help the sender scale back if the network gets busy [10–12].

Flow control. While slow start and congestion control are used to avoid network congestion, flow controls help in preventing the sender from overflowing the receiver with too much data. These controls are essential because the receiver drops packets when it is overloaded and those packets must be retransmitted, potentially increasing network congestion, and reducing system performance [10–12].

Error control. It is the responsibility of transport layer to provide error-free data transmission. For error control, three simple tools (Checksum, ACK and Timeout) are used. Error control includes mechanism for detecting lost, damaged, or unordered and also duplicated segments. After errors are detected, it includes methods for correcting them [12].

Reliability services. These services are used to retransmit corrupt, lost, and dropped packets. Positive acknowledgements confirm to the sender that the recipient actually received a packet (failure to transmit this acknowledgement means "resend the packet"). Sequencing is used to number packets so that packets can be put back in order and lost packets can be detected. Error checking detects corrupted packet [10–12].

4.4 Transport Layer Protocols

The User Datagram Protocol (UDP). The UDP is described in [RFC 768]. It is a transport layer protocol defined for use with the IP network layer protocol. It provides a best-effort datagram service to an End System (IP host). The basic operation of UDP protocol can be briefly described by the following steps [3, 13]:

(1) UDP gets the messages from the application layer.
(2) UDP then attaches source and destination port number fields for the multiplexing/demultiplexing service.
(3) It adds two fields for length of data and checksum.
(4) It passes the resulting segment to the network layer.
(5) The network layer encapsulates the resulting segment into an IP datagram and then makes a best-effort attempt to deliver the segment to the receiving host.
(6) At the receiving host, UDP uses the destination port number to deliver the segment's data to the correct application process (demultiplexing process).
(7) UDP has no handshaking between sending and receiving transport layer entities before sending a segment (so UDP is a connectionless protocol).

UDP does not provide any communications security. Applications that need to protect their communications against eavesdropping, tampering, or message forgery, therefore, need to separately provide security services using additional protocol mechanisms [14].

UDP data structure

The UDP header has four fields. Each of them consists of two bytes as shown in Fig. 8. The description of each field in detail is as follows [3, 11, 13]:

Fig. 8 The UDP segment structure RFC [768]

```
0         7 8      15 16     23 24     31
+--------+--------+--------+--------+
|     Source      |   Destination   |
|      Port       |      Port       |
+--------+--------+--------+--------+
|                 |                 |
|     Length      |    Checksum     |
+--------+--------+--------+--------+
|
|          data octets ...
+---------------- ...
```

(a) Source port address: This field indicates the port of the sending process which sends the datagram to perform multiplexing process.
(b) Destination port address: It indicates the port of the destination process to which the datagram is to be sent to perform demultiplexing process.
(c) Length: This field specifies the number of bytes in the UDP segment which includes the header plus data.
(d) Checksum: This field is an optional 16-bit one's complement of the one's complement sum of a pseudo-IP header, UDP header, and UDP data. The pseudo-IP header contains the source IP address and destination IP address, protocol and UDP length as shown in Fig. 9. The pseudo-header is used only for this calculation and is then discarded. It is not actually transmitted. The UDP software in the destination device creates the same pseudo-header when calculating its checksum to compare it with the one transmitted in the UDP header. Computing the checksum over the regular UDP fields protects against bit errors in the UDP message itself. Adding the pseudo-header allows the checksum to also protect against other types of problems as well, most notably the accidental delivery of a message to the wrong destination [15].

Here we present an example of UDP datagram [16]:

Datagram (Data + Header): 04 89 00 35 00 2C AB B4 00 01 01 00 00 01 00 00 00 00 00 00 04 70 6F 70 64 02 69 78 06 6E 65 74 63 6F 6D 03 63 6F 6D 00 00 01 00 01

Fig. 9 UDP pseudo-header format RFC [768]

```
               15 16     23 24     31
           -----+--------+--------+
         :urce address            |
           -----+--------+--------+
        :tination address         |
           -----+--------+--------+
       protocol|    UDP length    |
           -----+--------+--------+
```

UDP Header: 04 89 00 35 00 2C AB B4

Data: 00 01 01 00 00 01 00 00 00 00 00 00 04 70 6F 70 64 02 69 78 06 6E 65 74 63 6F 6D 03 63 6F 6D 00 00 01 00 01

Source Port	04 89
Destination Port	00 35
Length	00 2C
Checksum	AB B4
Data	DNS Message

Applications of UDP

There are various applications that use UDP as their transport protocol such as [11]:

(1) Routing Information Protocol (RIP).
(2) Simple network management protocol (SNMP).
(3) Dynamic host configuration protocol (DHCP).
(4) Voice and video traffic.

The Transmission Control Protocol (TCP). TCP is a transport layer protocol which is connection oriented. It provides a reliable byte stream to the upper layer (application layer). TCP has a mechanism based on positive acknowledgements and also provides a congestion avoidance mechanism to reduce the transmission rate when the network becomes overloaded [11, 17]. We can say that TCP is a time-tested transport layer protocol that provides several features such as reliability, flow control, and congestion control [11]. TCP is designed to dynamically adapt to different network conditions and to be robust in the face of many kinds of failures [13]. TCP is defined in RFC [793], RFC [1122], RFC [1323], RFC [2018], and RFC [2581].

The TCP protocol basic operation can be described briefly by the following steps [3]:

(1) Before one application process can begin to send data to another application process, the two processes must first "handshake" with each other.
(2) In three-way handshake, the first two segments carry no payload (no application layer data) and the third of these segments may carry a payload.
(3) Once a TCP connection is established, the two application processes can send data to each other.
(4) The client process passes a stream of data through the socket.
(5) Once the data passes through the socket, the data is in the hands of TCP running in the client.
(6) TCP directs this data to the connection's send buffer (one of the buffers that are set aside during the initial three-way handshake).
(7) From time to time, TCP grabs chunks of data from the send buffer and passes the data to the network layer.

TCP segment structure

When we compare a TCP segment format with a UDP segment format, we can observe the key difference between the two protocols. Since the UDP has a smaller header frame size, it is faster than TCP but it lacks reliability [12]. Figure 10 shows the format of the TCP segment format. A brief description of each field is as follows:

(a) Source port address: The 16-bit source port number, which specifies the sender of the segment to perform the multiplexing process [11, 18].
(b) Destination port address: The 16-bit destination port number used by sender to send the segment to the receiver to perform the demultiplexing process [11, 18].
(c) Sequence number: It is the 32-bits sequence number of the first data octet in the segment (except when SYN is present) [11, 18].
(d) Acknowledgement number: It is of 32-bits. If ACK control bit is set, this field indicates the value of the next sequence number of the segment to be received. This is always required to be sent once a connection is established [11, 18].
(e) Reserved: It is of 6-bits, which is reserved for future use and must be kept zero [11], RFC [793].
(a) Control bits: It is of 6-bits, which are control bits listed below [11], RFC [793]:
URG: Urgent Pointer field
ACK: Acknowledgement field
PSH: Push Function
RST: Reset the connection
SYN: Synchronize sequence numbers
FIN: No more data from sender

```
 0                   1                   2                   3
 0 1 2 3 4 5 6 7 8 9 0 1 2 3 4 5 6 7 8 9 0 1 2 3 4 5 6 7 8 9 0 1
+-+-+-+-+-+-+-+-+-+-+-+-+-+-+-+-+-+-+-+-+-+-+-+-+-+-+-+-+-+-+-+-+
|          Source Port          |       Destination Port        |
+-+-+-+-+-+-+-+-+-+-+-+-+-+-+-+-+-+-+-+-+-+-+-+-+-+-+-+-+-+-+-+-+
|                        Sequence Number                        |
+-+-+-+-+-+-+-+-+-+-+-+-+-+-+-+-+-+-+-+-+-+-+-+-+-+-+-+-+-+-+-+-+
|                    Acknowledgment Number                      |
+-+-+-+-+-+-+-+-+-+-+-+-+-+-+-+-+-+-+-+-+-+-+-+-+-+-+-+-+-+-+-+-+
|  Data |           |U|A|P|R|S|F|                               |
| Offset| Reserved  |R|C|S|S|Y|I|            Window             |
|       |           |G|K|H|T|N|N|                               |
+-+-+-+-+-+-+-+-+-+-+-+-+-+-+-+-+-+-+-+-+-+-+-+-+-+-+-+-+-+-+-+-+
|           Checksum            |         Urgent Pointer        |
+-+-+-+-+-+-+-+-+-+-+-+-+-+-+-+-+-+-+-+-+-+-+-+-+-+-+-+-+-+-+-+-+
|                    Options                    |    Padding    |
+-+-+-+-+-+-+-+-+-+-+-+-+-+-+-+-+-+-+-+-+-+-+-+-+-+-+-+-+-+-+-+-+
|                             data                              |
+-+-+-+-+-+-+-+-+-+-+-+-+-+-+-+-+-+-+-+-+-+-+-+-+-+-+-+-+-+-+-+-+
```

Fig. 10 The TCP Segment Structure RFC [793]

(b) Window: Its size is 16-bits. It specifies the number of data octets beginning with the one indicated in the acknowledgement field which the sender of this segment is expecting to receive [11].
(c) Checksum: It is of 16-bits in size. This field is the 16-bit one's complement of the one's complement sum of all the 16-bit words contained in the header and data [11], RFC[793].
(d) Urgent pointer: Its size is also of 16-bits. This field communicates the current value of the urgent pointer. It indicates the sequence number of the octet following the urgent data RFC [793].
(e) Options: Options occupy space at the end of the TCP header. It is multiple of 8-bits in length and occupy the space at the end of the header. All options are included in the checksum also RFC [793].

Here we present an example of TCP segment [19]:

TCP segment (data + header): 7E 21 45 00 00 4B 57 49 40 00 FA 06 85 77 C7 B6 78 0E CE D6 95 50 00 6E 04 9F 74 5B EE A2 59 9A 00 0E 50 18 24 00 E3 2A 00 00 2B 4F 4B 20 50 61 73 73 77 6F 72 64 20 72 65 71 75 69 72 65 64 20 66 6F 72 20 61 6C 65 78 75 72 2E 0D 0A 67 B2 7E

Start	7E
SEP	21
IP Header	45 00 00 4B 57 49 40 00 FA 06 85 77 C7 B6 78 0E CE D6 95 50
TCP Header	00 6E 04 9F 74 5B EE A2 59 9A 00 0E 50 18 24 00 E3 2A 00 00
Data	2B 4F 4B 20 50 61 73 73 77 6F 72 64 20 72 65 71 75 69 72 65 64 20 66 6F 72 20 61 6C 65 78 75 72 2E 0D 0A
FCS	67 B2
Stop	7E

TCP Header:

SRC_PORT = 110,
DEST_PORT = 1183,
SEQ = 745BEEA2 ACK = 599A000E
DTO = 5, FLG = 18, WIND = 9216,
TCP_SUM = E32A, URP = 0000, (No Options)

Data: +OK Password required for alexur\r\n

Control Flags (FLG = 18):

FLG = 00011000
Urgent Pointer URG = 0
Acknowledgement ACK = 1
Push Function PSH = 1
Reset connection RST = 0

Synchronization SYN = 0
Finished data FIN = 0
State = ACK-PSH

Applications of TCP
The following is a list of applications that use TCP [11]:

(1) File transfer protocol (FTP).
(2) Hypertext transfer protocol (HTTP).
(3) Simple mail transfer protocol (SMTP).
(4) Interactive mail access protocol (IMAP): which allows the clients to access e-mail messages and mailboxes over the network.
(5) Post office protocol (POP) which allows the clients to read and remove e-mails which are residing on a remote server. It is also used in e-mail applications.
(6) Remote login (Rlogin) which allows the remote login capability to the network.
(7) Secure shell (SSH) which allows the remote access to computers as well as data encryption.

5 Network Layer

5.1 Definition of Network Layer

The main aim of this layer is to deliver packets from source to destination across multiple links (networks) [13, 20]. It routes the signal through different channels to the other end and acts as a network controller. It also divides the outgoing messages into packets and assembles the incoming packets into messages for higher layers [20]. The network layer has some responsibilities or functions to do its job [3]:

(a) Forwarding: When a packet arrives at a router's input link, the router must move the packet to the appropriate output link.
(b) Routing: The network layer must determine the route or path taken by packets as they flow from a sender to a receiver. The algorithms that calculate these paths are referred to as routing algorithms.
(c) Connection Setup: Some networks require the routers along the chosen path from source to destination to handshake with each other in order to set up state before network layer data packets, within a given source to destination connection, begin to flow.

5.2 Principles of Network Layer

Figure 11 shows a simple network with two hosts, (H1) and (H2), and several routers on the path between (H1) and (H2). Suppose that H1 is sending information to H2. We can briefly describe the role of the network layer in the following steps [3]:

(1) At the sending end, the network layer in H1 takes segments from the transport layer in H1, encapsulates each segment into a datagram (network layer packet).
(2) Then it sends the datagrams to its nearby router R1.
(3) The receiving host (H2), the network layer receives the datagrams (network layer packets) from its nearby router R2.

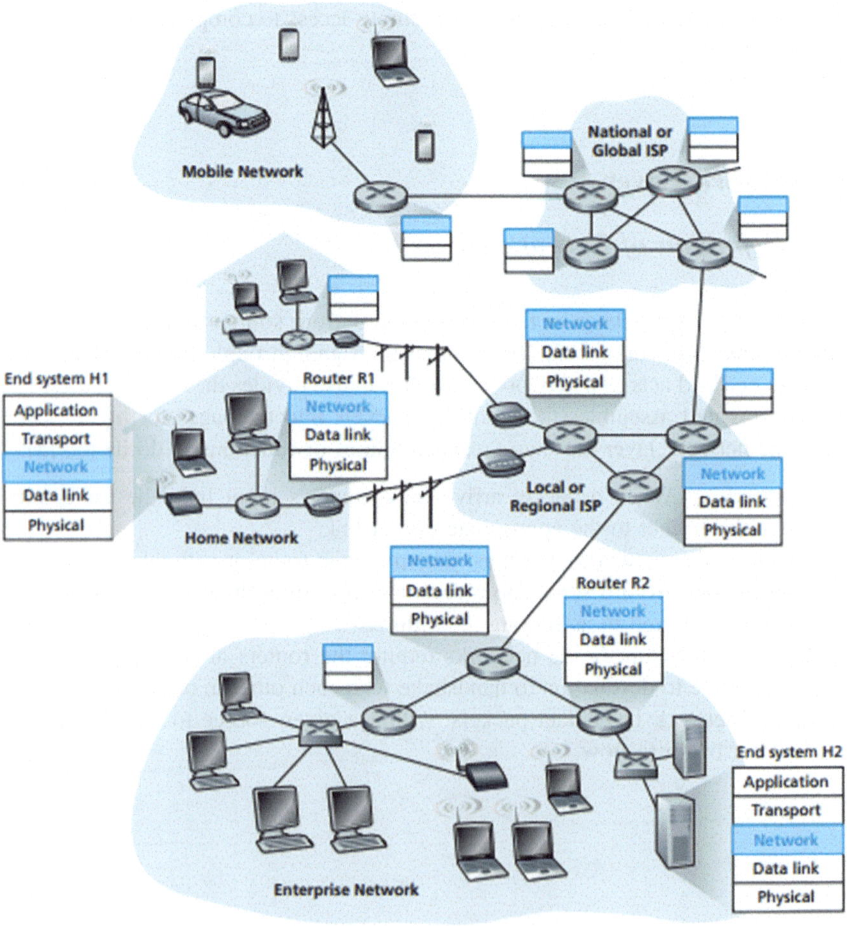

Fig. 11 Simple Network Example [3]

(4) It then extracts the transport layer segments and delivers the segments up to the transport layer at (H2). The primary role of the routers is to forward datagrams from input links to output links.

5.3 *The Network Service Model*

The network service model defines the characteristics of end-to-end transport of packets between sending and receiving end systems. In the sending host, when the transport layer passes a packet to the network layer, specific services that could be provided by the network layer include [3]:

(1) Guaranteed delivery: This service guarantees that the packet will eventually arrive at its destination [3].
(2) Guaranteed delivery with bounded delay: This service not only guarantees delivery of the packet, but delivery within a specified host-to-host delay bound (for example, within 100 ms) [3].
(3) In order packet delivery: This service guarantees that packets arrive at the destination in the order that they were sent [3].
(4) Guaranteed minimal bandwidth: This network layer service emulates the behavior of a transmission link of a specified bit rate (for example, 1 Mbps) between sending and receiving hosts. As long as the sending host transmits bits (as part of packets) at a rate below the specified bit rate, then no packet is lost and each packet arrives within a prespecified host-to-host delay (for example, within 40 ms) [3].
(5) Guaranteed maximum jitter: This service guarantees that the amount of time between the transmission of two successive packets at the sender is equal to the amount of time between their receipt at the destination [3].
(6) Security services: Using a secret session key known only by a source and destination host, the network layer in the source host could encrypt the payloads of all datagrams being sent to the destination host. The network layer in the destination host would then be responsible for decrypting the payloads. With such a service, confidentiality would be provided to all transport layer segments between the source and destination hosts [3].

5.4 The Network Layer Protocols

Here we take the Internet as a reference to show the mechanism of network layer protocols in it. The Internet's network layer has three major components [3]:

(1) The Internet protocol (IP), which has two versions of IP in use nowadays, IPV4 and IPV6 [3].
(2) The routing component, which determines the path a datagram follows from source to destination such as the Routing Information Protocol (RIP), Open Shortest Path First (OSPF), and the Border Gateway Protocol (BGP) [3, 21].
(3) A facility or a tool to report errors in datagrams and respond to requests for certain network layer information such as the Internet Control Message Protocol (ICMP) [3].

The Internet Protocol (IP). The Internet Protocol (IP) is defined in RFC [791]. It is designed for use in interconnected systems of packet-switched computer communication networks. Such a system has been called a "catenet". The Internet protocol provides transmitting blocks of data called datagrams from sources to destinations, where sources and destinations are hosts identified by fixed length addresses. The Internet protocol also provides fragmentation and reassembly of long datagrams, if necessary, for transmission through "small packet" networks [22], RFC [791]. There are two versions of IP in use nowadays, IPV4 and IPV6 [3].

Internet Protocol version 4 (IPV4). Internet Protocol Version 4 (IPv4) is the fourth revision of the Internet Protocol and a widely used protocol in data communication over different kinds of networks. IPv4 is a connectionless protocol used in packet-switched computer networks such as Ethernet. It provides the logical connection between network devices by providing identification for each device. There are many ways to configure IPv4 with all kinds of devices including manual and automatic configurations depending on the network type. IPv4 is based on the best-effort model. This model guarantees neither delivery nor avoidance of duplicate delivery; these aspects are handled by the upper layer transport [23].

A Datagram Format

IPv4 datagram format is shown in Fig. 12. The key fields in the IPv4 datagram are the following [3]:

In the following, we describe each of the IPV4 format components:

(1) Version number: These 4-bits specify the IP protocol version of the datagram [3]. For IPv4, the number is 4. The purpose of this field is to ensure compatibility between devices that may be running different versions of IP. In general, a device running an older version of IP will reject datagrams created by newer implementations, under the assumption that the older version may not be able to interpret the newer datagram correctly [24].
(2) Header length: Specifies the length of the IP header, in 32-bit words. This includes the length of any options fields and padding. The normal value of this field when no options are used is 5 (5 32-bit words = 5*4 = 20 bytes) [24].

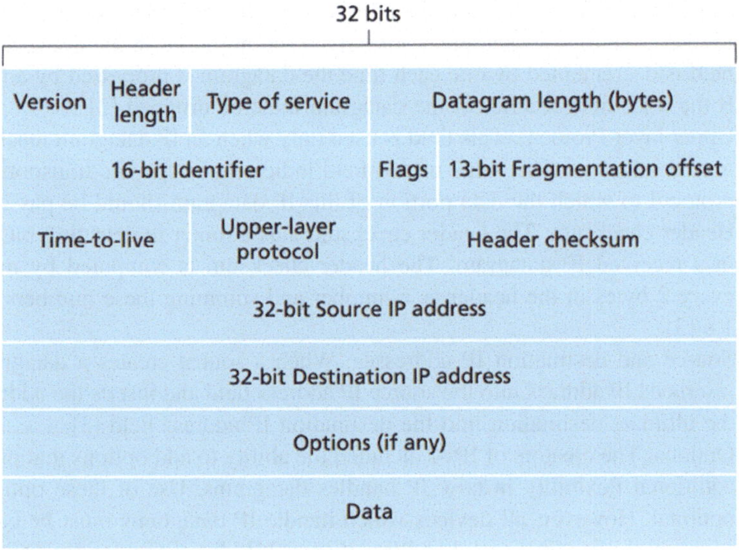

Fig. 12 IPV4 datagram format [3]

(3) Type of service: The type of service (TOS) bits is included in the IPv4 header to allow different types of IP datagrams to be distinguished from each other. For example, it might be useful to distinguish real-time datagrams such as those used by an IP telephony application from non-real-time traffic such as FTP [3].

(4) Datagram length: Specifies the total length of the IP datagram in bytes (header + data). Since this field is 16-bits wide, the maximum length of an IP datagram is 65,535 bytes, though most are much smaller [24].

(5) Identifier: This field contains a 16-bit value that is common to each of the fragments (will be discussed in the next section) belonging to a particular message. For datagrams originally sent unfragmented, it is still filled in, so it can be used if the datagram must be fragmented by a router during delivery. This field is used by the recipient to reassemble messages without accidentally mixing fragments from different messages. This is needed because fragments may arrive from multiple messages mixed together since IP datagrams can be received out of order from any device [24].

(6) Flags: Three control flags, two of which are used to manage fragmentation and one that is reserved [3, 24].

(7) Fragmentation offset: When a message is fragmented, this field specifies the offset, or the position, in the overall message where the data in this fragment goes. It is specified in units of 8 bytes (64-bits). The first fragment has an offset of 0 [24].

(8) Time-to-live: This field is included to ensure that datagrams do not circulate forever (due to, for example, a long-lived routing loop) in the network. This field is decremented by one each time the datagram is processed by a router. If the TTL field reaches 0, the datagram must be dropped [3].
(9) Upper layer Protocol: This field is used only when an IP datagram reaches its final destination. The value of this field indicates the specific transport layer protocol to which the data portion of this IP datagram should be passed [3].
(10) Header checksum: The header checksum aids a router in detecting bit errors in a received IP datagram. The header checksum is computed by treating every 2 bytes in the header as a number and summing these numbers using 1 s [3].
(11) Source and destination IP addresses: When a source creates a datagram, it inserts its IP address into the source IP address field and inserts the address of the ultimate destination into the destination IP address field [3].
(12) Options: The creators of IPv4 included the ability to add options that provide additional flexibility in how IP handles datagrams. Use of these options is optional. However, all devices which handle IP datagrams must be capable of properly reading and handling them [24]. Data (payload): This field contains the transport layer segment (TCP or UDP) to be delivered to the destination.

IP datagram fragmentation
Not all link layer protocols can carry network layer packets of the same size. Some protocols can carry big datagrams, whereas other protocols can carry only little packets [3]. So when an IP datagram is too large for the maximum transmission unit (MTU) of the underlying data link layer protocol used for the next leg of its journey, it must be fragmented before sending it across the network. The higher layer message to be transmitted is not sent in a single IP datagram but rather broken down into two or one IP datagrams called fragments. Each of these smaller IP datagrams is encapsulated in a separate link layer frame and is sent over the outgoing link separately [3, 24].

Internet protocol version 6 (IPV6). IP version 6 (IPv6) is defined in RFC [2460]. The Internet Engineering Task Force began an effort to develop the IPv4 protocol. A prime motivation for this effort was the realization that the 32 bit IP address space was beginning to be used up, with new subnets and IP nodes being attached to the Internet (and being allocated unique IP addresses) at a breathtaking rate. To respond to this need for a large IP address space, a new IP protocol, IPv6, was developed. The designers of IPv6 also took this opportunity to tweak and augment other aspects of IPv4, based on the accumulated operational experience with IPv4 [3, 25]. While the basic use of datagrams hasn't changed since IPv4, many modifications were made to their structure and format when IPv6 was created. The increase in the size of IP addresses from 32-bits to 128-bits adds a whopping extra 192-bits, or 24 bytes of information to the header. This in turn led to an effort to remove fields that weren't strictly necessary, to compensate for the necessary increase in size. However, changes were also made to IPv6 datagrams to add features to them and to make them better suit the needs of modern Internet working [25]. The following is a list of the most significant overall changes to datagrams in IPv6 [3, 25]:

(1) Multiple Header Structure: Rather than a single header that contains all fields for the datagram, the IPv6 datagram supports a main header and then extension headers for additional information when needed.
(2) Streamlined Header Format: Several fields are removed from the main header to reduce its size and increase efficiency. Only the fields that are truly required for pretty much all datagrams remain in the main header; others are put into extension headers and used as needed. Some are removed because they are no longer needed, such as the Internet Header Length field; the IPv6 header is of fixed length.
(3) Renamed Fields: Some fields have been renamed to better reflect their actual use in modern networks.
(4) Greater Flexibility: The extended headers allow a great deal of extra information to accompany datagrams when needed and options are also supported in IPv6.
(5) Elimination of Checksum Calculation: In IPv6, a checksum is no longer computed on the header. This saves both the calculation time spent by every device that packages IP datagrams (hosts and routers) and the space the checksum field took up in the IPv4 header.
(6) Improved Quality of Service Support: A new field, the Flow Label, is defined to help support the prioritization of traffic.

IPV6 datagram format

The format of the IPv6 datagram is shown in Fig. 13.

A comparison of Figs. 12 and 13 reveals the simpler, more streamlined structure of the IPv6 datagram. The following fields are defined in IPv6 [3, 26]:

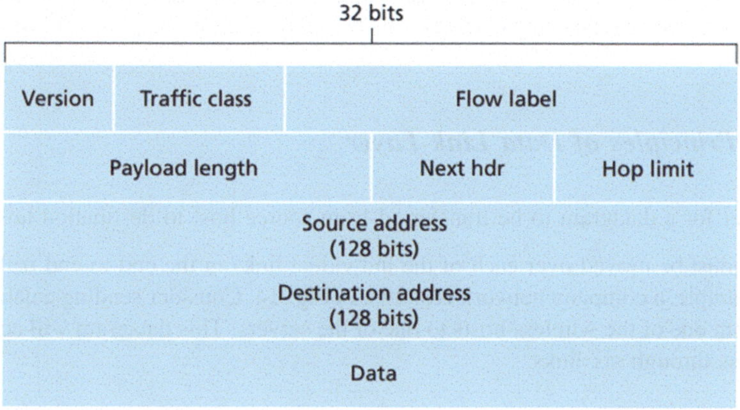

Fig. 13 IPV6 datagram format [3]

(1) Version: This 4-bit field identifies the IP version number. Not surprisingly, IPv6 carries a value of 6 in this field.
(2) Traffic class: This 8-bit field is similar to the TOS field in IPv4.
(3) Flow label: This 20-bit field is used to identify a flow of datagrams.
(4) Payload length: This 16-bit value is treated as an unsigned integer giving the number of bytes in the IPv6 datagram following the fixed length, 40-byte datagram header.
(5) Next header: This field identifies the protocol to which the contents (data field) of this datagram will be delivered (for example, to TCP or UDP).
(6) Hop limit: The contents of this field are decremented by one by each router that forwards the datagram. If the hop limit count reaches zero, the datagram is discarded.
(7) Source address: The 128-bit IP address of the originator of the datagram. As with IPv4, this is always the device that originally sends the datagram [26].
(8) Destination address: The 128-bit IP address of the intended recipient of the datagram; unicast, anycast, or multicast [26].
(9) Data: This is the payload portion of the IPv6 datagram. When the datagram reaches its destination, the payload will be removed from the IP datagram and passed on to the protocol specified in the next header field.

6 The Data Link Layer

6.1 Definition of Data Link Layer

The data link layer or layer-2 is the layer that transfers data between adjacent network nodes in a Wide Area Network (WAN) or between nodes on the same Local Area Network (LAN). The data link layer provides the functional and procedural means to transfer data between network entities and might provide the means to detect and possibly correct errors that may occur in the physical layer [27].

6.2 Principles of Data Link Layer

In order for a datagram to be transferred from source host to destination host [3]:

(1) It must be moved over each of the individual links in the end-to-end path. For example, a company network is shown in Fig. 14. Consider sending a datagram from one of the wireless hosts to one of the servers. This datagram will actually pass through six links:

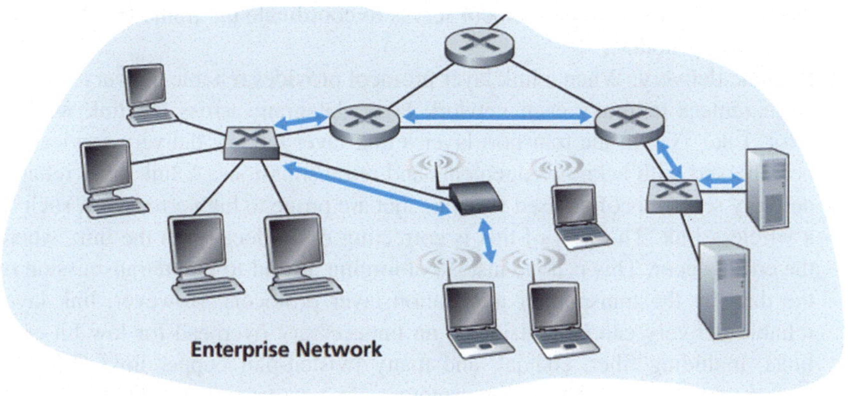

Fig. 14 Six link layer hops between wireless host and server [3]

(a) WiFi link between sending host and WiFi access point.
(b) Ethernet link between the access point and a link layer switch.
(c) Link between the link layer switch and the router.
(d) Link between the two routers.
(e) Ethernet link between the router and a link layer switch
(f) Ethernet link between the switch and the server.

(2) At any given link, a transmitting node encapsulates the datagram in a link layer frame and transmits the frame into another link.

6.3 *The Services Provided by the Link Layer*

The basic service of any link layer is to move a datagram from one node to an adjacent node over a single communication link. However, the details of the provided service can vary from one link layer protocol to the next. Possible services that can be offered by a link layer protocol include [3]:

(1) Framing: Almost all link layer protocols encapsulate each network layer datagram within a link layer frame before transmission over the link. A frame consists of a data field, in which the network layer datagram is inserted, and a number of header fields. The structure of the frame is specified by the link layer protocol [3].
(2) Link access: Medium Access Control (MAC) protocol specifies the rules by which a frame is transmitted from link to another. For point-to-point links that have a single sender at one end of the link and a single receiver at the other end of the link, the MAC protocol is simple or nonexistent. Here the sender can send a frame whenever the link is idle. The more interesting case is when multiple nodes share a single broadcast link which called multiple access

problem. Here, the MAC protocol serves to coordinate the frame transmissions of the many nodes [3].

(3) Reliable delivery: When a link layer protocol provides reliable delivery service, it guarantees to move each network layer datagram across the link without error. Like TCP in the transport layer a link layer reliable delivery service can be achieved with acknowledgements and retransmissions. A link layer reliable delivery service is often used for links that are prone to high error rates, such as a wireless link. The goal of this is correcting errors locally on the link, where the errors occur. This is done instead of forcing an end-to-end retransmission of the data by the transport or application layer protocols. However, link layer reliable delivery can be considered an unnecessary overhead for low bit-error links, including fiber, coaxial, and many twisted-pair copper links. For this reason, many wired link layer protocols do not provide a reliable delivery service [3].

(4) Error detection and correction: The link layer hardware in a receiving node can incorrectly decide that a bit in a frame is a zero when it was transmitted as a one, and vice versa. Such bit errors are introduced by signal attenuation and electromagnetic noise. Because there is no need to forward a datagram that has an error, many link layer protocols provide a mechanism to detect such bit errors. This is done by having the transmitting node include error detection bits in the frame, and having thc receiving node perform an error check. Error detection in the link layer is usually more sophisticated and is implemented in hardware. Error correction is similar to error detection, except that a receiver not only detects when bit errors have occurred in the frame but also determines exactly where in the frame the errors have occurred and then corrects these errors [3].

6.4 Elementary Data Link Layer Protocols

A utopia simplex protocol. As an initial example, we will consider a protocol that is as simple as it can be because it does not worry about the possibility of anything going wrong. In this protocol [13, 28, 29]:

(1) Data are transmitted in one direction only.
(2) Both the transmitting and receiving network layers are always ready.
(3) Processing time can be ignored.
(4) Infinite buffer space is available.
(5) The communication channel between the data link layers never damages or loses frames.

The Protocol components

The protocol consists of two distinct procedures, a sender and a receiver. The sender runs in the data link layer of the source machine, and the receiver runs in the data link layer of the destination machine. Let's describe these procedures in the following steps:

(1) The sender is in an infinite while-loop just pumping data out onto the line as fast as it can. The body of the loop consists of three actions [13]:

    (a) Go fetch a packet from the network layer.
    (b) Construct an outbound frame.
    (c) Send the frame on its way.

(2) The receiver is equally simple:

    (a) Let us assume that we are at state wait_for_event, it waits for something to happen which is the arrival of an undamaged frame.
    (b) When the frame arrives and the procedure wait_for_event returns, with event set to frame_arrival.
    (c) It receives the frame by the call to from_physical_layer, extract data.
    (d) The data portion is passed on to the network layer, and the data link layer settles back to wait for the next frame, effectively suspending itself until the frame arrives [13].

A simplex stop/wait protocol for error-free channels. In this protocol, we assume that [28]

(1) Data are transmitted in one direction only.
(2) No errors occur (perfect channel).
(3) The receiver can only process the received information at a finite rate.

These assumptions imply that the transmitter cannot send frames at a rate faster than the receiver can process them. The problem here is how to prevent the sender from flooding the receiver. A general solution to this problem is to have the receiver provide some sort of feedback to the sender. The process could be as follows:

(1) The receiver sends an acknowledge frame back to the sender telling the sender that the last received frame is processed and passed to the host; permission to send the next frame is granted.
(2) The sender, after having sent a frame, must wait for the acknowledge frame from the receiver before sending another frame [28]. As in the previous protocol:

    (a) The sender fetches a packet from the network layer, using it to construct a frame, and sending it on its way.
    (b) The sender waits until an acknowledgement frame arrives before looping back and fetching the next packet from the network layer. The only difference between the receiver of the previous protocol and this one is that after delivering a packet to the network layer, the receiver of this protocol sends an acknowledgement frame back to the sender before entering the wait loop again [10].

A simplex stop/wait protocol for noisy channels. In this protocol the unreal "error-free" assumption in the previous protocol is eliminated. Frames may be either damaged or lost completely. We assume that transmission errors in the frame are detected by the hardware checksum. One suggestion is that the sender would send a frame, the receiver would send an ACK frame only if the frame is received correctly. If the frame is in error the receiver simply ignores it. Then the transmitter will time out and retransmit it [13, 28]. One fatal flaw here is that if the ACK frame is lost or damaged, duplicate frames are accepted at the receiver without the receiver knowing it. Imagine a situation where the receiver has just sent an ACK frame back to the sender saying that it correctly received and already passed a frame to its host. However, the ACK frame gets lost completely, the sender times out and retransmits the frame. There is no way for the receiver to tell whether this frame is a retransmitted frame or a new frame, so the receiver accepts this duplicate and transfers it to the host. So, the protocol fails in this aspect. To overcome this problem, it is required for the receiver to be able to distinguish a frame that it received for the first time from a retransmitted one. One way to achieve this

(1) The sender puts a sequence number in the header of each frame it sends.
(2) The receiver then can check the sequence number of each arriving frame to see if it is a new frame or a duplicate to be discarded.
(3) The receiver needs to distinguish only two possibilities: a new frame or a duplicate; so, it only needs a 1-bit sequence number.
(4) At any instant, the receiver expects a particular sequence number. Any wrong sequence numbered frame arriving at the receiver is rejected as a duplicate.
(5) A correctly numbered frame arriving at the receiver is accepted, passed to the host, and the expected sequence number is incremented by 1 [13, 28].

7 The Physical Layer

7.1 Definition of the Physical Layer

The physical layer is the foundation on which the network is built. It defines the electrical, timing and other interfaces by which bits are sent as signals over channels [3]. It concerns with the physical characteristics of the network, such as the types of cables connecting devices, the types of connectors, the cables lengths, and so on [4].

7.2 Operation of the Physical Layer

The physical layer transmits raw bits over a communication channel. It defines the electrical and physical specifications of the data connection. It also concerns with the relationship between a device and a physical transmission medium (e.g., a copper or a fiber optical cable). The design processes make sure that if one side sends a 1-bit, the other side receives it as a 1-bit and not as a 0-bit. The physical layer also specifies characteristics of the electrical signals used to transmit data over cables from a network node to another. Transmission media may be roughly grouped into two types. The first type is the guided media, such as copper wire and fiber optics. The second type is the unguided media, such as terrestrial wireless, satellite, and lasers. There are three basic forms of network media on which data is represented

(1) Copper cable: the signals are patterns of electrical pulses.
(2) Fiber: the signals are patterns of light.
(3) Wireless: the signals are patterns of radio transmissions.

When the physical layer encodes the bits into the signals for a particular medium, it must also distinguish where one frame ends and the next frame begins. Indicating the beginning of frame is often a function of the Data Link layer. However, in many technologies, the physical layer may add its own signals to indicate the beginning and end of the frame. Figure 15 shows the representation of signals in the physical layer.

7.3 *Physical Layer Standards*

The physical layer consists of hardware which is designed and/or developed by engineers. This hardware may include electronic circuitry, media, and connectors.

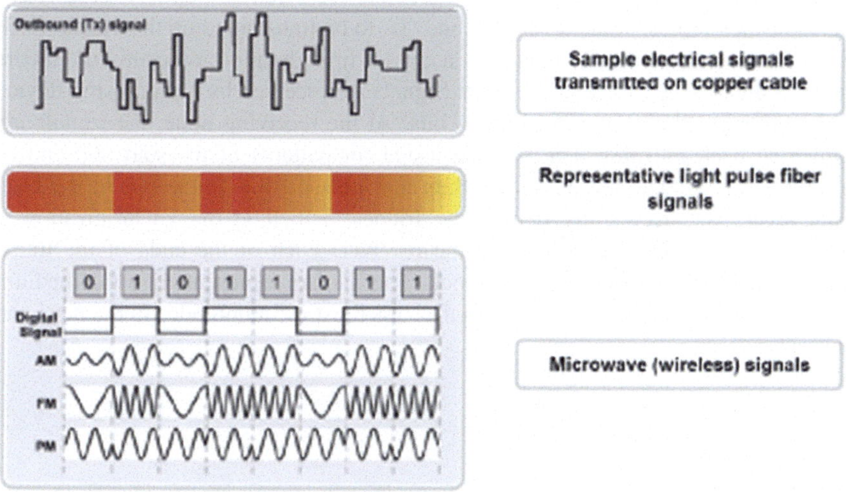

Fig. 15 Representations of signals on the physical media [4]

Thus, the standards governing this hardware should be defined by the relevant electrical and communications engineering organizations. These technologies include four areas of the physical layer standards:

(1) Physical and electrical properties of the media
(2) Mechanical properties of the connectors (materials, dimensions, pinouts).
(3) Bit signal representation (encoding)
(4) Definition of control information signals

Examples of hardware components may include: network adapters, Network Interface Cards (NICs), connectors and interfaces, cable materials, and cable designs. All these components are specified in standards associated with the physical layer.

7.4 Physical Layer Principles

There are three fundamental functions of the physical layer which are the physical components, the data encoding, and the signaling [29].

Encoding. Signaling is a method to convert a stream of data bits into a predefined code. This is done to

(1) Provide a predictable pattern that can be recognized by both the sender and the received.
(2) Distinguish data bits from control bits and provide better media error detection.
(3) Provide codes for control purposes such as identifying the beginning and end of a frame.

Signaling. The physical layer must generate the electrical, optical, or wireless signals which represent the levels “1” and “0” to be transmitted on the media. Each signal is transmitted on the media has a certain time duration to occupy the media and this duration is called the bit time. Signals are received by a processing device and returned to their representation as bits. At the receiving node, the signals are converted back into bits. The bits are tested for patterns of the start and end of complete frames. The physical layer then transmits the bits frame bits to the Data Link layer. To successfully deliver the frames, a method of synchronization is needed between the transmitter and the receiver. Each of the ends of the transmission medium maintains its own clock. The bits are represented on the medium by changing one or more of the signal characteristics: amplitude, frequency, and phase.

Signaling Methods:

Non-Return to Zero (NRZ) Signaling Method

The bitstream is transmitted as a series of voltage two values which a low voltage value represents a logical 0 and a high voltage value represents a logical 1. This representation has the following:

(1) The NRZ is used for only slow-speed data links.
(2) The NRZ inefficiently uses the bandwidth and it is susceptible to electromagnetic interference.
(3) When long strings of 1 s or 0 s are consecutively transmitted, the boundaries between individual bits can be lost. Then, no voltage transitions are detectable on the media.

Manchester Encoding Method

Bit values are represented as voltage transitions from a low voltage to a high voltage (a bit value of 1) and from a high voltage to a low voltage (a bit value of 0). Although at high signaling speeds, the Manchester Encoding inefficient, it is employed by 10BaseT Ethernet (i.e., Ethernet running at 10 Megabits per second).

Encoding (Grouping Bits)

Encoding is grouping bits before being presented to the media to

(1) Improve the efficiency at higher speed data transmission.
(2) Detect errors.
(3) Represent more data across the media by transmitting fewer bits.

The stream of transmitted signals needs to start in such a way such that the receiver recognizes the beginning and end of the frame. One way to provide the frame detection is to begin and end each frame with a certain pattern of signals representing bits. Figure 16 gives the characteristics of the physical layer.

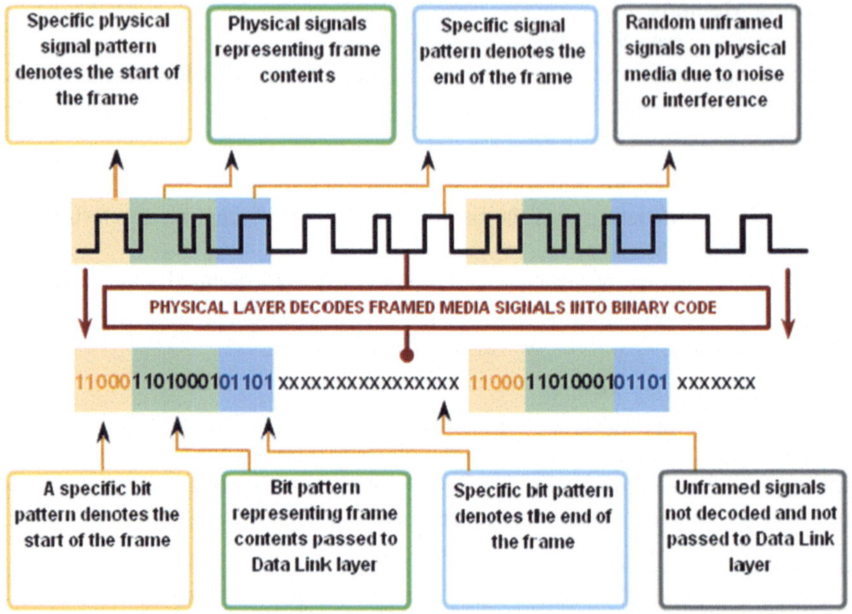

Fig. 16 Characteristics of the physical layer [4]

8 Conclusions

This paper aimed to introduce a survey of the famous and important protocols and processes of the TCP/IP network layers. The layers are generally introduced and then the important protocols and processes of each layer are explained. The layer is generally defined, and its function is explained and then the protocols are given. Examples are also explained when it is needed.

References

1. Parziale, L. 2016. TCP/IP tutorial and technical overview, 8th edn. IMB Corporation.
2. https://docs.oracle.com/javase/tutorial/networking/sockets/definition.html. Accessed 11 July 2019.
3. Kurose, James F. and Keith W. Ross. 2013. Computer networking: Atop-Down approach, 6th edn. London: Pearson Education Inc.
4. http://cnp3book.info.ucl.ac.be/1st/html/application/application.html. Accessed 10 July 2019.
5. http://cnp3book.info.ucl.ac.be/1st/html/application/app-protocols.html. Accessed 10 July 2019.
6. Ravali, P. 2015. A comparative evaluation of OSI and TCP/IP models. *International Journal of Science and Research* 4 (7): 514–521.
7. http://cnp3book.info.ucl.ac.be/1st/html/application/principles.html. Accessed 10 July 2019.

8. http://docs.oracle.com/javase/7/docs/technotes/guides/net/http-keepalive.html. Accessed 10 July 2019.
9. http://www.corenetworkz.com/2007/06/what-is-role-of-transport-layer-in-osi.html. Accessed 10 July 2019.
10. www.linktionary.com/t/transport.html. Accessed 10 July 2019.
11. Kumar, Santosh, and Sonam Rai. 2012. Survey on transport layer protocols: TCP & UDP. *International Journal of Computer Applications* 46 (7): 20–25.
12. Khan, Inam Ullah, and Muhammad Abul, Hassan. 2016. Transport layer protocols and services. *IJRCCT* 5(9): 490–493.
13. Tanenbaum, Andrew S., and David J. Wetherall. 2011. *Computer Networks, 5th edn.*. London: Pearson Education, Inc.
14. http://www.erg.abdn.ac.uk/users/Gorry/course/inet-pages/udp.html. Accessed 10 July 2019.
15. http://www.tcpipguide.com/free/t_UDPMessageFormat-2.htm. Accessed 9 July 2019.
16. http://www.netfor2.com/udp.htm. Accessed 9 July 2019.
17. Rahmani, Mehrnoush, Andrea, Pettiti, Ernst, Biersack, Eckehard, Steinbach, and Joachim, Hillebrand. 2008. A comparative study of network transport protocols for in-vehicle media streaming. In *2008 IEEE International Conference on Multimedia and Expo*. Germany: IEEE.
18. Forouzan, Behrouz A. 2003. Data communications and networking, 2nd edn. New York: Mc Graw hill.
19. http://www.netfor2.com/tcp.htm. Accessed 8 July 2019.
20. http://www.studytonight.com/computer-networks/osi-model-network-layer. Accessed 9 July 2019.
21. https://www.techopedia.com/definition/25927/routing-protocol. Accessed 10 July 2019.
22. https://en.wikipedia.org/wiki/Internet_Protocol. Accessed 10 July 2019.
23. https://www.techopedia.com/definition/5367/internet-protocol-version-4-ipv4. Accessed 10 July 2019.
24. http://www.tcpipguide.com/free/t_IPDatagramGeneralFormat.htm. Accessed 10 July 2019.
25. http://www.tcpipguide.com/free/t_IPv6DatagramOverviewandGeneralStructure.htm. Accessed 10 July 2019.
26. http://www.tcpipguide.com/free/t_IPv6DatagramMainHeaderFormat.htm. Accessed 10 July 2019.
27. https://en.wikipedia.org/wiki/Data_link_layer. Accessed 10 July 2019.
28. http://www.dsi.unive.it/~franz/reti/dll/Protocolli.html. Accessed 10 July 2019.
29. https://uomustansiriyah.edu.iq/media/lectures/5/5_2017_12_24!10_44_25_PM.pdf. Accessed 11 July 2019.

A Trust-Based Ranking Model for Cloud Service Providers in Cloud Computing

Alshaimaa M. Mohammed and Fatma A. Omara

Abstract With the rapid growth of Cloud services, many Cloud Service Providers (CSPs) offer similar service functionalities. Hence, guaranteeing the available CSPs having a trust degree would increase the performance of the cloud environment. Therefore, selecting the trusted CSP whose services satisfy the Cloud Service Consumers' (CSC) requirements becomes a challenge. According to the work in this paper, a ranking model for CSPs has been introduced based on a combination of the trust degree for each CSP and the similarity degree between the CSPs' parameters and the CSCs' requested parameters. The proposed model consists of four phases; Filtrating, Trusting, Similarity, and Ranking. In the Filtrating phase, the existing CSPs in the system will be filtered based on their parameters. The CSPs trust values are calculated in the Trusting phase. Then, the similarity between CSCs' requirements and CSPs' services is calculated. Finally, the ranking of CSPs will be performed. To evaluate the performance of the proposed CSPs Ranking model, a comparative study has been done among the proposed CSP Ranking model and the most up to date four models using two QoS case studies and Armor data set. According to the comparative results, it is found that the proposed CSPs Ranking model outperforms the existing models with respect to execution time, time complexity, and precision of the system.

Keywords Cloud service provider · Cloud service consumers' request · Ranking · Trust · Similarity · Fuzzy controller · Dynamic adaptive particle swarm optimization

---

A. M. Mohammed (✉)
Faculty of Science, Computer Science & Mathematics Department, Suez Canal University, Ismailia, Egypt
e-mail: Shaimaa_mostafa@science.suez.edu.eg

F. A. Omara
Faculty of Computer Science and Information, Computer Science Department, Cairo University, Giza, Egypt
e-mail: f.omara@fci-cu.edu.eg

A. Z. Ghalwash et al. (eds.), *Internet of Things—Applications and Future*,
Lecture Notes in Networks and Systems 114,
https://doi.org/10.1007/978-981-15-3075-3_22

1 Introduction

Nowadays, Cloud Computing is considered one of the most challenging emerging technologies [1, 2]. It provides a pool of IT computing services (i.e., CPU, Networks, Storage, and applications) that offer a range of dynamic, elastic, and on-demand services to the Consumer on the basis of usages under "pay-as-you-go" pricing model [3]. These opportunities offer many advantages such as reducing the cost of the resources' scalability, self-service, location independence, and rapid deployment [4]. The main feature of the Cloud Computing is that self-service provisioning is provided, which allows the users to deploy their own sets of computing resources [5]. The core of providing the required services in the Cloud is on-demand manner, where the Cloud Service Consumers (CSCs) request specific services (e.g., computation, storage, memory, etc.) from the Cloud Service Provider (CSP). The CSP must provide the service(s) that satisfies the CSCs' requests in terms of its Quality of Service (QoS) [6]. On the other hand, there are many CSPs that offer the same service(s) with different parameters [7]. Therefore, selecting the proper CSP to provide the requested service(s) has recently attracted considerable attention from the industry and academia and it is considered one of the most critical and strategic problems in the Cloud environments. From the CSCs' point of view, selecting the proper CSP is essential to assure future performance and maintain compliance with laws, policies, and rules [8].

By increasing the number of CSPs who provide similar services, ranking them becomes the most challengeable issue [9]. From the security point of view, the increasing number of CSPs causes the cloud environment to be more competitive day by day. One of the most important factors of security in the cloud is the CSPs' trust parameter which plays a vital role to make the cloud business grows and the CSP can get more profit. To make cloud computing more attractive, CSPs' trust must be addressed where CSCs can interact with the CSPs and use their services safely [10], Therefore, to rank CSPs, the trust parameter of the CSPs must be taken into consideration as it plays a significant role for prioritizing CSPs. The CSP's trust is defined as the expectation that CSP can be relied on, predictable, and act fairly [11]. By making the CSP trustable, it helps the CSC to interact with proper CSP [12]. Therefore, most existing selection researches based on CSCs' requirements only [2, 8, 13], and others consider only the trust parameter of CSP [10]. According to the work in this paper, a CSPs Ranking model has been introduced to provide a list of the most ranking CSPs to the CSC. This model based on CSPs trust degree and the similarity of their resources with respect to the CSCs' requirements. On our point of view, the reliability of the CSPs ranking model would be increased by considering the trust degree of CSPs.

The proposed CSPs Ranking model consists of four phases; Filtrating, Trusting, Similarity, and Ranking. As the number of the available CSPs is increased, the available CSPs are filtrated in the Filtrating phase to prevent unwanted CSPs. In the Trusting phase, CSPs' trust degrees will be calculated for the accepted CSPs. In the similarity phase, the similarity value between the CSCs' requests and the parameters of the accepted CSPs will be calculated. In the Ranking phase, the CSPs ranking degree will be determined based on the CSPs' trust degrees and their similarity value.

The paper is organized as follows, related work of the CSPs ranking is introduced in Sect. 2. Section 3 is dedicated to illustrating the principles of the proposed CSPs' Ranking model. The performance evaluation of the proposed CSPs' Ranking model relative to the most up to date models (i.e., hypergraph computational (HGCM), Hypergraph-Binary Fruit Fly Optimization Algorithm (HBFFOA), security risk, and E-TOPSIS algorithms) is discussed in Sect. 4. Finally, conclusions and future work are presented in Sect. 5.

2 Related Work

In [14], a ranking model has been introduced to rank the CSPs based on Hypergraph Computational model (HGCM) using the Service Measurement Index (SMI) metrics of parameters [15]. The model represents the relations between the SMI metrics in the form of Hyper-edges, where the intersection is the interrelations between them. A minimum distance algorithm is used to arrange the attributes in a hierarchical order. Neighborhood relations between the Hyper-edges have been established by fixing a threshold value using the minimum distance algorithm. The recursive using Helly property influences the order of CSPs to select the proper CSP. The time complexity of this model is determined as follows [14]:

$$O\left(n^2 + NP\right) \quad (1)$$

where n is the number of sub-attributes, N is the number of attributes, and P is the number of CSPs. The computation overhead is considered the main drawback of this model, especially, by increasing the number of CSPs.

A CSPs ranking model using checkpoint-based load balancing has been introduced in [16]. This model uses integration between checkpoint and load balancing algorithm to increase the availability of the requests of consumers in the real-time scheduling environment [17]. Initially, the user will access the system based on the obtained ranks according to the services that have been accessed earlier and the preferred value is determined by subtracting the service's ranking. Then, the priority degree for each CSP is determined by adding the preferred values for all services provided by the CSP. Finally, the CSPs are sorted based on their priority degrees in ascending order. According to this model, a simple technique is used to calculate CSPs ranks based on the opinion of the CSCs only.

A CSP ranking framework, called security risk, has been proposed to find out the secured CSPs [18]. According to this framework, a model based on a security risk assessment approach has been developed to determine the vulnerabilities and define the risks related to CSPs. The vulnerable CSPs are determined based on the stochastic process for the security risk measurement for each CSP. Then, it ranks these CSPs based on the Markov chain. After ranking the CSPs, the best CSP is selected based on the reliability and security parameters only.

An Extended TOPSIS (E-TOPSIS) model has been proposed for ranking CSPs [19]. This model is based on seven criteria; accountability, agility, assurance, financial, performance, security and privacy, and usability. It uses Minkowski distance ($\mathrm{d}(X, Y)$) to measure distances between the solutions using the following equation:

$$\mathrm{d}(X, Y) = \left(\sum |X_i - Y_i|^p\right)^{1/p} \tag{2}$$

The model generates solutions by varying the value of p for a specific interval $[l, h]$ and the step value of p which determines the number of iterations. For each iteration, it outputs a list of CSPs ranking. Finally, the most ranking arrangement is defined. By increasing the number of iterations, the execution time will be increased. This is considered the main drawback of this model.

Recently, an approach based on the Hypergraph-Binary Fruit Fly Optimization (HBFFO) Algorithm for cloud service ranking has been presented [20]. This approach consists of three phases; Filtering, Selecting, and Ranking phases. The role of the Filtering phase is to filter CSPs based on the user's requirements using the Hypergraph technique [14]. Then, the time-varying mapping function and Helly property have been used to identify the trustworthy CSPs in the Selecting Phase. In the Ranking phase, the HBFFO algorithm has been used to rank these CSPs based on their trustworthy, credibility, and the user QoS requirements. This model provides a service selection model, in which the service selection middleware consists of many service repositories with one repository for each service type. According to the service type in the CSC's request, the CSC could forward directly to the repository of this service. The experimental analysis of this approach is done using a dataset contains two parameters only (Response Time and Throughput). The complexity of this approach is defined using Eq. (3):

$$\ll \mathrm{O}\left(m^3\right); \text{where } m \text{ is the number of CSPs} \tag{3}$$

According to the related work, it is found that some ranking models define CSPs ranking based on the CSCs' requests only. Other models define CSPs ranking based on their parameters only. Our proposed CSP Ranking model concerns both (i.e., CSCs requirements and CSPs parameters).

3 The Proposed Cloud Service Provider (CSP) Ranking Model

The main function of the proposed Cloud Service Providers (CSP) Ranking model is that when a new request from a CSC is received, all CSPs in the cloud will be ranked based on their trust degrees and the similarity between the requests' parameters and the parameters of each CSP. Again here, the proposed CSP Ranking model consists of four phases; Filtrating, Trusting, Similarity, and Ranking.

According to our previous work, a CSP Trusting model has been developed based on SLA parameters (i.e., Reputation, Availability, Turnaround Time, data Integrity, authorization, Data Recovery, Reliability) and CSCs parameters which concern about CSPs' feedbacks such as Users' Access Frequency, and Service Success Rate (for more details see [21]). As the proposed CSP Ranking model consists of four phases; Filtrating, Trusting, Similarity, and Ranking. The main function of the Filtrating phase is to prevent unwanted CSPs to be passing to the Trusting phase. This phase has been developed using Fuzzy Controller System [22]. In the Trusting phase, the trust degree of the accepted CSPs has been determined using Dynamic Adaptive Particle Swarm Optimization (DAPSO) technique [23].

According to the work in this paper, a CSP Ranking model has been introduced and developed based on the previously developed CSP Trust model. The proposed CSP Ranking model consists of four phases; Filtrating, Trusting, Similarity, and Ranking. The framework of the proposed CSP Ranking model is illustrated in Fig. 1.

The Data flow of the proposed CSP Ranking model is illustrated in Fig. 2. According to Fig. 2, the data for all existing CSPs will be normalized and passed to the Filtrating phase to determine the accepted and unaccepted CSPs using Fuzzy Controller technique. The data of the accepted CSPs will be sent to the Trusting Phase to determine their trust degree using Dynamic Adaptive Particle Swarm Optimization (DAPSO) technique. When the CSC sends his request for a service, the data of the request will be normalized. The Similarity phase will define the similarity between the CSC's request and the data of the accepted CSPs using Cosine Similarity technique. Finally, the accepted CSPs will be ranked based on their trusting and similarity values.

3.1 Filtration Phase

In this phase, a Fuzzy Controller System (FCS) is used to filter the available CSPs and define the CSPs to be considered in the trusting phase [24]. The decision-makers use the FCS as an intelligent technique to support the decision-making processes [22, 25]. So, it is considered as a powerful and effective controller and predictive tool. In many clouds research work, the FCS is used as a

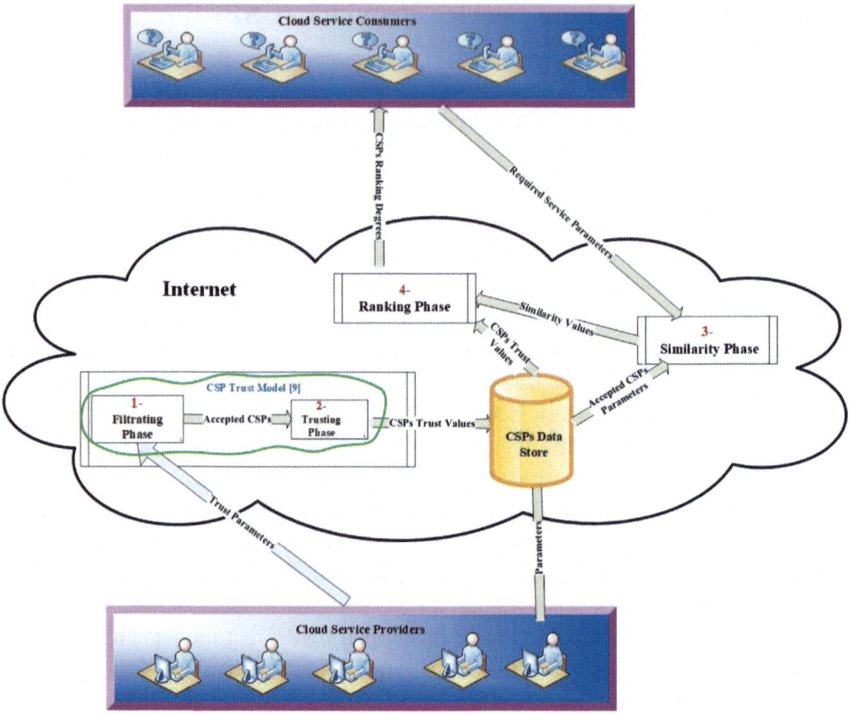

Fig. 1 The proposed CSP ranking model

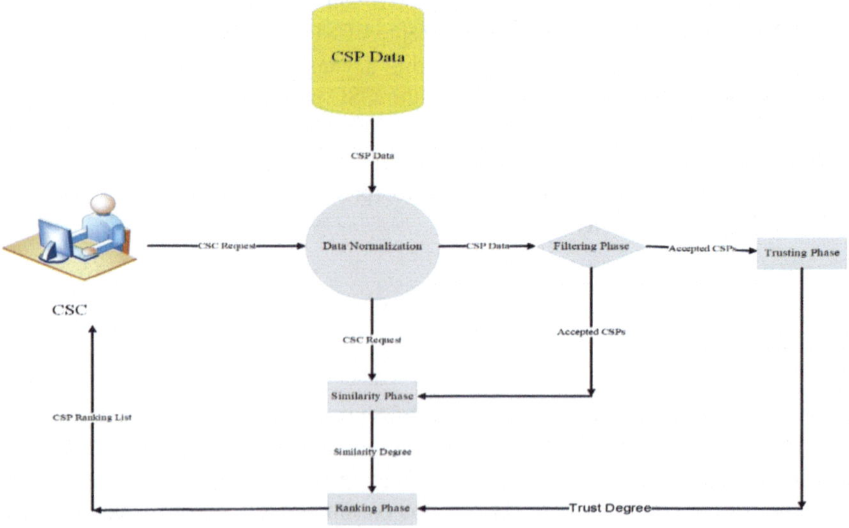

Fig. 2 Data flow diagram for CSP ranking model

predictive tool to predict the degree of the provider's security and trust based on IF-Then rules [26].

For the proposed model, the filtrating phase uses the FCS to prevent unwanted providers to be involved in the Trusting, Similarity, and Ranking phases. So, the execution time is decreased, and the performance of the system is improved. Namely, for each provider, the FCS takes the values of the nine parameters (i.e., SLA and users' parameters) of this provider as input and produces an output value which indicates whether this provider will be involved in the trust metric stage or not. The output of the FCS is one of two parameters; GOOD or POOR. The CSPs and the values of their nine parameters could be represented as a matrix, A(n, m), where n is the number of the CSPs and m is the number of parameters to be concerned (i.e., the above nine parameters). Each column in matrix A represents the values of a specific parameter for all CSPs, and row represents the values of the nine parameters for each CSP. Therefore, each CSP with his jth parameter is presented as a(i, j), where i is the CSP identification (CSPi) and j is the value of the jth parameter:

$$A(n,m) = \begin{bmatrix} a_{11} & a_{12} & \cdots & a_{19} \\ a_{21} & a_{22} & \cdots & a_{29} \\ \vdots & \vdots & \cdots & \vdots \\ a_{n1} & a_{n2} & \cdots & a_{n9} \end{bmatrix}$$

Therefore, an If-Then rule has been introduced to determine the FCS decision (i.e., GOOD or POOR). According to this rule, the decision for each CSP is determined as follows:

If-Then Rule: For each CSP_i, i = 1, 2 … n, where n is the number of CSPs

$$\left.\begin{array}{c} \text{if} \sum_{j=1}^{9} F(P_j, op_j) \text{ is GOOD, Then } CSP_i \text{ is GOOD} \\ \text{Otherwise, } CSP_i \text{ is POOR} \end{array}\right\} \quad (4)$$

where P_j indicates the parameter j (j = 1, 2 … 9), and op_j is the logical AND/OR operator associated with parameter j. By considering AND operator only for all nine parameters, the number of the CSPs that pass to the trust metric stage decreases (i.e., discarding the important ones). In the opposite, if the OR operator is only used, the number of CSPs increases, including less important ones. Therefore, the operator of each parameter could be AND or OR depending on its features. Namely, the used operators are determined according to Eq. (5):

$$op_j = \begin{cases} \text{AND}, & \text{if mean value of} \left(P_j, \text{for all CSPs}\right) \geq \text{mean value of all parameters} \\ \text{OR}, & \text{otherwise} \end{cases} \quad (5)$$

i.e., AND operator will be associated with parameter j in If-Then rule when the mean value of this parameter for all CSPs is greater than or equal to the mean value of all nine parameters' values for all CSPs. Otherwise; OR operator will be applied.

After determining the operators, they will be substituted in Eq. (4) to filter the CSPs. We must notice that nine operators will be produced using Eq. (5), while Eq. (4) needs only eight operators. This can be done by using the conjunction relation between the operators of two consequence parameters as follows:

$$\left.\begin{array}{l}\mathrm{If}\left(op_i == op_j\right), \mathrm{set\,Cop}_{i,j} = op_i \\ \mathrm{Otherwise;\ set\,Cop}_{i,j} = \mathrm{XOR}\,op\end{array}\right\} \quad (6)$$

where $\mathrm{Cop}_{i,j}$ is the conjunction operator between op_i and op_j of the consequence parameters.

After that, the value of each element a_{ij} in matrix A will be ranked GOOD or POOR depending on its value according to Eq. (7):

$$\mathrm{Rank}(a_{ij}) = \begin{cases}\mathrm{GOOD}, & \mathrm{if\ mean}\left(\mathrm{A}_j\right) < a_{ij} \leq \max\left(\mathrm{A}_j\right) \\ \mathrm{POOR}, & \mathrm{otherwise}\end{cases} \quad (7)$$

where mean (A_j) and max (A_j) are the mean and the maximum values of parameter j for all CSPs, respectively.

Finally, the results of Eqs. (6) and (7) for each CSP will be applied to If-Then rule (i.e., Eq. (4)) to determine that this CSP will be accepted to be involved in the trust metric stage (i.e., CSP is GOOD) or not (i.e., CSP is POOR).

3.2 Trusting Phase

In the Trusting phase, the trust degree of the accepted CSPs for all services they provided will be defined using the Particle Swarm Optimization technique (PSO) [27–30].

The PSO is considered one of the most commonly used optimization techniques in many areas such as function optimization, artificial neural network training, fuzzy system control, and other areas [28].

By developing Trusting phase using PSO technique, v_i^k and x_i^k denote the velocity and the position of particle i in iteration k, respectively. The velocity v_i^{k+1} is computed using Eq. (8) [31]:

$$v_i^{k+1} = w_i^{k+1} * v_i^k + c_1 * rand_1 * \left(pbest_i^k - x_i^k\right) + c_2 * rand_2 * \left(gbset^k - x_i^k\right) \quad (8)$$

where w_i^{k+1} is the inertia weight, c_1, c_2 are constants, $rand_1, rand_2$ are random numbers between 0 and 1, *pbest* is the best position that each particle reached, and *gbest* is the best position of the group of particles.

Position x_i^{k+1} of particle i is calculated based on its velocity v_i^{k+1} and the previous position x_i^k as in Eq. (9) [31].

$$x_i^{k+1} = x_i^k + v_i^{k+1} \tag{9}$$

Alam [32] has claimed that the common initial value of PSO velocity is between 10 and 90% of position value, the common number of iterations is between "500" and "10000", and the common value of inertia weight is in the range from 0.4 to 0.9.

A Dynamic Adaptive Particle Swarm Optimization (DAPSO) technique is developed to determine the value of inertia weight (w_i^{k+1}) for the PSO algorithm dynamically [23]. It uses a dynamic adaptive inertia factor in the PSO algorithm to adjust its convergence rate and control the balance of global and local optima. It determines the inertia weight as in Eqs. (10) and (11).

$$w_i^{k+1} = w_{\min} + (w_{\max} - w_{\min}) \sin\left(\frac{\beta_i(k)\pi}{2}\right) \tag{10}$$

where $\mathrm{w}_{\max}, \mathrm{w}_{\min}$ are the maximum and minimum values of w (i.e., 0.4, 0.9), respectively.

$$\beta_i(k) = \frac{f_i(k) - f_g(k)}{f_w(k) - f_g(k)} \tag{11}$$

where $f_i(k)$ is the fitness function of the ith particle in kth iteration, and $f_g(k), f_w(k)$ are the best and worst fitness values of the swarm in the kth iteration, respectively.

The Trusting phase has been implemented using DAPSO technique. In addition, a CSPs data store has been introduced to store the services' features of the accepted CSPs and their trust degrees.

To determine the CSPs' trust degrees, DAPSO technique is used with considering nine parameters, SLA and users' parameters, (i.e., Reputation, Availability, Turnaround Time, data Integrity, authorization, Data Recovery, Reliability, Users' Access Frequency, and Service Success Rate) for each accepted CSP as PSO initial population. The model supposed that the fitness function for each CSP is calculated as in Eq. (12):

$$F_i = 1 - X_i; i = 1, 2, \ldots 9 \tag{12}$$

where X_i is the parameter value (i) for each CSP.

The pseudocode of the used DAPSO technique is depicted as follows:

Algorithm DAPSO

i. Initialize a population of particles having a position ($X_i^1 = T_i$), and velocity ($V_i^1 = 70\,\%\, of\, X_i$).
// $T_i; i = 1, \ldots, 9$ are the values of the proposed parameters for each accepted provider
ii. Set parameters of PSO ($c_1 = c_2 = 2$).
iii. Set iteration $k = 1$
iv. Calculate Fitness function $F_i^k = 1 - X_i^k, \forall i$
v. Set $Pbest_i^k = X_i^k, \forall i$ and $Gbest^k = \max(X_i)$
vi. Set $k = k + 1$
vii. Update the inertia weight as in Eqs. (10) and (11)
viii. $\mathrm{f_g(k)} = Gbest^k, \mathrm{f_w(k)} = \min\left(F_i^k\right)$
ix. Update velocity and position of particles

$$V_i^{k+1} = w \times V_i^k + c_1 \times rand(1) \times \left(Pbest_i^k - X_i^k\right) + c_2 \times rand(2) \times \left(Gbest^k - X_i^k\right); \forall i$$

i.

$$X_i^{k+1} = X_i^k + V_i^{k+1}; \forall i$$

x. Evaluate Fitness function $F_i^{k+1} = 1 - X_i^{k+1}, \forall i$
xi. Update $Pbest\, \forall i$

$$if\ F_i^{k+1} \le F_i^k \text{ then } Pbest_i^{k+1} = X_i^{k+1} \text{else } Pbest_i^{k+1} = Pbest_i^k$$

xii. Update $Gbest\, \forall i$

$$\text{if} \max\left(F_i^{k+1}\right) \le \max\left(F_i^k\right) \text{ then } Gbest^{k+1} = \max\left(X_i^{k+1}\right) \text{else } Gbest^{k+1} = Gbest^k$$

xiii. If $k \neq 100$ then go to step (vi) else go to step (xii)
xiv. Output the solution as $Gbest^{k+1}$

3.3 *Similarity Phase*

According to the proposed CSP Ranking model, the CSC defines his requested parameters (i.e., CPU Utilization, Response Time, Cost, Availability, Usability, Flexibility, Security Management, etc. [13, 33]) and their requested values about these parameters. Therefore, the similarity phase has been developed to determine the similarity value between the parameters of the CSC's required service and the

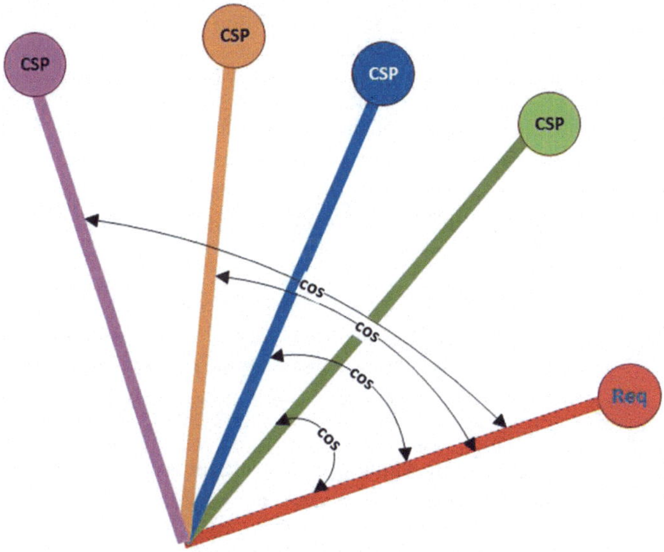

Fig. 3 Cosine similarity between parameters of CSC's required service and CSP's parameters

associated parameters for each accepted CSP that could provide the required service. To evaluate the similarity degree, the proposed model uses Cosine Similarity measure by considering the angle between two sequences of values, where a greater similarity implies a smaller angel [34]. Therefore, as the angle between the parameters of the CSC's required service and associated published parameters of the accepted CSPs is small, the similarity between them is large (see Fig. 3).

The Cosine Similarity for each accepted CSP is calculated using Eq. (13) [34]:

$$S[i] = \cos(\alpha) = \frac{\sum_{j=1}^{m} P_j[i]P_j[\text{req}]}{\sqrt{\sum_{j=1}^{m}(P_j[i])^2}\sqrt{\sum_{j=1}^{m}(P_j[\text{req}])^2}} \tag{13}$$

where $P_j[i]$ is the parameter value (*j*) of the accepted CSP_i, $P_j[\text{req}]$ is the parameter value (j) for the CSC' required service, and $j = 1, 2, \ldots, m$ is the number of parameters in the CSC's required service.

3.4 Ranking Phase

In this phase, the ranking value for CSPs will be determined by combining CSPs' trust degrees, and the similarity values which have been produced from the Similarity phase. The ranking value for each CSP is calculated using Eq. (14).

$$\text{Rank}[\text{CSP}_i] = w_1 * T[i] + w_2 * S[i], i = 1, 2, \ldots, n \tag{14}$$

where

- w_1, and w_2 describe the weight of the trust degree and similarity value respectively. These values are defined by calculating the standard deviation of the trust degrees and similarity values for all trusted CSPs
- $T[i]$ and $S[i]$ are the trust degree and similarity value for CSP_i respectively, and
- n is the number of accepted CSPs.

w_1, and w_2 are normalized as the sum of weight values equal 1. For example, suppose the standard deviation for trust degree and similarity value are $S_T = 0.1$, $S_S = 0.3$, respectively. These values are normalized by dividing them by their sum as $sum = 0.1 + 0.3 = 0.4$ then $w_1 = \frac{S_T}{sum} = \frac{0.1}{0.4} = 0.25$, $w_2 = \frac{S_S}{sum} = \frac{0.3}{0.4} = 0.75$ where $w_1 + w_2 = 0.25 + 0.75 = 1$.

The CSP with higher ranking value (i.e., the higher trust degree and similarity value) is the most candidate one for providing the service to the CSC.

3.5 Data Normalization

Since the model is based on QoS parameters of CSPs and CSC requirement parameters, and these parameters have their different ranges and units. Therefore, it is necessary to normalize this data to a uniform range [0, 1]. Suppose that $[QoS]^{\text{CSP}} \leftarrow (A_1, A_2, \ldots A_n)$, represents the data provided by the CSPs, and $[QoS]^{\text{CSC}} \leftarrow (R_1, R_2, \ldots R_m)$, represents the CSC request parameters. These data are normalized according to Eqs. (15) and (16), respectively:

$$\text{norm}(A_i)^{\text{CSP}} \leftarrow \frac{[A_i - \text{Min}(QoS)^{\text{CSP}}]}{[\text{Max}(QoS)^{\text{CSP}} - \text{Min}(QoS)^{\text{CSP}}]} \tag{15}$$

$$\text{norm}(R_i)^{\text{CSC}} \leftarrow \frac{[R_i - \text{Min}(QoS)^{\text{CSP}}]}{[\text{Max}(QoS)^{\text{CSC}} - \text{Min}(QoS)^{\text{CSP}}]} \tag{16}$$

where $\text{Min}(QoS)^{\text{CSP}}$, $\text{Min}(QoS)^{\text{CSC}}$ are the minimum values of QoS parameters of CSPs and CSC requirement parameters, and $\text{Max}(QoS)^{\text{CSP}}$, $\text{Max}(QoS)^{\text{CSC}}$ are the maximum values of QoS parameters of CSPs and CSC requirement parameters, respectively.

4 Performance Evaluation of the Proposed CSP Ranking Model

The implementation environment and the time complexity of the proposed CSP Ranking model will be discussed. Then, a comparative study will be done to evaluate the performance of the proposed CSP Ranking model relative to the existing models (i.e., hypergraph computational (HGCM) [14], Hypergraph-Binary Fruit Fly Optimization Algorithm (HBFFOA) [20], security risk [18], and Extended TOPSIS (E-TOPSIS) [19]).

4.1 The Implementation Environment

To evaluate the performance of our proposed CSP ranking model, HP ProBook computer with core i5@2.5 GHz processor, 6 GB RAM, under Windows Ten platforms are used. The experiments have been done using C# language using Microsoft Visual Studio 2010.

Armor Dataset is used to evaluate the proposed CSP Ranking model. This dataset contains the values of QoS parameters for 7334 CSPs [35]. According to Armor dataset, it is found that 6991 out of 7334 providers with no data. Therefore, only 343 providers out of 7334 will be used to evaluate the proposed model. Because the Filtering and Trusting phases of our proposed model are based on nine parameters (i.e., Reputation, Availability, Turnaround Time, data Integrity, authorization, Data Recovery, Reliability, Users' Access Frequency, and Service Success Rate), the Turn Around Time in Armor dataset is considered as the Response Time, Reputation as the Feature, Integrity as Ease of Use, Reliability as Technical Support, and Authorization as Customer Service. The parameters which do not exist in Armor dataset (i.e., Data Recovery, Service Success Rate, Users' Access Frequency) are generated randomly in the range of [0, 1].

4.2 Comparative Study

This section provides the comparative study to evaluate the performance of our proposed CSP Ranking model with respect to the existing models (hypergraph computational (HGCM) [14], Hypergraph-Binary Fruit Fly Optimization Algorithm (HBFFOA) [20], security risk [18], and Extended TOPSIS (E-TOPSIS) [19]). The comparison is carried out using three steps; Validation, Time Complexity, and Performance Evaluation of the proposed model.

4.2.1 Our Proposed Model Validation

To validate our proposed model, our proposed model, and HGCM model have been implemented with considering the QoS case study which is presented in HGCM model [14]. This QoS case study is used in two different cases. The first case study consists of three IaaS CSPs (Amazon EC2, Windows Azure and Rackspace), and the second one consists of five IaaS CSPs (Amazon EC2, Windows Azure, Rackspace Private cloud using OpenStack, and Private cloud using Eucalyptus) (see Table 1).

With considering the first case study (i.e., three CSPs (CSP1, CSP2, CSP3, and one CSC), our proposed model produces the same results as HGCM model where the CSPs are ranked as CSP3 < CSP1 < CSP2 with the CSP3 is the most suitable provider. The execution time of implementing the existing HGCM model and our proposed CSP Ranking model are 1.0448, 0.6798 ms, respectively. Therefore, our proposed model produces the same result as the HGCM model with an execution time reduction of 34.9%.

By considering the second case study (i.e., five CSPs (CSP1, CSP2, CSP3, CSP4, and CSP5) and one CSC), the implementation of the existing HGCM and our proposed CSP Ranking models produce the same result with CSP3 is the most suitable provider. The execution time of implementing the existing HGCM model and the proposed CSP Ranking model is 3.363, 0.4138 ms, respectively. Therefore, the proposed CSP Ranking model outperforms the existing HGCM model by execution time reduction of 87.69%.

Table 1 QoS dataset of Amazon EC2 (CSP1), Windows Azure (CSP 2), Rackspace (CSP 3), private cloud using OpenStack (CSP 4), and private cloud using Eucalyptus (CSP 5)

Parameter	Case study 1			Case study 2		CSC requirement
	CSP1	CSP2	CSP3	CSP4	CSP5	
Accountability	4	8	4	5	5	4
CPU	9.6	12.8	8.8	12.8	9.6	6.4 GHZ
Memory	15	14	15	16	14	10 GB
Disk	1690	2040	630	500	400	500 GB
Availability (%)	99.95	99.99	100	99.99	99.90	99.9
Cost	0.68 $/h	0.96 $/h	0.96 $/h	0.95 $/h	0.66 $/h	<1 $/h
Data inbound	10	10	8	10	10	100 GB/Month
Data outbound	11	15	18	11	11	200 GB/Month
Storage	12	15	18	12	10	1000 GB
Service response time	120	780	200	180	180	120 s
Security	4	8	4	5	5	4

4.2.2 Time Complexity Comparison

In this section, the time complexity of our proposed model is calculated and compared with the existing models (hypergraph computational (HGCM) [14], Hypergraph-Binary Fruit Fly Optimization Algorithm (HBFFOA) [20]).

As our proposed model consists of four phases (Filtrating, Trusting, Similarity, and Ranking), we calculate its time complexity by calculating the time complexity of Trusting phase, Similarity phase, and Ranking phase).

Time Complexity of the Trusting Phase

In the Trusting phase, the trust values for all accepting CSPs will be calculated using the PSO technique. It should be noted that this phase is executed only once throughout the proposed CSP Ranking model. But, when a new CSP is included in the system or even the features of the existing CSP(s) are changed, the Trusting phase will be reactivated. Therefore, the time complexity for calculating the trust value for CSPs in the Trusting phase is determined in Eq. (17):

$$O[S(kn + k + n)] \simeq O(Skn) \tag{17}$$

where S is the number of the accepted CSPs, k is the number of iterations for the PSO technique, and n is the number of parameters in the trust calculation.

Time Complexity of the Similarity Phase

When a new request from CSC is sent to the system, the proposed model will pass it to the Similarity phase to calculate the similarity degree between this request and the parameters of the accepted CSPs. Therefore, the time complexity of the Similarity phase is determined using Eq. (18):

$$O(Sn) \tag{18}$$

where n is the number of parameters in the CSC's requested service and S is the number of the accepted CSPs.

Time Complexity of the Ranking Phase

In the Ranking phase, the ranking value for each CSP is determined using Eq. (14). Therefore, the time complexity for this phase is determined using Eq. (19):

$$O(n \log S) \tag{19}$$

where n is the number of parameters in the CSC's requested service and S is the number of the accepted CSPs.

It should be noticed that When the CSC sends his request, our proposed model retrieves CSPs trust values without recalculating it. Then, the model performs the similarity and ranking phases only.

According to the time complexity Eqs. (1), (3) and (19), the time complexity of the proposed CPS Ranking model is less than the time complexity of the HGCM, and the BHFFOA models [14, 20]. Also, the number of CSPs in our proposed CPS Ranking model is less than the number of CSPs in the existing HGCM and BHFFOA models because of using Filtrating phase.

In case of all CSPs are not filtered and passed to the Trusting phase, the number of CSPs in our developed CSP Ranking model will be equal to the number of CSPs in the existing HGCM, and BHFFOA models. This is considered the worst case of our developed CSP Ranking model. In this worst case, our proposed CSP ranking model still has low time complexity.

4.3 Performance Evaluation Using Armor Dataset

In this section, the comparative study of our proposed model is determined with respect to the existing models (hypergraph computational (HGCM) [14], security risk [18], and Extended TOPSIS (E-TOPSIS) [19]). This comparative study is based on defining two performance parameters; Execution Time and Precision. Armor dataset is used to evaluate these two performance parameters by considering the number of available CSPs 20, 40 … 340, respectively [35].

4.3.1 Execution Time

Figures 4 and 5 illustrate the execution time of our proposed and existing models (hypergraph computational (HGCM), security risk, and Extended TOPSIS (E-TOPSIS)) respectively. According to Fig. 4, we notice that the execution time of our proposed model and E-TOPSIS model are very close because of time scales. Therefore, Fig. 5 is introduced for these two models to identify the difference between them using a small range of time. Figure 6 presents the average execution time of all models.

According to the results in Figs. 4, 5, and 6, the execution time of our proposed model is smaller than those existing models. From Fig. 6, it is noticed that the average execution time of the existing models; hypergraph computational (HGCM), security risk, and Extended TOPSIS (E-TOPSIS) and the proposed CSP Ranking models are 2.89, 2.49, 0.099, and 0.058 ms, respectively. So, the proposed CSP

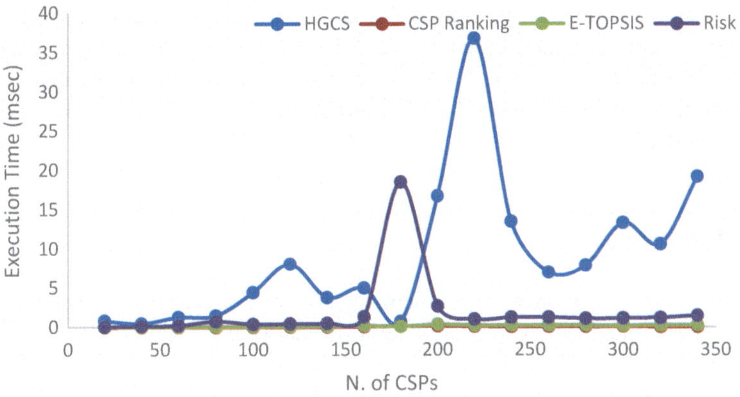

Fig. 4 Execution time for CSPs ranking and existing models

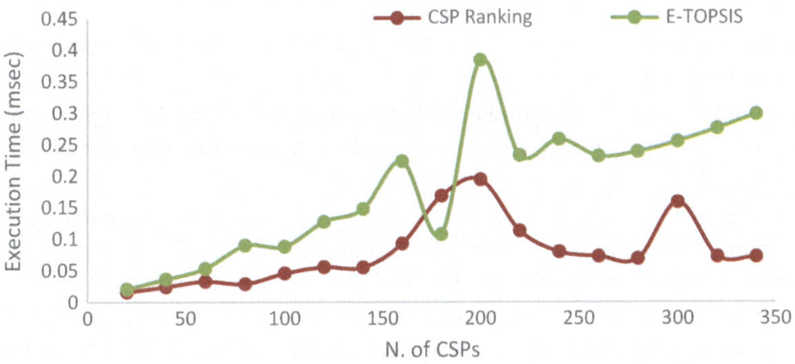

Fig. 5 Execution time for CSPs ranking and E-TOPSIS models

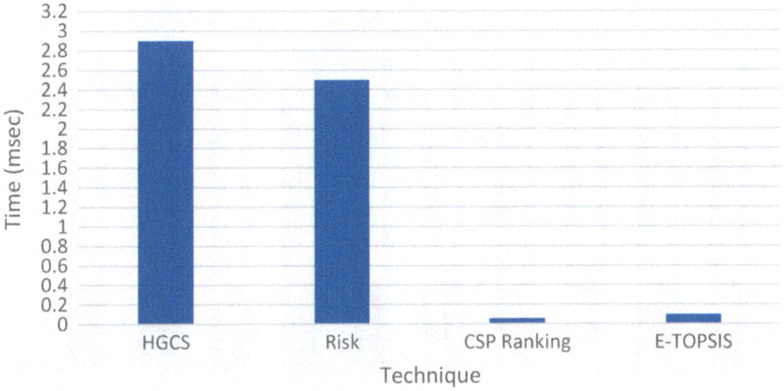

Fig. 6 Average execution time for CSPs ranking models

Ranking model outperforms the existing models (hypergraph computational (HGCM), security risk, and Extended TOPSIS (E-TOPSIS)) by reducing execution time by 97.99%, 97.67%, and 41.4%, respectively, and the average reduction of the execution time of our proposed CSP Ranking model is 79.02% relative to the existing models.

4.3.2 Precision

Precision is defined as how experiment results are close to each other [36]. It is usually measured as the average deviation of data (see Eq. (20)) [37].

$$\text{Precision} = \frac{\sum_{i=1}^{n} |x_i - \mu|}{n} \tag{20}$$

where μ is the mean value for data, and n is the data numbers.

We evaluate the precision of our proposed CSP Ranking model with respect to the existing models for 100 CSPs (see Fig. 7), and the average of Precision for 340 CSPs is shown in Fig. 8.

According to Fig. 7, it is found that our proposed model has higher precision values with respect to CSPs variation (i.e., 10, 20, …, 100 CSPs) relative to the existing models.

From Fig. (8), it is noticed that the average precision of the existing models (hypergraph computational (HGCM), security risk, and Extended TOPSIS (E-TOPSIS)) and the proposed model are 0.22, 0.26, 0.13, and 0.34 respectively. So, our proposed CSPs Ranking model outperforms the existing models (hypergraph computational (HGCM), security risk, and Extended TOPSIS (E-TOPSIS)) with respect to the average precision by 35.29%, 23.53%, and 61.76% respectively, and the average increase of the precision is 40.19%.

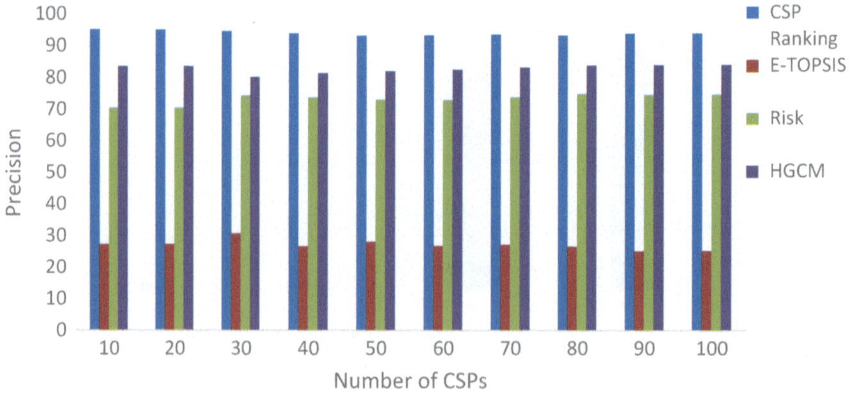

Fig. 7 Precision evaluations for 100 CSPs

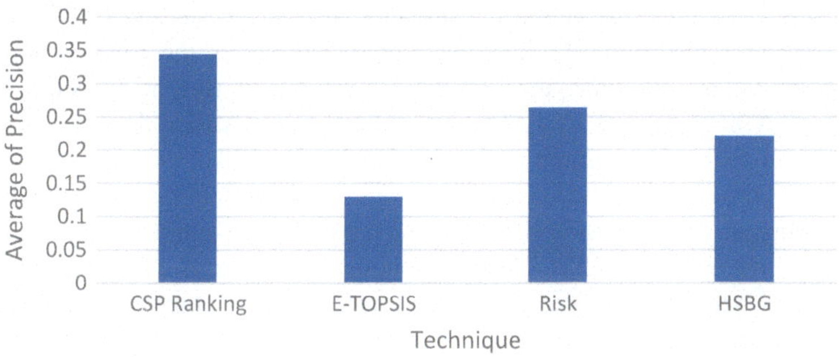

Fig. 8 Average of precision for 340 CSPs

Generally, the proposed CSP Ranking model is considered the proper model to define CSPs' ranking degree due to the following four reasons:

1. The model has been developed in module manner using four phases (i.e., Filtrating, Trusting Similarity, and Ranking). Therefore, the model can easily be extended to include extra functions.
2. In most cases, the Filtrating and Trusting phases are activated only once.
3. The execution time of the model is decreased especially when the number of CSPs in the system has become large.
4. The precision of the model is increased by increasing the number of CSPs in the system.

5 Conclusions and Future Work

Ranking CSPs becomes one of the most important challenges in the cloud environment because it helps CSCs to select the nearest CSP that attains their requirements in suitable execution time. According to the work in this paper, a ranking model has been introduced. The proposed CSP Ranking model is based on the trust parameters of the existing CSPs, and the similarity between the parameters of CSPs, and the CSCs' requested services. The CSP Ranking model consists of four phases; Filtrating, Trusting, Similarity, and Ranking. In the Filtrating phase, the CSPs will be filtered by rejecting the unwanted CSP's based on their parameters. In the Trusting phase, the trust degrees of the accepted CSPs will be determined. The results of the first and second phases are stored in a CSPs' data store. In the Similarity phase, the similarity percentage between the parameters of the CSC's requested service and the CSPs' parameters is determined using the Cosine Similarity Measure. In the Ranking phase, the ranking values of CSPs are determined by combining CSPs' trust degrees and the similarity values between the services' features of the accepted CSPs offering the requested services and the

CSC's service features to define the proper CSP that attains the CSC's requirements. Because the proposed Ranking model determines the trust value of the accepted providers only once, the search space is reduced, and the execution time is decreased. A comparative study has been done using Armor dataset among the existing models (hypergraph computational (HGCM), security risk, and Extended TOPSIS (E-TOPSIS)) and the developed CSP Ranking model to validate the performance of the proposed model. According to the comparative results, it is found that the proposed CSP's Ranking model outperforms the existing models (i.e., HGCM, security risk, and E-TOPSIS) with respect to the execution time, the time complexity and the precision.

In the future, the proposed CSP Ranking model will be used in the services composition process to select the best providers.

References

1. Mulholland, Andy, Jon Pyke, and Peter Fingar. 2010. Enterprise cloud computing FAQ. TechTarget, Dec 2010. http://whatis.techtarget.com/definition/Enterprise-Cloud-Computing-FAQ. Accessed April 2017.
2. Hasan, Samer, Vatsavayi Valli Kumari. 2017. Numerical similarity algorithms for cloud service discovery and selection system. *International Journal of Intelligent Engineering and Systems* 10 (3): 226–234.
3. Kumar, Rakesh Ranjan, and Chiranjeev Kumar. A multicriteria decision-making method for cloud service selection and ranking. *Advances in computer and computational sciences,* vol. 2, 139–147.
4. Ashford, W. 2010. Security in the cloud: Top nine issues in building users' trust. TechTarget. http://www.computerweekly.com/feature/Security-in-the-cloud-Top-nine-issues-in-building-users-trust. Accessed May 2017.
5. Aboulnaga, Ashraf, Kenneth Salem, and Ahmed A. Soror, Umar Farooq Minhas, Peter Kokosielis, and Sunil Kamath. Deploying database appliances in the cloud. *IEEE Data(base) Engineering Bulletin* 32 (1): 13–20.
6. Attar, Noha El., Wael Awad, and Fatma A. Omara. RPOA WLB: Resource provisioning optimization approach based on RPOA with load balance. *International Journal of Computer Applications* 105 (7): 34–41.
7. Baranwal, Gaurav, Deo Prakash Vidyarthi. 2014. A framework for selection of best cloud service provider using ranked voting method. In *2014 IEEE international advance computing conference* (*IACC*), Gurgaon, India.
8. Lang, Michael, Manuel Wiesche, and Helmut Krcmar. 2018. Criteria for selecting cloud service providers: A Delphi study of quality-of-service attributes. *International Journal of Information Systems Theories and Applications* 55 (6): 746–758.
9. Preethi, J., N. Sujaudeen, T.T. Mirnalinee, and P. Venugopal. 2015. Cloud service ranking and selection using linear programming. *International Journal of Computer Applications* 124 (3): 39–43.
10. Gholami, Atoosa, and Mostafa Ghobaei Arani. 2015. A trust model for resource selection in cloud computing environment. In *2nd international conference on knowledge-based engineering and innovation*, Iran.
11. Firdhous, Mohamed, Osman Ghazali, and Suhaidi Hassan. 2011. Trust management in cloud computing: A critical review. *International Journal on Advances in ICT for Emerging Regions* 4 (2): 24–36.

12. Randhawa, Sukhchandan, and Sushma Jain. Trust models in cloud computing: A review. *International Journal Wireless and Microwave Technologies* 4: 14–27.
13. Ezenwoke, Azubuike, Olawande Daramola, and Matthew Adigun. 2018. QoS-based ranking and selection of SaaS applications using heterogeneous similarity metrics. *Journal of Cloud Computing: Advances, Systems and Applications* 7 (15): 1–12.
14. Somu, Nivethitha, Kannan Kirthivasan, and Shankar Sriram. 2017. A computational model for ranking cloud service providers using hypergraph based techniques. *Future Generation Computer Systems* 68: 14–30.
15. Garg, Saurabh Kumar, Steve Versteeg, and Rajkumar Buyya. SMICloud: A framework for comparing and ranking cloud services. In *2011 fourth IEEE international conference on utility and cloud computing*, Australia.
16. Belgaum, Mohammad Riyaz, Safeeullah Soomro, Zainab Alansari, and Muhammad Alam. 2016. Cloud service ranking using checkpoint based load balancing in real time scheduling of cloud computing. In *International conference on advanced computing and intelligent engineering (ICACIE2016)*, India.
17. Singh, Dilbag, Jaswinder Singh, and Amit Chhabra. 2015. Evaluating overheads of integrated multilevel checkpointing algorithms in cloud computing environment. *International Journal Computer Network and Information Security* 5: 29–38.
18. Talbi, Jamal, and Haqiq, Abdelkrim. 2015. A novel framework for ranking cloud service providers using security risk approach. In *International conference on big data cloud and applications*, Morocco.
19. Rădulescu, Constanţa Zoie, and Iulia Cristina Rădulescu. 2017. An extended TOPSIS approach for ranking cloud service providers. *Studies in Informatics and Control* 26 (2): 183–192.
20. Somu, Nivethitha, M.R. Gauthama Raman, Kannan Kirthivasan, and V.S. Shankar Sriram. 2018. A trust centric optimal service ranking approach for cloud service selection. *Future Generation Computer Systems* 86: 234–252.
21. Mohammed, Alshaimaa M., Ehab Morsy, and Fatma A. Omara. 2017. Trust model for cloud service providers using fuzzy controller. *International Journal of Computer Science and Information Security (IJCSIS)* 15 (1): 373–387.
22. Bezděk, V. 2014. Using fuzzy logic in business. *Procedia - Social and Behavioral Sciences* 371–380.
23. Zhu, Hao, Hu Yumei, and Weidong Zhu. 2019. A dynamic adaptive particle swarm optimization and genetic algorithm for different constrained engineering design optimization problems. *Advances in Nonlinear Dynamics and Vibrations on Mechanical Systems* 11 (3): 1–27.
24. Guanrong, Chen, and Trung Tat Pham. 2001. *Introduction to fuzzy sets, fuzzy logic, and fuzzy control systems*. Florida: CRC Press LLC.
25. Alvaré, A.M.D. 2015. *Fuzzy logic as an instrument to perform security analysis*. United States: ProQuest LLC.
26. Prasath, V., Nithya Bharathan, N.P. Neetha, N. Lakshmi, and M. Nathiya. 2013. Fuzzy logic in cloud computing. *International Journal of Engineering Research & Technology* (*IJERT*) 2 (3).
27. Grover, Garima, and Ila Chaudhary. Implementation of particle swarm optimization algorithm in VHDL for digital circuits optimization. *International Journal Information Engineering and Electronic Business* 5: 16–21.
28. Zhang, Yudong, Shuihua Wang, and Genlin Ji. 2015. A comprehensive survey on particle swarm optimization algorithm and its applications. *Mathematical Problems in Engineering* 2015: 1–38.
29. Marini, Federico, and Beata Walczak. 2015. Particle swarm optimization (PSO). A tutorial. *Chemometrics and Intelligent Laboratory Systems* 149: 153–165.
30. Chavan, S.D., and Nisha P. Adgokar. 2013. An overview on particle swarm optimization: Basic concepts and modified variants. *International Journal of Science and Research* (*IJSR*) 4 (5): 255–260.

31. Bachelorarbeit, Miljenko Jakovljevic. 2011. *Particle swarm optimization for generating input data in measurement based worst-case execution time analysis*. Wien: Vienna University of Technology.
32. Alama, M.N. 2016. Particle swarm optimization: Algorithm and its codes in MATLAB. ResearchGate, 1–10.
33. Hung, Tran Cong, Nguyen Xuan Phi. 2016. Study the effect of parameters to load balancing in cloud computing. *International Journal of Computer Networks & Communications* 8 (3): 33–45.
34. Chen, Jiqian, Jun Ye, and Du Shigui. 2017. Vector similarity measures between refined simplified neutrosophic sets and their multiple attribute decision-making method. *Symmetry* 9 (153): 1–13.
35. Noor, Talal H., and Quan Z. Sheng. 2011. Trust as a service: A framework for trust management in cloud environments. In *International conference on web information systems engineering*, Berlin, Heidelberg, 314–321.
36. Anderson, P. 2013. Accuracy and precision. Sophia Learning. https://www.sophia.org/tutorials/accuracy-and-precision-3. Accessed Oct 2018.
37. Al-Saleh, Mohammad Fraiwan, and Adil Eltayeb Yousif. 2009. Properties of the standard deviation that are rarely mentioned in classrooms. *Austrian Journal of Statistics* 38 (3): 193–202.

A Survey for Sentiment Analysis and Personality Prediction for Text Analysis

A. Genina, Mariam Gawich and Abdel Fatah Hegazy

Abstract In the last few years' researchers started to pay attention for social media using personality prediction and sentiment analysis but due to the high costs and the privacy of these datasets this paper will present a comparative study on Sentiment Analysis and Personality Prediction through Social Media. First, what is meant by Social Media? It is a place which is designed to allow people (share, comment, post and …) for their beliefs and thoughts through websites and applications. The data gathered form social media users are using to represent the user behavior based on the different situations of everyday life. With machine learning techniques can be used to understand the mood. For that we can predict the human behavior through social media using machine learning techniques (clustering techniques, classification techniques). In this paper, clustering techniques will be discussed such as K-means clustering, hierarchical clustering, DBSCAN clustering. The personality assessment will help the market place and organizations to find the suitable person to a specific job and vice versa. For this reason, the sentiment analysis techniques can facilitate the personality assessment through the texts and the emotions that he has expressed on the social media websites.

Keywords Social media · Personality prediction · Sentiment analysis · Machine learning and NLP

A. Genina (✉) · M. Gawich
French University in Egypt, Cairo, Egypt
e-mail: alaa.genina@gmail.com

M. Gawich
e-mail: mariamjawich@gmail.com

A. F. Hegazy
Arab Academy for Science, Technology & Maritime Transport, Sheraton, Cairo, Egypt
e-mail: abdheg@hotmail.com

A. Z. Ghalwash et al. (eds.), *Internet of Things—Applications and Future*, Lecture Notes in Networks and Systems 114,
https://doi.org/10.1007/978-981-15-3075-3_23

1 Introduction

Social media is an area where users symbolize themselves to the world. The accounts of social media are private and personal then it can describe the personal life of the users. There are a lot of bounces such as posting, commenting, and updating status can uncover personal information. Text commented or posted by users can be analyzed to procure information [1].

Personality is substantive as a rightly fixed feature of an individual which mentions individual's priorities. Personality has been shown to be pertinent to many types of interactions such as social relationships, predicting songs preferences, and linking between the job tasks and the personality [2].

The data gathered from social media users are used to represent the user behavior based on the different situations of everyday life. Machine learning techniques can be used to understand the mood [3].

Sentiment Analysis also known as Opinion Mining is the process of defining the emotional mode behind a series of words (blogs, user comments, etc.) using natural language processing to determine whether it is negative, positive, or neutral and to examine the problem of studying texts [4].

Sentiment Analysis can be performed by three approaches: Machine Learning, Lexicon-Based, and Hybrid approach. For machine learning, it uses diversified features to construct a classifier that can characterize text that expresses sentiment from the supervised and unsupervised data. For the lexicon-based approach, it uses a diversity of words suspended by polarity score, to decide the general assessment score of a given content. The powerful asset of this technique is that it does not require any training data. Examples of these tools are the Wordnet [5], SentiWordNet [6], BabelNet [7], and W-WSD [8]. For the hybrid: It is a combination between machine learning and lexicon-based approaches and it is not commonly used.

The social media is a trended area and it is controlling the whole world around us. According to Thought for your Penny's website, they gathered some statistics that show how much the world is using the social media and its importance [9].

The personality assessment is a complex process that takes into account several dimensions to determine the type of the personality. The personality assessment will help the market place and organizations to find the suitable person for a specific job and vice versa. For this reason, the sentiment analysis techniques can facilitate the personality assessment through the texts and the emotions that he has expressed on the social media websites.

One of the biggest challenges in social media is the understanding of the informal language that users use on the social media specially the slang terms and the emoticon expressions.

2 Related Work

Text mining is the process of analyzing and scout about large amounts of unstructured text data supported by software that can distinguish concepts, patterns, topics, keywords, and other attributes in the data. It is also known as text analytics. Now the data scientists and users can analyze huge sets of unstructured data because of the development of big data platforms [10]. The human beings can understand the unstructured data and make it structured due to the linguistic techniques on the contrary with the computers it is a difficult task for them. Most of the available data in any organization are appeared in text format, so if we make a comparison between data mining and text mining then text mining is more significant. But as text mining is utilized for structuring the unstructured text data then this assignment is more demanding as compared to data mining. In general, the data concerning to social media sites is not gathered for the research aim, it is compulsory to change the structure of the data coming from the social media. 80% of the available text on the web is unstructured while only 20% is structured as shown in Fig. 1 [11].

2.1 Text Mining Versus Data Mining

A study by [11, 12] pointed out that text mining is much more complicated than data mining because of the irregular and unstructured data patterns, while data mining is dealing with the structured sets of data. The tools that were used in data mining were only dealing with structured data [11, 13].

Text mining is similar to an intelligence system which is extracting appropriate words or sentences from the unappropriate words and then mutates those words into the particular suggestions. Text mining is a new field having the main aim of data recovery, machine learning, information mining, and computational linguistics [11, 14].

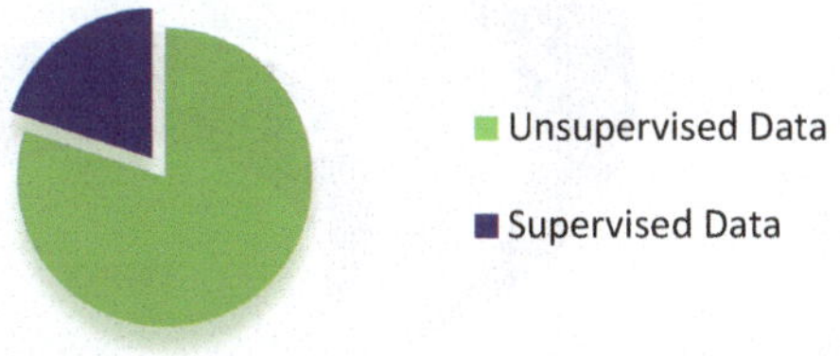

Fig. 1 Text available on web [11]

2.2 Sentiment Analysis

For sentiment analysis, it could be analyzed using those three different methods: SentiWordNet [6], BabelNet [7], and WSD [8] using natural language processing to determine is it negative, positive, or neutral.

Sentiment analysis and opinion mining were introduced in the early 2000s as manners to recognize and analyze opinions and feelings [15].

To define feelings through texts, humans utilize their language knowledge; on the contrary, the computers needs to be told which phrases or words change the polarity of texts. In sentiment analysis, the unit of analysis can be from the document level to the word level, as shown in Fig. 2 [15].

At the document level, the sentiment of the full document is taken to locate the overall sentiment of each individual document. At the sentence level, the same decision occurs, but each sentence is deemed individually when calculating sentiment. Feature-level sentiment is measured on the attribute level, particularly when applying sentiment analysis to customer or product feedback and reviews [15].

2.3 Lexicon Approach

To specify a sentiment value to an unprecedented text the lexicon approach utilizes before scored words and word phrases. Each value of word or phrase given by lexicon must be matched with the corresponding word or phrase. For the full text, the values are then summed [15].

SentiWordNet is a lexicon that utilizes WordNet, a lexical database. To assign numeric sentiment score measures between −1 for negative sentiment tokens to +1

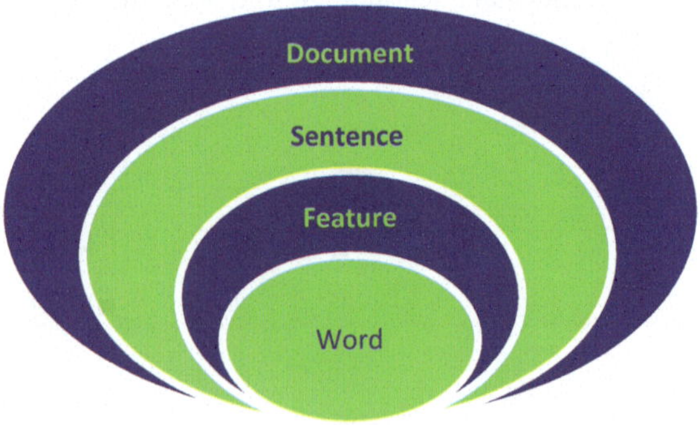

Fig. 2 Levels of sentiment analysis [15]

for positive sentiment tokens. SentiWordNet contains words and phrases, or n-grams, with the largest phrases consist of six terms. The AFINN lexicon scores words between −5 and 5, with negative numbers indicating negative polarity and positive numbers representing positive polarity. For both of the lexicons providing numerical sentiment values, the absolute value of the sentiment score indicates the strength of its polarity [15].

2.4 Machine Learning Approach

The machine learning approach to sentiment analysis constructs a classifier, on a dataset with categorized sentiments. For this purpose, the datasets from news articles, headlines, movie reviews, and product reviews have been annotated. Domain-specific datasets perform best when utilized to unseen data from that domain [15].

First, the classifier is trained on a training set. Then, the model is applied to the testing data to classify the text's sentiment. Some common machine learning and classification analysis methods applied to sentiment analysis are support vector machines (SVM), naïve Bayes (NB), k-nearest neighbors (kNN), and logistic regression [15].

2.5 Personality Prediction

There are so many different types of personality prediction assessments such as (MBTI, Big Five, Disc Assessment, The Winslow Personality Profile, Process Communication Model, The Holtzman Inkblot Technique, etc.). There are many machine learning algorithms that can be applied on the dataset. These algorithms will be used to cluster the dataset. Examples of these algorithms are the K-means, TFIDF, DBSCAN, and Hierarchical Clustering.

2.6 Clustering Techniques

Clustering is a method of unsupervised learning and is a combined technique for statistical data analysis applied in numerous domains. In Data Science, to obtain some valuable ideas from our data we can apply clustering analysis by seeing what groups the data points fall into when we use a clustering algorithm [16].

2.7 K-Means Clustering

K-Means is the most popular clustering algorithm. It's taught in a lot of inceptive data science and machine learning courses because of it's easiness to understand and implement in code [16].

First, we pick out a number of classes/groups to utilize and randomly initialize their respective center points to detect the number of groups to utilize. Every data point is classified by counting the space between that point and each group center, after that classifying the point to be in the group whose center is closest to it. Based on these classified points, you have to recompute the group center by taking the mean of all the vectors in the class. Finally, you have to iterate these steps for a set number of iterations or until the group centers do not change much between iterations [16].

K-Means has a linear complexity O(n), so it has the advantage of being very fast as all we do is calculating the distance between points and group centers. But it also has a couple of disadvantages. First, you have to pick out how many classes or groups there are. This is not always ideal with the clustering algorithms, also it starts with a random choice of cluster centers thus it may yield various clustering results in different runs of the algorithm. Therefore, the outcomes may not be repeatable and lack consistency [16].

2.8 Hierarchical Clustering

Hierarchical clustering is an alternative to prototype-based clustering algorithms. It will discover the specific number of clusters by itself. In addition, it enables the plotting of dendrograms. Dendrograms are visualizations of a binary Hierarchical clustering. There are two approaches to this type of clustering: Agglomerative or Bottom-up hierarchical clustering and divisive or top-down hierarchical clustering [17].

Divisive: Also known as top-down Hierarchical Clustering this method starts to split the cluster iteratively into smaller ones until each one of them consists of only one sample. This procedure is applied recursively until each document is in its own singleton cluster [18].

Agglomerative: This method starts with each sample being a different cluster and then merging them by the ones that are closer to each other until there is only one cluster [17].

2.9 Density-Based Spatial Clustering of Applications with Noise (DBSCAN)

DBSCAN is the abbreviation for a density-based clustered algorithm similar to mean-shift [16, 17], it is also like the Hierarchical clustering the specific number of clusters do not need to be given. The flexibility is very high in the forms and sizes that the clusters may adopt. The noise data and outliers could be identified and deal with them very usefully [17].

3 Framework Proposal for Comparative Study

The comparative study between the sentiment analysis approaches is based on the objective of each Approach, the use of sentimental analysis, the use of personality, the applied method, and the dataset (Table 1).

There are several approaches that are applied in the domain of sentiment analysis and personality prediction on the social media the criteria used for comparison between approaches:

Across the MBTI dimensions, they have extracted user activity features and a number of linguistic and perform a preliminary feature analysis. They have evaluated a number of benchmark models as a supervised machine learning task they framed personality prediction, but as a disadvantage, they did not use sentimental analysis for slang terms or emoticon expressions [22].

It focuses on the adoption of various sentiment analyzers with machine learning algorithms to determine the approach with the highest accuracy rate for learning about election sentiment [19].

With the use of machine learning algorithms for learning about election sentiment analyzers to determine the approach with the highest accuracy rate they focused on the adoption of various sentiments but they did not take in consideration the slang terms or the emoticon expressions and either they did not understand the behavior of the persons based on any personality test or model [19].

Based on the big five models they discussed the techniques performed to predict the student's personality and in order to get the most significant classifier they made a comparison for the performance but only the extraversion dimension was considered in this study and they didn't use the sentiment analysis [2].

By the use of the morphological sentence pattern model, they proposed an automated sentiment analysis using the NLP and the Jsoup HTML Parser, but the model could not analyze the data neither detecting the pattern and they did not use the sentiment analysis for the slang terms or the emoticon expressions [20].

Table 1 The comparison between the sentimental analysis approaches [2, 19–21]

Approach	Objective	Sentiment analysis	Emoticon processing	Personality prediction	Method	Dataset
Reddit: a gold mine for personality prediction [21]	Using the MBTI dimension they extracted user activity feature and number of linguistic	No	No	Yes	• SVM • Regularized logistic regression (LR), and • Three-layer multilayer perceptron (MLP)	9 k
Machine learning-based sentiment analysis for Twitter accounts [19]	Determine the approach with the highest accuracy rate for learning about election sentiment	Yes	No	No	• Machine Learning (Naïve Bayes, SVM) • Sentiment lexicons (W-WSD, SentiWordNet (with the use of python code) • Tweepy API • Weka tool	Tweets gathered from public accounts were 100,000 in number
A comparative study of different classifiers for automatic personality prediction [2]	Discussed several techniques to predict student's personality	No	No	Only extraversion dimension	Several classifiers provided in WEKA such as Bayes, Functions, rules, tree, and meta	From Twitter, they extracted undergraduate student's data
Sentiment analysis on social media using morphological sentence pattern model [20]	Proposed an automated sentiment analysis using the morphological sentence	Yes	No	No	• NLP (Natural Language Processing) • Jsoup HTML parser	1,000 documents from movie review sites

4 Conclusion

According to the comparison established between the approaches, will figure that no one has taking into consideration the emoticon expressions or the slang terms they have discussed only the sentiment analysis or personality prediction except Reddit: A Gold Mine for Personality Prediction discussed both sentiment analysis and personality prediction but without taking into consideration the emoticon expressions. Social media is an area where users symbolize themselves to the world. Text commented or posted by users can be analyzed to procure information. The data gathered from social media users are using to represent the user behavior based on the different situations of everyday life. Machine learning techniques can be used to understand the mood. For that as a future work, a model that will incorporate between emoticon expressions the sentiment analysis and personality prediction will be proposed.

References

1. Pratama, B. Yudha, and R. Sarno. 2015. Personality classification based on twitter text using Naive Bayes, KNN and SVM. In *International conference on data and software engineering (ICoDSE)*. Yogyakarta, Indonesia.
2. Ngatirin, N. Rahayu, Z. Zainol, and T. Lee Chee Yoong. 2016. A comparative study of different classifiers for automatic personality prediction. In *6th IEEE international conference on control system, computing and engineering (ICCSCE)*. Batu Ferringhi, Malaysia.
3. Lima, Ana Carolina. E.S., and Leandro N. de Castro. 2016. Predicting temperament from twitter data. In *2016 5th IIAI international congress on advanced applied informatics*. IEEE.
4. https://medium.com/retailmenot-engineering/sentiment-analysis-series-1-15-min-reading-b807db860917. Last accessed 21 Sept 2018.
5. https://wordnet.princeton.edu/download. Last accessed 9 Sept 2018.
6. http://sentiwordnet.isti.cnr.it/. Last accessed 21 Sept 2018.
7. https://babelnet.org/download. Last accessed 21 Sept 2018.
8. Pal, A. Ranjan, and D. Saha. 2015. Word sense disambiguation: a survey. *International Journal of Control Theory and Computer Modeling* 5(3): 1–16.
9. http://thoughtforyourpenny.com/technology/interwebz/100-social-media-stats-know-2018/?doing_wp_cron=1537651716.4822630882263183593750. Last accessed 22 Sept 2018.
10. https://searchbusinessanalytics.techtarget.com/definition/text-mining. Last accessed 15 June 2019.
11. Salloum, S.A., M. Al-Emran, A. Abdel Monem, and K. Shaalan. 2017. A survey of text mining in social media: facebook and twitter perspectives. *Advances in Science, Technology and Engineering Systems* 2(1): 127–133.
12. A. Akilan. 2015. Text mining: challenges and future directions. In *Electronics and communication systems (ICECS), 2nd, international conference*, 1679–1684. IEEE.
13. S.B. Navathe, and E. Ramez. 2000. Data warehousing and data mining. *Fundamentals of Database Systems*: 841–872.
14. M. Sukanya, and S. Biruntha. 2012. Techniques on text mining. In *Advanced Communication Control and Computing Technologies (ICACCCT), 2012 IEEE International Conference on*, 269–271. IEEE.

15. Anandarajan, M., C. Hill, and T. Nolan. 2018. Text Sentiment: Learning and Lexicon Models. *Modeling*.
16. https://towardsdatascience.com/the-5-clustering-algorithms-data-scientists-need-to-know-a36d136ef68. Last accessed 17 June 2019.
17. https://towardsdatascience.com/unsupervised-machine-learning-clustering-analysis-d40f2b34ae7e. Last accessed 17 June 2019.
18. Manning, C.D., P. Raghavan, and H. Schütze. 2019. *Introduction to Information Retrieval*. Divisive clustering.
19. Hasan, A., S. Moin, A. Karim, and S. Shamshirband. 2018. Machine learning-based sentiment analysis for Twitter accounts. *Mathematical and Computational Applications* 23(11): 1–15.
20. Han, Y., and K. Ko Kim. 2018. Sentiment analysis on social media using morphological sentence pattern model. In *15th international conference on software engineering research, management and applications (SERA)*. London, UK.
21. Reddit: A gold mine for personality prediction
22. https://www.reddit.com/. Last accessed 9 Sept 2018.

An Effective Selection for ERP Modules Using Data Mining Techniques

Eslam Abo Elsoud, Mariam Gawich and Abdel Fatah Hegazy

Abstract Enterprise resources planning "ERP" is an integrated system that improves productivity that drives large-sized companies to study and adopt it in their business processes. There is a little emphasis on factors that predict the implementation of ERP System will success or fail depends on some factors such as Company size, Industry Sector, Budget, IT infrastructure, Group affiliation, Top management support, Clear goals and objectives, Availability of skilled stuff and Internationalization. The major techniques focused on the decision to implement the ERP system in the organization but they ignore the recommendation of ERP modules. Therefore, this paper will propose a model that helps the ERP vendor to decide to implement the ERP system and it recommends suitable ERP modules according to the mentioned factors.

Keywords Enterprise resources planning · ERP system · Critical ERP success factors · Data mining · ERP implementation outcome · Prediction

1 Introduction

Enterprise resource planning (ERP) systems are basically software packages that organize the enterprise business data into an integrated database that facilitate the organization with useful information generated from the transformation of those business data [1]. It also improves productivity that drives large-sized companies to study and adopt it in their business processes. Once the companies implement the

E. A. Elsoud (✉) · A. F. Hegazy
Arabic Academy for Science Technology & Maritime Transport, Alexandria, Egypt
e-mail: Eslam_aboelsoud@ymail.com

A. F. Hegazy
e-mail: abdheg@hotmail.com

M. Gawich
French University in Egypt, Cairo, Egypt
e-mail: mariam.gawish@ufe.edu.eg

A. Z. Ghalwash et al. (eds.), *Internet of Things—Applications and Future*,
Lecture Notes in Networks and Systems 114,
https://doi.org/10.1007/978-981-15-3075-3_24

ERP system, each department will use the same database and share the same data that immediately improve the communication between the different departments as a good start.

To implement the ERP systems, it must be taken into consideration the variety of business processes for the companies to complete the implementation processes in large companies that can take a long time. A good estimation of the implementation time could be a few weeks for small and medium business companies and three months up to one year or more for larger companies.

ERP system is very expensive; it needs huge investment in the ERP applications. Furthermore, the re-engineering of the ERP system in a way that reflects the modification of the business processes has several disadvantages; limitations of the customization, risk of losing the sensitive data, and the need for training courses for end users. Also, the infrastructure of the companies plays an important role to complete the implementation process.

There are many challenges to implement ERP Systems such as the reputation of the ERP vendor's in the market and the understanding of the company infrastructure. For the reputation of ERP vendor, it means the willing to support the customer throughout the product lifecycle. For the company infrastructure, it is important that the ERP vendor determines the capability of the users to learn the new system, the availability of a suitable infrastructure to build the system.

One of the solutions that predict thc success or failure implementation and make an effective selection ERP module is Machine learning (ML) which allows outcomes without being explicitly programmed. The principle of machine learning is to build algorithms that can receive input data and use statistical analysis to predict an output while updating outputs as new data becomes available. This paper is organized as follows.

2 Related Work

This section will show the related work including the objective, method, advantages, disadvantages, future work, conclusion, and the dataset for each technique.

An approach [2] is deployed to overcome ERP application failures of the ERP implementation and increase the success rate of ERP projects. The success of ERP systems depends on many factors, User satisfaction is one of the important factors among them. It used hybrid of ANFIS (Adaptive Neuro-Fuzzy Inference System) and KNN (K-Nearest Neighbor). The approach get that the prediction accuracy can increase and the error level decrease more comparatively than existing methods. This method of prediction can be used as a decision-making tool to support the management of organizations when making decisions regarding the implementation of ERP.

KNN might be very easy to implement but as the dataset grows efficiency or speed of algorithm declines very fast.

Another conceptual model [3] is published to predict User Satisfaction by looking at human, Technological, and Organizational factors as predictors.

It used K-Nearest Neighbor (KNN) Classification method to predict the user satisfaction.KNN gives high prediction accuracy in most of the cases and low average error (0.25) in comparison with ANFIS (0.3378) and ANN (0.6053) methods.

It also used 14 CSF's taken from an earlier study which is good.

Although KNN is very easy to implement but as the dataset grows efficiency or speed of algorithm declines very fast. KNN works well with a small number of input variables but as the numbers of variables grow KNN algorithm struggles to predict the output of new data point. So, after comparing KNN with ANFIS and ANN. Also, calculated average error for all test cases and demonstrate that KNN gives high predication accuracy in most of the cases and low average error (0.25) in comparison ANFIS (0.3378) and ANN (0.6053) methods. So this approach is novel and using KNN, prediction accuracy can be further improved for US to make successful ERP.

Another approach [4] applies both the fuzzy logic analytical model and an expert judgment method support vector machines (SVM) classifier to predict whether the ERP software implementation project succeeds or fails using fuzzy logic to predict the key ERP outcome and SVM; which tells about the prediction of good and bad performances of ERP project. The missing point here is that the model presented does not consider all possible factors and parameters that might be associated with ERP success level acquisition.

Also, there is a Comparative study [5] of Institute based ERP based on ANFIS, ANN, and MLRA which identifies the antecedents of end-user satisfaction with an Enterprise Resource Planning (ERP) system, in the context of educational institute using (ANFIS) Adaptive Neuro-Fuzzy Interference System, (ANN) Artificial Neural Network, and (MLRA) Multiple Linear Regression Analysis.

The study established the efficiency of ANFIS as a good predictor of project risk of ERP implementation measured to evaluate overall IS success but the model presented factors such as Staff, Team Work, and Technical as predictor variables and does not consider some factors and parameters that might be associated with ERP success level acquisition.

The results establish that ANFIS is able to predict outcomes well with an error (RMSE) of 0.2945 and outperforms ANN and MLRA with errors of 0.85 and 0.86, respectively.

Another research established by Rouhani and Ravasan [6] predicting ERP post-implementation success based on organizational profiles has been discussed using Artificial Neural Network (ANN) and Expert system. This research attempted to link the organizational profiles to the ERP success level. Which is no one has ever used the capability of ANNs in articulating to bridge this gap between the nature and profile of organizational factors. The trained expert system predicts, with an average correlation coefficient of 0.744, which is respectively, high. Besides, the total correct classification rate of 0.685 indicates good prediction power.

3 Proposed Model for ERP Selection

For ERP selection there are many techniques that can predict the implementation of ERP System will success or fail depends on some factors using methods like K-Means, EM algorithms, Naive Bayes, Multilayer Perceptron and K-star. All of the previous techniques were focused on the decision to implement the ERP system in the organization but they have ignored the recommendation of other ERP modules. There is a little emphasis on other factors such as Company size, Industry Sector, Budget, IT infrastructure, Group affiliation, Top management support, Clear goals and objectives, Availability of skilled staff and Internationalization. Therefore, the proposed model will help the ERP vendor to decide to implement the ERP system and it recommends suitable ERP modules according to the mentioned factors. As Fig. 1 shows the input of the model here is the initial data that the ERP consultant should collect them from the companies which want to apply the system, collecting these initial data is not a simple task but it can be a big issue, collecting the data will be processed to clean the data by filling in missing values, smoothing noisy data, identifying or removing outliers, and resolving inconsistencies, this is the preprocessing phase after that data mining technique (Prediction Algorithm) is applied to determine if the customer will buy the ERP system or not, in case the decision is yes (the customer will implement the ERP system), another Algorithm

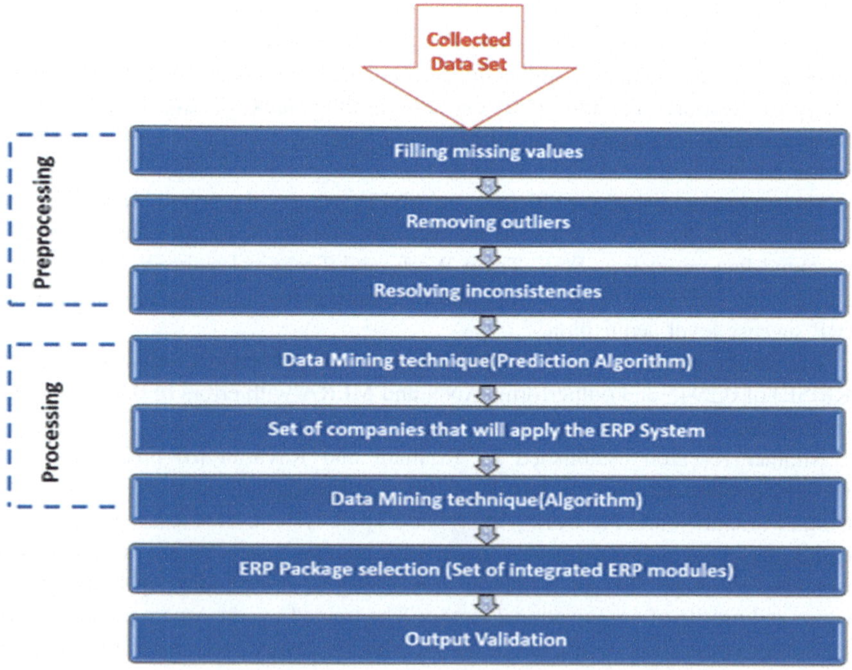

Fig. 1 Proposed model for ERP selection

will describe what the customer will buy/implement exactly, this is the processing phase, so the output is a set of suitable ERP modules according to the enterprise needs to improves its productivity.

The Components of the model are explained in the following points.

3.1 Dataset

It consist of the following factors

- **Company size**: large, medium or small enterprise.
- **Industry sector**: to determine the main modules.
- **Budget**: to know the possibilities of the customer/enterprise.
- **IT infrastructure**: it's important to know if the IT infrastructure in the enterprise is suitable to implement the ERP system or not.
- **Group affiliation**: it's one of the critical factors because if the key user has no affiliation it can lead to problems.
- **Top management support**: to ensure if there is a problem from the employees, the top management will solve it immediately.
- **Clear goals/objectives**: what you want what you get.
- **Availability of skilled staff**: to train them on the system after implementation.
- **Internationalization**: to know the financial transactions they do and the currency they trade with.
- **Well define business processes**: to achieve the enterprises' goals with a smart way
- **Numbers of transactions per day**: the performance must take into considerations
- **Numbers of key users**: to compare the numbers with the budget.

3.2 Preprocessing

Clean the data by filling in missing values, smoothing noisy data, identifying or removing outliers, and resolving inconsistencies.

3.3 Data Mining Technique (Prediction Algorithm)

Prediction algorithm is applied to determine whether the customer will buy the ERP system or not. Once the decision is yes, another algorithm will be applied.

3.4 Data Mining Algorithm

Data Mining Algorithm describes what the customer will buy exactly. It is designed to operate on databases containing transactions (for example, collections of items bought by customers, or details of a website frequentation).

3.5 Validation (Output)

The output of each algorithm involves a set of rules that determines the decision to buy the ERP system, as well as the recommendation of suitable ERP modules based on the company factors. At the end, there would be a presentation of results in the forms of charts or diagrams which describe the results and its accuracy of each algorithm will be validated by ERP Consultants.

4 Proposed Framework for Comparative Analysis

Table 1 shows the criteria of comparison which are the input, processing, and output.

Table 1 Comparison between the ERP selection techniques

Approach	Critical success factors	ERP Selection	ERP success of failure prediction
ERP user satisfaction by hybrid of ANFIS and KNN classification	No Used user satisfaction only	No	Yes
User satisfaction prediction in ERP using KNN classifier for high prediction accuracy	No Used user satisfaction only	No	Yes
Prediction of ERP outcome measurement and User satisfaction using adaptive Neuro-fuzzy inference system and SVM Classifiers approach	No This paper considers the Human, Technological and Organizational factors only	No	Yes
ERP success prediction: an artificial neural network approach	Yes	No	Yes
Proposed model	Yes	Yes	Yes

There are many techniques that can predict the implementation of ERP System will success or fail depends on some factors using methods like K-Means, EM algorithms, Naive Bayes, Multilayer Perceptron and K-star. All of the previous techniques were focused on the decision to implement the ERP system in the organization but they ignored the following

- The recommendation of other ERP modules.
- There is a little emphasis on other factors such as Company size, Industry Sector, Budget, IT infrastructure, Group affiliation, Top management support, Clear goals and objectives, Availability of skilled staff, Internationalization, Business Process, Number of transactions per day, and Number of key users.

5 Conclusion

After collecting these initial data and is not a simple task but it can be a big issue, collecting the data will be processed to clean the data by filling in missing values, smoothing noisy data, identifying or removing outliers, and resolving inconsistencies, the model will be implemented by the use of data mining techniques and output involves a set of rules that determines the decision to implement the ERP system and the recommendation of suitable ERP modules based on the enterprise critical success factors at the end it will be validated by ERP Consultants.

References

1. Wanas, G.S., A.H. El-Bastawissy, and M.A. Kadry. 2015. Decreasing ERP Implementation Failure in Egypt. In *The 25th international conference on computer theory an applications (ICCTA)*.
2. Kumawat, P. 2018. High prediction accuracy and low error for ERP user satisfaction by hybrid of ANFIS and KNN classification. *Sandip Foundation's International Journal on Emerging Trends in Technology (IJETT)* 5(2). ISSN: 2455–0124 (Online), September 2018.
3. Kumawat, P. 2018. User satisfaction prediction in ERP using KNN classifier for high prediction accuracy. *Sandip Foundation's International Journal on Emerging Trends in Technology (IJETT)* 5(2). ISSN: 2455–0124 (Online), September 2018.
4. Kumawat, P., G. Kalani, and N.K. Kumawat. 2015. Prediction of ERP outcome measurement and user satisfaction by using adaptive neuro fuzzy inference system and SVM classifiers approach. *Proceedings of the International Congress on Information and Communication Technology (ICICT)* 1: 229–237. https://doi.org/10.1007/978-981-10-0767-5_25.
5. Gupta, R., S. Chouhan, and N. Bhuria. 2014. Comparative study of Institute based ERP based on ANFIS, ANN and MLRA. *International Journal of Engineering and Technical Research (IJETR)* 2(5). ISSN: 2321-0869.
6. Rouhani, S., and A.Z. Ravasan. 2013. ERP success prediction: an artificial neural network approach. *Elsevier B.V Scientia Iranica* 20(3): 992–1001.

Knowledge Representation: A Comparative Study

Mariam Gawich

Abstract The intelligent computer system relies on the existence of knowledge. Knowledge refers to a set of facts and rules that describe the domain of interest. The encoded knowledge enables the computer system to interpret and analyze the received information. It also enables the computer system to infer new knowledge from the current facts. The encoded knowledge can include a set of rules that can be captured from the human experts to solve a complex problem such as the disease diagnosis. By this way, the intelligent computer system stimulates the human reasoning that is used to solve problems in an automated way. The process of the symbolization of knowledge in a formal manner in order to be understood by the intelligent systems called knowledge representation. The objective of this paper is the investigation of the major knowledge representation techniques. Furthermore, the paper presents a framework proposal for a comparative study to clarify the difference between the knowledge representation techniques.

Keywords Artificial intelligence · Knowledge engineering · Knowledge representation

1 Introduction

Knowledge representation [1, 2] is a field of Artificial intelligence domain. John McCarthy [3] assumes that the creation of artificial intelligence systems depends on the use of the common sense. The common sense refers to the essential facts concerning the events, objects, and object properties. According to this point of view, the artificial intelligence is associated with the ways to represent knowledge to the computer systems. Elaine Rich and Kevien Knight defined the knowledge representation into two notions. The first notion is the set of facts which represent the things which are pertinent to the domain of interest. The second notion is the

M. Gawich (✉)
Université Française En Égypte (UFE), Cairo, Egypt
e-mail: mariam.gawish@ufe.edu.eg

A. Z. Ghalwash et al. (eds.), *Internet of Things—Applications and Future*, Lecture Notes in Networks and Systems 114,
https://doi.org/10.1007/978-981-15-3075-3_25

representation of facts which refer to the ways to represent knowledge in a formal manner to be understood by the computer systems.

There are four kinds of knowledge [4]; the simple relational knowledge, the inheritable knowledge, the inferential knowledge, the procedural knowledge. The simple relational knowledge involves the representation of the declarative facts that depict the domain of interest. They can be represented in terms of entities and predefined relationships. In other words, the database management system encompasses this kind of knowledge. It is difficult to infer new fact using the formalism due to the lack of the semantic representation between the entities. The Inheritable knowledge refers to the representation of knowledge in terms of a hierarchy of classes, subclasses, attributes, and values that describe the domain of interest. The hierarchical structure is also called slot and filler structure. The inheritance property between the parent class and its subclasses facilitates the acquisition of new fact by the use of inference mechanism. The inferential knowledge refers to the representation of the knowledge in a way that facilitates the inference of a new fact from the current fact. The knowledge can be represented in a logical form that enables the execution of mathematical deduction and the use of inference procedures to obtain new knowledge. The inference procedures involve the forward and backward chaining. Procedural knowledge refers to the operational knowledge that determines the actions that will be executed based on the realization of particular conditions.

The objective of this paper is to present a comparative study of the major knowledge representation techniques and exploring their capabilities. The research paper is organized as follows: Sect. 2 investigates the major knowledge representation techniques. Section 3 presents a framework proposal to compare the knowledge representation techniques. Section 4 provides a conclusion.

2 Knowledge Representation Techniques

2.1 Predicate Logic

Predicate logic [5, 6] is used to encode the facts provided by sentences. Predicate logic enables the knowledge engineer to represent sentences in terms of objects (variables), object properties, and relations between the objects. Furthermore, the knowledge engineer can use the universal and the existential quantifiers to represent the quantified variables. The Fig. 1 presents examples of English sentences and their equivalent logical sentences encoded with the predicate logic. The universal quantifier ($\forall$) is used to express the universe of an object such as the sentence "Every man likes cars". The existential quantifier ($\exists$) is used to express that there exists some predicate in the universe such as the sentence "Anoine has at least one car". The predicate logic allows the knowledge engineer to represent the class subsumption such as the

English sentences	English sentences encoded with predicate logic
• Antoine likes dogs	• L(Antoine, dogs)
• Every man likes vehicle	• ∀ x: man(x)-→ vehicle (x)
• Antoine has at least one car	• ∃ x: car (x,Antoine)
• The car is a vehicle	• isa (car, vehicle)
• Antoine is a man	• instance (Antoine, man)

Fig. 1 Examples of English sentences encode with predicate logic

sentence that denotes the car is a vehicle. The class membership can be also presented such as the sentence that denotes Antoine is a man.

Although it is a powerful technique to represent the relational knowledge, it is suitable to represent simple general statements not the ambiguous statements. Furthermore, it is considered as a semi decidable technique; it doesn't own a decision procedure. For this reason, the inference rules are used to conclude new facts.

2.2 *Rule Based*

Rule Based is a technique of knowledge representation [7, 8]; its objective is to simulate the mental process of the knowledge experts applied to determine the possible solutions for a given problem. For this reason, the rule based is also called expert system. The rule-based system enables the knowledge engineer to represent the encoded knowledge in terms of IF and THEN rules which are also called production rules. The IF presents group of assertions and THEN presents group of actions that are related to the assertions. Figure 2 demonstrates the structure of rule with an example. It involves two main sections; The IF section and THEN section. IF section presents the condition which is also called the premise or the antecedent. The THEN section presents the action which is also called the conclusion or the consequent.

In the rule-based system, the logical expression in the premise section will be tested by its comparison with the initial facts. The IF section can involve one or more premises that are attached by the logical operators (AND-OR). The THEN section can have one or more clauses (actions).

IF	antecedent1		
AND	antecedent 2	IF	the season is winter
⋮		AND	the temperature is <0 degrees
OR	antecedentN	OR	the weather is windy
THEN	consequent	THEN	it is cold

Fig. 2 Example of production rules [8]

The rule-based system involves the following main elements; a group of facts, a group of rules, and a termination criterion. The Group of facts is the current declarative assertions that present the initial facts of the domain knowledge. These facts are stored in a database. The group of rules contains all the possible conditions and the related actions. They are stored in a knowledge base. Both group of facts and group of rules are presented by the knowledge engineer. The termination criterion is used to decide the existence or the inexistence of possible actions.

There are two inferences reasoning chaining that can be applied, the forward chaining and the backward chaining. The forward chaining is executed by the matching between the group of facts and the group of rules to acquire new facts. The backward chaining is applied when there is a goal and the user asks the rule-based system to prove it. Therefore, all the THEN section for all rules will be verified until the system finds the clause that matches the required goal.

2.3 Semantic Network

The notion of the semantic network [9, 10] was proposed by a philosopher called porphyry. He supposed the usage of the semantic network to model the Aristotle's method. In the artificial intelligence domain, the semantic network [11] is considered as a group of nodes (classes) that are interlinked together through labeled arcs. The semantic network is a technique to represent knowledge in an isa hierarchy. Figure 3 shows a semantic network that describes the main concepts of the lung cancer disease. The top superclass is the lung cancer that is associated with subclasses with isa relations. The class can have properties and instances.

There are various kinds of semantic network [9]: the definitional network, assertional network, implicational network, executable network, learning network, and the hybrid networks. The definitional network focuses on the use of the subsumption relations; by this way, the inheritance of the properties can be applied between a supertype and its subtypes. It is called definitional network because it is supposed that the presented knowledge is right. The definition is considered as a declarative fact. Example: Paris is the capital of France. The assertional network focuses on the representation of assumption that can be true. The assumption is considered to be true if it is attached with a model operator (quantifier). On the other hand, the assertional network is considered as propositional semantic network; it is used to represent the semantic of the sentences. The implicational semantic network is a propositional semantic network that focuses on the existence of implication between the nodes. The implicational network involves the causality network, the Bayesian and the belief network. The inference mechanism focuses only on the implication relations between the nodes. The executable network involves mechanisms that can update the semantic network itself. The mechanisms have data structure that can be modified by the change executed on external artifacts related to the semantic network. Example of these mechanisms: the message passing network, the attached procedures, and the graph transformations. The learning network

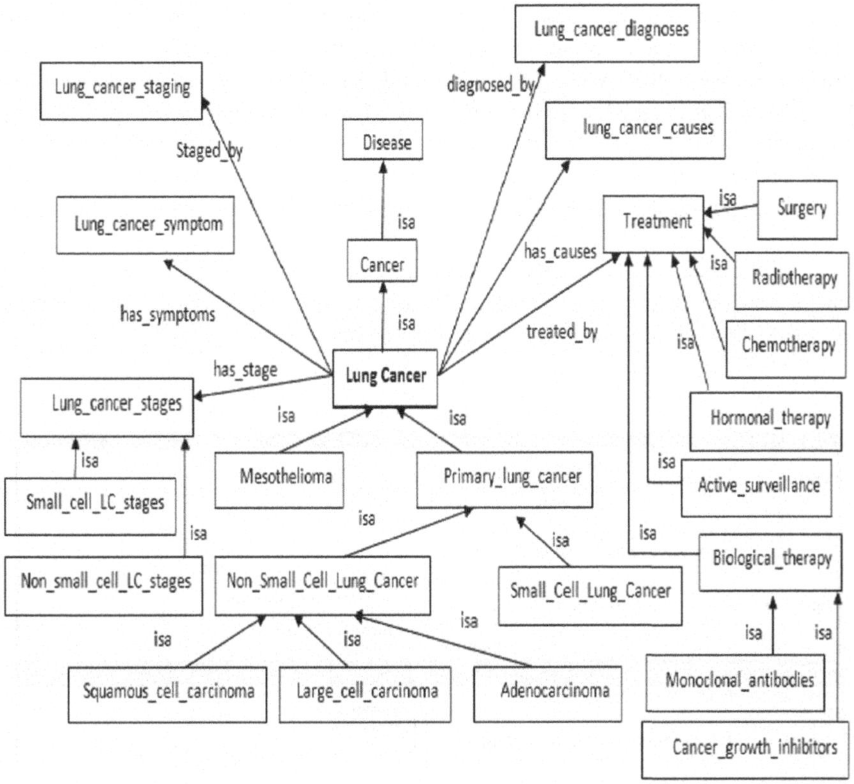

Fig. 3 Semantic network for lung cancer disease [12]

represents the gained knowledge acquired from a set of examples. The evolution of this network can be applied in several ways such as the rote memory and the weights modification. The neural network is the most common type of the learning network. The hybrid network involves the execution of a set of the mentioned networks.

2.4 *Frames*

Minsky [13] proposed the notion of frame as a knowledge representation technique. The frame [11] in the artificial intelligence is a set of slots and linked values that depict an object in the domain of interest. Richard Fikes and Tom KEHLER defined the frame [14] as a formal representation of an object or a collection of objects called class. In other words, a class can have one or more subclasses and it can be associated with one or more general classes. The frame enables the knowledge engineer to represent taxonomy of classes. Furthermore, the knowledge can encode some constraints on the values. Figure 4 demonstrates the frame taxonomy for transportation. The left side of the figure presents the AUTOMOBILE frame. It has three subclasses; SEDANS, COUPES, and STATION WAGONS. The class AUTOMOBILE is attached to the superclass VEHICLES and it owns specific slots that distinguish the AUTOMOBILE from other vehicle types. The right side of the figure presents an instance of SEDANS called CAR2. The production rules are applied on the frame to detect the generic deductive reasoning about the presented objects.

The difference [15] between the semantic network and the frame is that the later can present the data values of the attributes and it is more suitable to represent specific statements (facts) about the domain of interest.

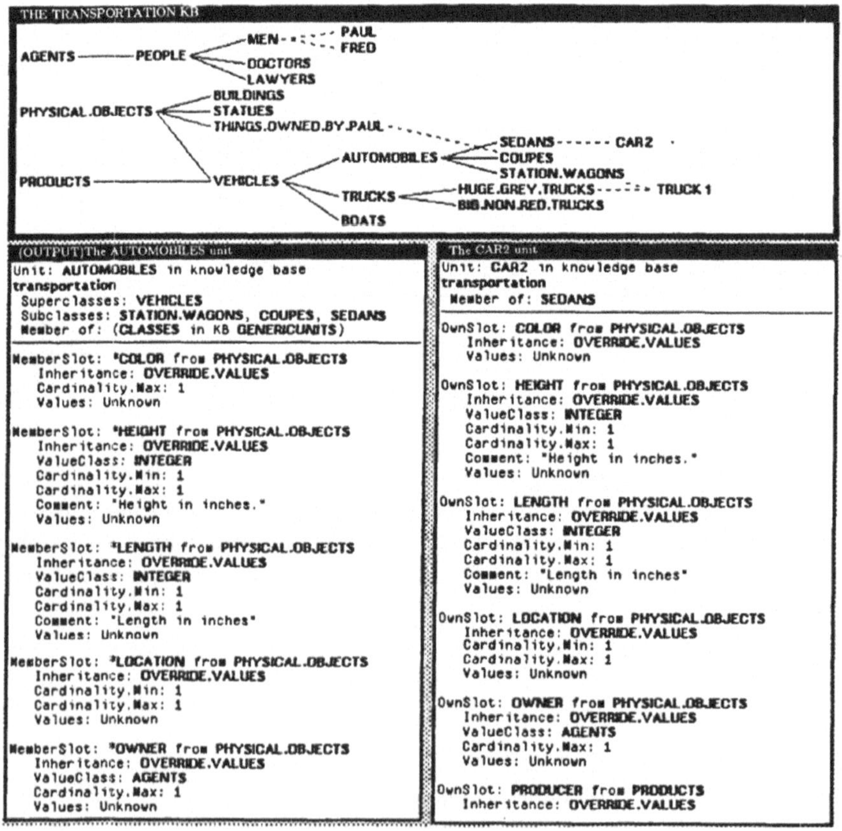

Fig. 4 Framework taxonomy [14]

2.5 *Ontology*

In the philosophy domain, ontology refers to the science of existence. In the artificial intelligence, ontology [16] refers to the formal explicit specification of shared conceptualization. Formal indicates that the ontology should be interpreted by the computer systems and shared means the ontology should be validated by a group of experts or a particular community. Conceptualization invokes that the domain of interest will be presented as a relational hierarchy that involves linked classes, subclasses, and individuals. Figure 5 demonstrates a hierarchical structure of the Rheumatoid disease ontology. Rheumatoid top superclass has three main subclasses the treatment, side effects, and the clinical finding. Each one of them includes subclasses.

The ontology enables the knowledge engineer to represent semantic relations between the classes, subclasses, and the individuals due to the expressiveness power of the ontology language OWL2 [18]. Table 1 demonstrates the semantic relations in terms of object properties.

Each object property has a domain and a range. The domain and range can include classes, subclasses, and individuals. Example: the "provoke" relation (object property) is created between two individuals that belong to the treatment class, Rituximab, and cytopenia.

The most common tool that is used to present a formal ontology is the protégé tool [19]. The protégé tool enables the knowledge engineer to represent all kinds of knowledge on the ontology. The Ontology structure enables the knowledge engineer can represent the simple relational knowledge between the classes in terms of relations that reflect declarative facts. He can also represent the inheritable knowledge due to the hierarchical structure of the ontology. Furthermore, the protégé tool enables the knowledge engineer to represent the inferential knowledge by the codification of a set of inferences rules on the ontology. The procedural knowledge can be represented by the use of IF-THEN rules.

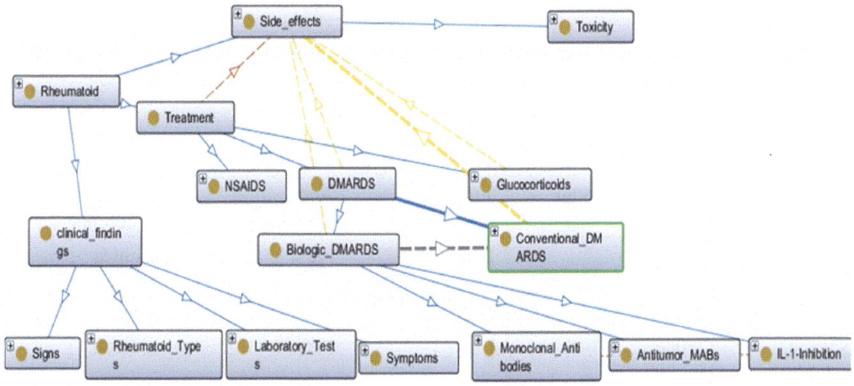

Fig. 5 Rheumatoid ontology [17]

Table 1 The object properties and data properties of the Rheumatoid ontology [17]

Object property			Data property
Name	Domain	Range	Name
Are classified	Clinical_findings	Symptoms, Laboratory_tests, Rheumatoid_types, Signs	Associated_treatment Dosage Initial_evaluation Monitoring Value
Can be combined with	Biologic_DMARDS	Conventional_DMARDS	
Causes	Treatment	Side_effects	
Determine	Laboratory_tests	Rheumatoid_types	
Provoke (sub-property of causes)	Rituximab (individual)	Cytopenia (individual)	
Cannot_be combined with	Il-1-Inhibition	Monoclonal_antibodies	

3 Framework Proposal for Comparative Study

The comparative study between knowledge representation techniques is based on the following criteria:

- Representational adequacy [4]: It refers to the ability of the technique to represent the expected knowledge with its different types.
- Inferential adequacy [4]: It refers to the manner to process the current knowledge in order to infer new knowledge.
- Acquisitional efficiency [4]: It refers to the capacity of the knowledge representation technique to obtain new knowledge in an automated way without the intervention of human.
- Semantic representation: It refers to the ability of the knowledge representation technique to represent the semantic relations between the elements of the represented knowledge.

Table 2 demonstrates the comparison between the knowledge representation techniques

Table 2 Comparison between knowledge representation techniques

	Representational adequacy	Inferential adequacy	Acquisition efficiency	Semantic representation
Predicate logic	False	True	False	Limited
Rule based	False	True	False	False
Semantic network	False	True	False	Limited
Frames	False	True	False	Limited
Ontology	True	True	True	True

The predicate logic represents inferential knowledge as facts and rules. Also, it can represent the relational knowledge and inheritable knowledge (the class membership and class inclusion; instance, and isa). For the inferential adequacy, the predicate logic describes the inferences that are needed. There is an existence of inference procedure. Some of them apply reason forward reasoning or backward reasoning. Although the predicate logic uses the resolution procedure, it is a semi-decidable technique which makes it inefficient to acquire new knowledge. For the semantic representation, the predicate logic enables the representation of the class inclusion in terms of isa relation which is considered as weak representation of the semantic of the relation.

The rule based enables to represent procedural knowledge, declarative knowledge, and inferential knowledge in terms of facts and rules except the hierarchical knowledge. Therefore, it is not fully adequate to represent all kinds of knowledge. For the inferential adequacy, the rule-based establishes the inferential adequacy through the use of Prolog tool that enables the application of backward reasoning to prove a goal and forward reasoning to infer new fact. For the acquisition efficiency, the rule-based uses prolog interpreter that has a fixed control strategy and the assertions define particular path but there is a difficulty to manipulate and update large pieces of codes. Therefore, the rule based cannot be considered as inefficient acquisition technique. For the semantic representation, it does not represent hierarchical structure which relates the facts together. Therefore, the rule-based does not support the semantic representation between facts.

Concerning the semantic network, it does not achieve the representational adequacy. It represents the relational declarative knowledge and the inheritable knowledge form in terms of the description of the nodes, the properties, and relations. Semantic network does not represent the procedural and inferential knowledge. For the inferential adequacy, the property inheritance is a useful form of inference. Moreover, the semantic network enables the use of inference mechanism to infer new fact. The semantic network has a difficulty to manipulate and update large numbers of nodes. Although the semantic network uses the hierarchical structure, the relation between the classes is limited to isa relation and there is no relation between instances.

For the frame, it represents the relational knowledge, the declarative and the inheritable knowledge form. It also enables the use of procedural knowledge but it ignores the representation of inferential knowledge. Therefore, it does not establish the representational adequacy. For the inferential adequacy, the property inheritance is a useful form of inference. Furthermore, the frame enables the use of production rule to infer new knowledge. For the acquisition efficiency, the technique has a difficulty to manipulate and update large numbers of frames. Although the frame uses the hierarchical structure, the relation between the classes is limited to membership relation and there is no relation between instances and subclasses.

For the ontology, it represents all kinds of knowledge. It represents explicitly the relational knowledge, the declarative and the inheritable knowledge form. Furthermore, it incorporates the procedural knowledge (IF and THEN) and the inferential knowledge by its use of the Semantic Web Rule Language (SWRL). For the inferential adequacy, the ontology supports the property inheritance which is a useful form of inference. It also enables the use of inference rules and procedural rules to infer new knowledge. Furthermore, it enables the execution of reasoners to recommend the addition of new relations between the ontology elements. For the Acquisition efficiency, the ontology tool such as protégé enables the acquisition of a new knowledge easily by the use of reasoner that does not need a human intervention. Furthermore, there is various researches [20–22] focus on the ability to evolve and update the ontology. For the semantic representation, the ontology enables the knowledge engineer the liberty to represent the name of the relation that is suitable to associate the ontology classes. Moreover, the relation between sibling classes and the relation between instances can be represented too.

4 Conclusion

The ontology is considered as the evolved form of knowledge representation technique that overcomes the drawbacks of the other previous techniques. What distinguishes the ontology from the other knowledge representation techniques is its adequacy to represent all kinds of knowledge. For this reason, the ontology can acquire new knowledge easier than any knowledge representation technique. Due to the rapid development of ontology tools and languages, the ontology is the only knowledge representation technique that focuses on the creation of semantic relationship between the instances. The semantic representation in the ontology can enhance the results of reasoning compared to the reasoning that will be applied with the other knowledge representation techniques.

References

1. Brachman, R., and H. Levesque. 1985. *Readings in knowledge representation*. San Mateo (Calif.): Morgan Kaufmann.
2. Van Harmelen, F., V. Lifschitz, and B. Porter. 2008. *Handbook of knowledge representation*, vol. 1. Elsevier.
3. McCarthy, J. 1989. Artificial intelligence, logic and formalizing common sense. In *Philosophical logic and artificial intelligence*, 161–190. Dordrecht: Springer.
4. Rich E., K. Knight. 1991. Knowledge representation issues. In *Artificial intelligence (Berlin, Germany)*, 105–129. New York: McGraw-Hill.
5. Rich E., and K. Knight. 1991. Using predicate logic. In *Artificial intelligence (Berlin, Germany)*, 131–169. New York: McGraw-Hill.
6. Grosan C., and A. Abraham. 2011. Knowledge representation and reasoning. In *Intelligent systems modern approach*, 17th ed, 139–145. Springer.
7. Abraham, A., P. Sydenham, and R. Thorn. Rule-based expert systems. In *Handbook of measuring system design*, 909–919. Wiley.
8. Grosan, C., and A. Abraham. 2011. Rule-based expert systems. In *Intelligent systems modern approach*, 17th ed, 149–185. Springer.

9. Sowa, J.F. 2012. Semantic networks. John_Florian_Sowa isi [2012-04-20 16: 51] > Author [2012-04-20 16: 51].
10. Sowa, J.F. 1991. Current issues in semantic networks. In *Principles of Semantic Networks: Explorations in the Representation of Knowledge*, 13–43. San Mateo, CA: Morgan Kaufmann Publishers.
11. Rich, E., and K. Knight. Weak slot-and-filler structures. In *Artificial intelligence (Berlin, Germany)*, 251–276. New York: McGraw-Hill.
12. Salem, A.-B.M., and M. Alfonse. 2008. Ontology versus semantic networks for medical knowledge. In *Proceedings of the 12th WSEAS International Conference on COMPUTERS*. Heraklion, Greece.
13. Minsky, M. 1975. A framework for representing knowledge. In *The psychology of computer vision*, ed. P. Winston, 211–277. New York: McGraw-Hill.
14. Fikes, R., and T. Kehler. 1985. The role of frame-based representation in reasoning. *Communications of the ACM* 28 (9): 904–920.
15. Stephan, G., H. Pascal, and A. Andreas. 2007. Knowledge representation and ontologies. In *Semantic web services*, 51–105. Berlin, Heidelberg: Springer.
16. Gruber, T. 1995. Toward principles for the design of ontologies used for knowledge sharing. *International Journal of Human-Computer Studies* 43 (5–6): 907–928.
17. Gawich, M., M. Alfonse, M.M. Aref, and A.-B.M. Salem. 2017. Development of rheumatoid ontology. In *Proceedings of IEEE eighth international conference on intelligent computing and information systems, ICICIS'17*, 421–426. Egypt.
18. https://www.w3.org/TR/owl2-primer/. Last accessed 23 July 2019.
19. https://protege.stanford.edu/. Last accessed 23 July 2019.
20. Wang, S., W. Wang, Y. Zhuang, and X. Fei. 2015. An ontology evolution method based on folksonomy. *Journal of Applied Research and Technology* 13 (2): 177–187.
21. Zablith, F. 2009. Evolva: a comprehensive approach to ontology evolution. In *European semantic web conference ESWC 2009: the semantic web: research and applications, part of the lecture notes in computer science book series (LNCS)*, vol. 5554, 944–948.
22. Sellami, Z., V. Camps, and N. Aussenac-Gilles. 2013. DYNAMO-MAS: a multi-agent system for ontology evolution from text. *Journal on Data Semantics* 2 (2–3): 145–161.

The Graphical Experience: User Interface Design Approach Based on User-Centered Design to Support Usability

Ibrahim Hassan

Abstract This paper introduces a visual design approach, which supplies UI researchers with testing methods that help them in solving the visual problems through all interaction process, and also guide them to know which part of the interface they need to refine it or introduce a different design solution, to meet the users' needs and goals.

Keywords Usability · Visual experience · User interface

1 Introduction

According to Ibrahim Hassan's model of communication, the designer (sender) examines the actual reality of the user (receiver) and considers his characteristics and objectives and needs, which are formulated and represented in a visual (representamen).

The designer creates the appropriate visual language (encoded) and then composes the visual message that serves the goals of the transmitter (designer) and fulfills the aspirations of the receiver (user).

Therefore, a good and actual study of the user reality enables the designer to determine the appropriate visual language for the user and proportional to his mental models and mental abilities, and even sometimes affect his feelings and internal impressions, to ensure the desired response.

We can limit the visual language that the designer uses through the user interface to four basic elements:

1. Color
2. Iconic and imagery elements (image and illustration, pictogram, and so on)
3. Typography and texts
4. Space (user interface formats and sizes).

I. Hassan (✉)
Faculty of Fine Arts University, Alexandria University, Alexandria, Egypt
e-mail: ibrahim_hassan@alexu.edu.eg

A. Z. Ghalwash et al. (eds.), *Internet of Things—Applications and Future*,
Lecture Notes in Networks and Systems 114,
https://doi.org/10.1007/978-981-15-3075-3_26

The role of the GUI designer is not limited to simply presenting information in an elegant way; the role of the GUI designer comes in three important ways:

The first concerns the aspect of understanding the design elements and how the user can interpret them correctly.
The second is the visual structure for these design elements, leading to unity in the design as a whole.
The third is how the visual perception of the design elements is achieved correctly by the user, so that the user can receive the visual message in the intended way and at the intended time, without the occurrence of any kind of visual diffraction, or visual noise.

According to Fig. 1, the research methodology is based on discussing the design elements in three important and pivotal issues, which leads to efficient user experience and achieves greater usability, namely:

1. Visual semiotics: It is intended to address the graphic elements in a manner consistent with the way human understanding.
2. Visual structure: It is the structural treatment of graphical interface elements.
3. Visual perception: The factors that help to recognize the elements of visual design correctly.

The research—for each of these three issues—determines the appropriate test methods that are performed by user samples to reach an acceptable visual language, which are largely in line with the characteristics and expectations of users (Fig. 2).

2 Visual Semiotics

The user interface can be viewed as a complex sign consisting of a small set of signs. If we assume that there is a special science of user interface, it will be the science of semiotics or semiology [21].

In addition, the semiotics or semiology is a theory looking at the meanings inherent in symbols and signs, whether photographed or written or spoken.

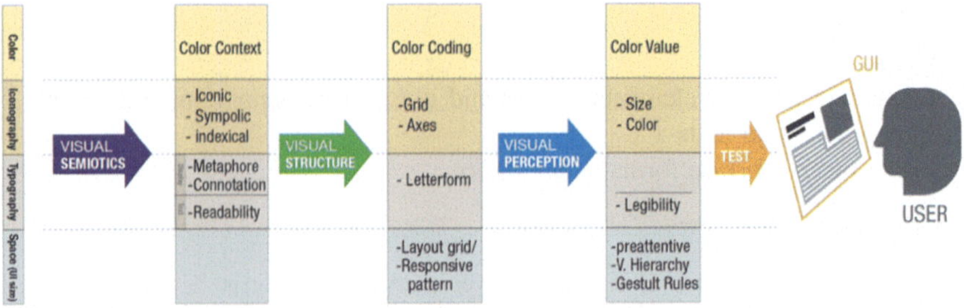

Fig. 1 GUI elements through the three visual concepts

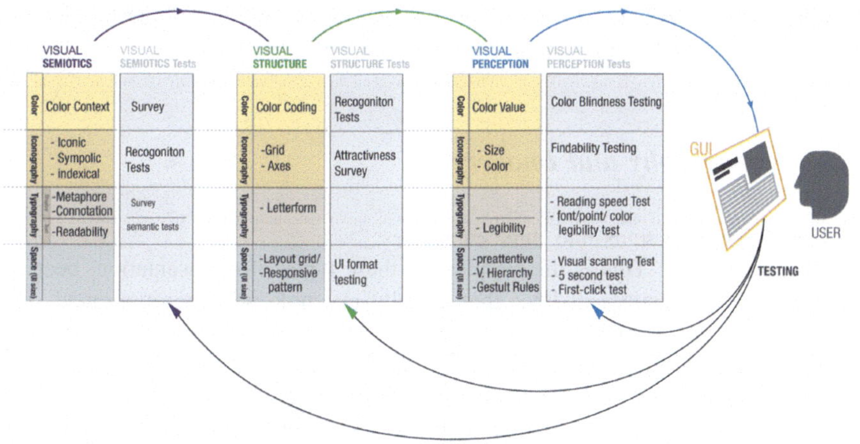

Fig. 2 A GUI Design Approach elements through the three visual concepts and the appropriate test methods for each stage

Andersen (1992) states that a designer builds a user interface. It can be used to tell people something, and from a semiotic point of view, the designer combines a variety of signs to build the user interface. To apply semiotics, the designers only need to take into account the effects of semiotics on everything they design [1, 20].

2.1 A Context and Its Impact on Color Connotation

Although many try to explain the color through the reference cultural or psychological or otherwise, but we can say that talking about color is just a path of speculation, which makes us in the end do not stand on solid ground that enable us to delve into the interpretation of applications of color accurately. It is important to know that the color meaning and connotations vary according to the context used, that the color connotations and meanings are different in different people, ages and specializations, so the designer should be careful when using color [6].

However, it must be recognized that there are many color indications that have a global agreement, if used in a specific context.

The process of interpreting the color meaning and connotations is completed only through the context in which it is used.

Color Survey. To reach the appropriate color group that matches the customer's experience and expectations, it is important to do a survey for a large segment of potential users. This survey can go in two ways.

A simple questionnaire can be asked by asking the target segment directly about the color that is expected to apply to a particular service. The other method is to display a single design through more than one color set, and a questionnaire is conducted on which color combination the user sees appropriate.

2.2 *Iconography and Imagery*

Pearce classify the signs into three categories:

The iconic sign: This is the simplest of these types of representation, because they consist of a pattern of lines that actually resemble what is represented. The icons display the characteristics and features of the body or object that it shows.

The indexical sign: This is related to time and place and directly relates to the concept assigned to it (such as smoke in its sign of fire).

The symbolic sign: Its meaning is based on a convention among a group of people or arbitrarily, and there is no relationship between the form of the mark and what it indicates.

Iconography and imagery tests. At this point we need to perform a test on the users, to see if the user understands what the icon represents? If images and image and illustrations are used, is the intended meaning reaches to the user easily? Can users correctly guess what will happen once they interact with the icon?

Testing for recognition is best done out-of-context: the icon is shown in isolation, in the absence of a text label or of other interface elements. Users presented with an icon must guess what that icon symbolizes. The purpose of this test is to make sure that **icons are recognizable**, and that people can easily deduce the object that it depicts.

In his paper "The importance of mobile interface icons on user interaction" Gatsou gives an example of older people who are likely to have less experience than other younger age groups with contemporary mobile devices and thus be less familiar with the icons displayed in the device or digital applications, so interpretation is difficult in such matters [9].

Gatsou made a comparison between a set of single-function icons for a number of different telephone systems, and measured the recognition rate for the visual representation of each icon (see Fig. 3).

In addition, Gatsou created a table of icons with a rate of 20:40% showing the reasons for the lack of understanding of these icons (Fig. 4), summarizing the research in several important results:

	iPhone	NOKIA		MOTOROLA		SAMSUNG		SONY ERICSSON	
Phone call									
Recognition rate	75%	50%	60%	20%	53.3%	21.7%	55.0%	60.0%	60.0%

Fig. 3 Shows the rate of recognition of the different visual representation of the "phone call" icon, [9]

Icons	R/r*	Original meaning	Possible reasons for misunderstanding
	40%	phone book	The graphic, intended to indicate a note book, is visually unclear.
	40%	internet	The use of a compass to denote the internet is unfamiliar to some users. Generally an icon of a globe is more effective.
	30%	phone book	The combination of head phone and notebook is confusing.
	30%	games	The addition of the word 'games' would probably add to clarity and effectiveness
	21.7 %	phone call	The combination of handset and an individual is ambiguous.
	20%	phone call	The resemblance to a notebook may cause confusion with the phone book icon.

*R/r = recognition rate

Fig. 4 A table of icons with a rate of 20: 40% showing the reasons for the lack of understanding of these icons, [9]

1. Complex or ambiguous graphic configurations reduce the ease of icon correct interpretation.
2. Using familiar metaphors increases the probability of correct interpretation.
3. Users have difficulty interpreting codes that use symbolic or abstract representation.
4. Icons that use concrete images are interpreted more.

2.3 Typography and Text

We can divide the fonts into two types: text fonts and display fonts.

Text font. It is used in writing multiple lines text, and is designed to be clear and readable in different sizes.

Display font. It is designed to attract attention and tighten the reader to the text, can contain more details, expressive and attractive appearance.

This distinction between the text fonts and displaying fonts is useful as it recognizes the possibility of different reception methods and priorities depending on the context in which the characters are received. We note that text fonts give priority to communicate the linguistic meaning, while the display fonts add a deeper energy than the delivery of the linguistic meanings, and it can add emotional feelings and connotations to the words [3, 18].

These different goals of the display fonts and text fonts enable us to conceptualize them on both ends of a continuum. The left side of the writing is the concept of Jean Cheold, which sees the main purpose of the typography as the transmission of language messages only, while on the other side of the display characters can be seen by the names of brands and advertising logotype that use distinctive font characters as part of its identity.

Van Leeuwen presents what he sees as two main principles for semiotics, which they share in the creation of the typographic meaning: connotations and metaphors [17].

Connotations and metaphors. According to van Leeuwen, the implicit connotations or concepts come from the importation of certain symbols and signs from a specific field, and then use them in creating the structural form for display fonts. These signs can be related to a historical period, a particular culture, profession, group, and so on, so that a new typeface is generated and the concepts and meanings associated with those signs are transmitted to it [14].

Van Leeuwen adds that it is not possible to understand every typeface through the connotation because many of the connotations cannot be applied to all typefaces, and it may be difficult to know where these are inspired from, and therefore difficult to understand.

In this case we can introduce another semiotic principle; it is a metaphor, or specifically the semiotics possibilities derived from the attributes of character's letterform.

The letter derives its meaning from structural properties and its formal features are produced by controlling the elements of weight, size, color, motion, and so on (Fig. 5).

Shaikh (2009) conducted a study to identify the personality and the emotional aspects of typefaces. Participants were asked about 40 typefaces of four main categories: serif, sans serif, display, and handwriting, using semantic differential measures [26].

He used the following differentials with Osgoods classic potency, evaluative, and activity dimensions:

- Potency reflects typefaces that are seen as having strength, power, or force.

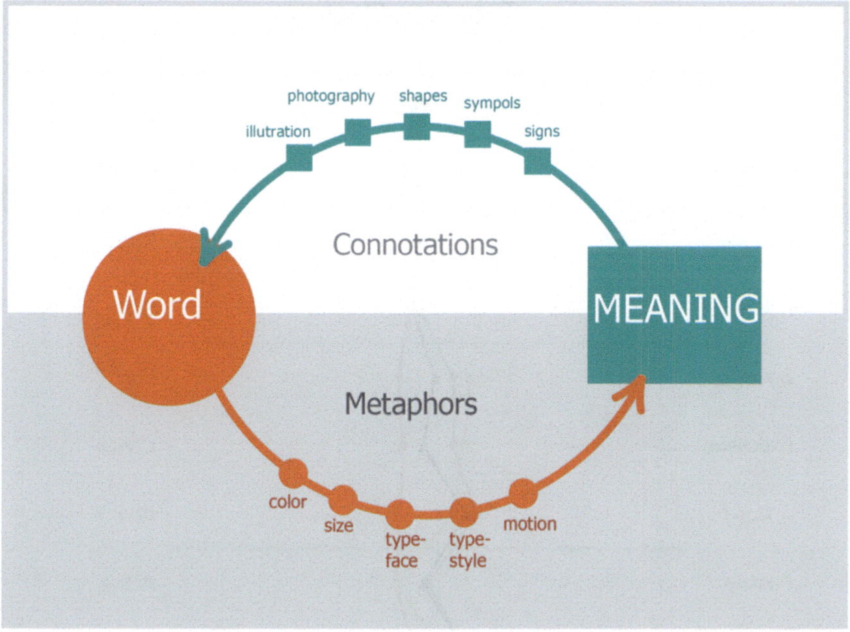

Fig. 5 Through connotation the meaning is achieved by adding external elements to the word, while the metaphor is achieved by modified the typographic characteristics of the word to achieve the meaning

- Evaluative reflects typefaces that are viewed as having value, worth, and importance.
- Activity reflects typefaces that are considered to be full of energy, movement, and action.

The items were presented in this order (factor dimension in parenthesis)

- Passive – Active (activity)
- Warm – Cool
- Strong – Weak
- Bad – Good (evaluative)
- Loud – Quiet (activity)
- Old – Young
- Cheap – Expensive (evaluative)
- Beautiful – Ugly (evaluative)
- Happy – Sad
- Delicate – Rugged (potency)
- Calm – Excited (activity)
- Feminine – Masculine (potency)
- Hard – Soft (potency)
- Fast – Slow (activity)
- Relaxed – Stiff (potency) (Fig. 6).

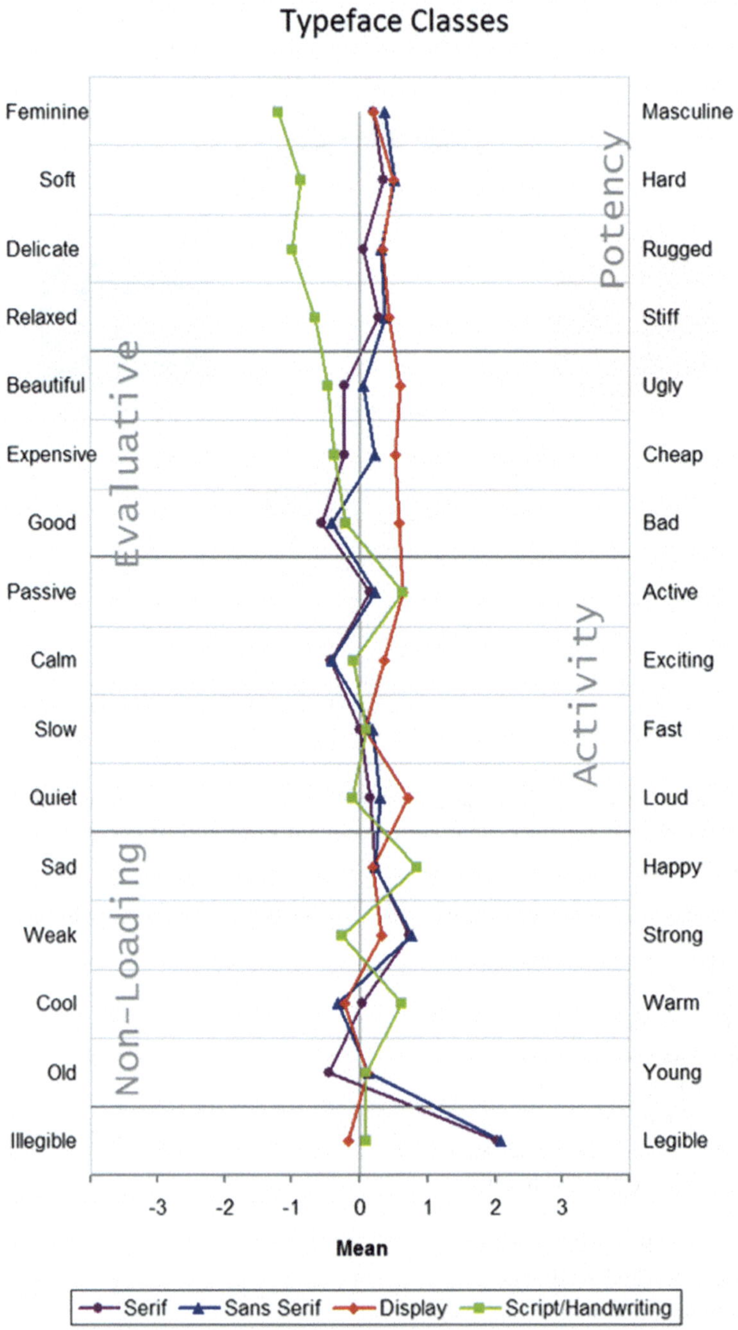

Fig. 6 Type classes on each semantic scale [15]

The 40 typefaces evaluated had unique personalities and ranks on the factors. Overall, the bolder, blockier, typefaces were seen as more potent, more evaluative, and less active. The scripted typefaces and the typefaces that were light in color due to the stroke weight were likely to be perceived as less potent, more evaluative, and less active. For the most part, the serif and sans serif typefaces were viewed as neutral on all factors. Additionally, such typefaces as Gigi, Curlz, and Kristen tended to be judged as active but not potent or evaluative. The three monospaced typefaces (Lucida Console, Courier New, and Consolas) all tended to score similarly on the three factors. These typefaces were seen as very potent and not evaluative; the scores on activity were either in the middle or at the low end.

The most commonly used typefaces for large blocks of text (serifs and sans serifs) differed minimally on the three factors. Arial was one of the most potent typefaces; Perpetua and Centaur were both very high on the evaluative factor; and Berlin Sans scored high in activity. The subtle differences in the ranks of the typefaces on each factor indicate that designers have a choice among the "common" typefaces when co-structuring documents. For example, a document that needs to be higher in potency and evaluative and neutral on activity might best be expressed in the typeface Georgia. On the other hand, a document that needs to be low in potency, high in evaluative, and neutral in activity might be best expressed using Century Gothic. By using the semantic differential charts for each typeface, desired personality traits on each of the 15 semantic scales could be used when choosing typefaces.

When collapsing across type classes (serif, sans serif, display, and script/handwriting) the results of this study indicate that participants perceive typefaces to be similar in personality based on the type class only when there are few variations among members of the type class as seen in serifs and sans serifs. Typefaces in the display and script/handwriting type classes varied widely within their respective classes. For the most part the serif and sans serif typefaces were also judged as more neutral in terms of personality, and they were reported to be more legible. These findings possibly explain why such typefaces are commonly used in "everyday" neutral settings with longer passages of text. The display and script/handwriting typefaces had "more" personality and were likely to be rated as less legible. Therefore, such typefaces are probably best suited for occasions when small amounts of text is needed to convey a strong message.

Effect of semiotics on the readability. Beginning in the 1950s, new developments transformed the study of readability, including a new test of reading comprehension and the contributions of linguistics and cognitive psychology. Researchers explored how the reader's interest, motivation, and prior knowledge affect readability. These studies in turn stimulated the creation of new and more accurate formulas. George Klare [16] defines readability as "the ease of understanding or comprehension due to the style of writing [9]." According to William S. Gray and Bernice Leary [11] "What Makes a Book Readable" and according to William Dubay "The Principles of Readability" there are four basic elements of reading ease: (1) content, (2) style, (3) structure, and (4) design [4] (Fig. 7).

Semantics and syntactic—which fall under the element number 2, that is style—play an important role in improving readability. Semantics and syntactic is a part of

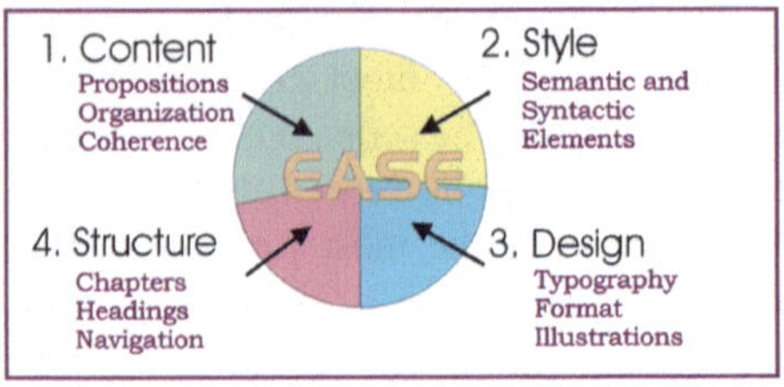

Fig. 7 The four basic elements of reading ease [4]

the study of semiotics, which is divided into three: "*pragmatics, semantics, and syntactic.*"

Semantics is an area of semiotics in which the researchers attempt to determine the significance of signs within and throughout various cultures. Syntactic is the study of the ways signs are combined with each other to form complex messages.

Teenie Matlock in his search "Experimental Semantics" has put a methodology to test the word semantic. He notes that "Some experiments use a between-subjects design, in which participants are engaged in only one experimental condition. Other experiments use a within-subjects design, in which participants are in two or more conditions."

"Experimental semantics uses 'online' methods and 'offline' methods. Online studies measure behavior in real time, often in milliseconds [27].

For example, in an online lexical decision task, participants might view items such as 'hand,' 'thub,' and 'cherry' on a computer screen, and quickly indicate whether they are real words or not by pressing a 'yes' or 'no' button. In an online narrative comprehension task, they might read a sentence and quickly decide whether it relates to a short story that they had just read. Response times generally differ for many reasons (because of a participant's familiarity with a specific word, because of a word's frequency etc.), but in a controlled experiment where a linguistic phenomenon is systematically manipulated while other things are kept constant, response times can be used to draw inferences on linguistic processing."

> In offline studies, behavior is not measured in real time. In that case, participants might read a paragraph and then leisurely answer questions, or read a sentence and generate a drawing that reflects their understanding of its meaning. Both offline and online approaches are informative, and both have advantages and disadvantages. Offline experiments are simple to set up and run, but incapable of pinpointing when and how processing unfolds in time. Online studies can be work intensive and require other technical expertise, but they are well suited for investigating the dynamics of language processing.

3 The Visual Structure

The designer's attention to the structural aspect of the design process greatly helps to avoid many of the problems of visual perception and content understanding through the user interface [13].

The visual structure provides the user with guides on how to do something, when designing user interfaces with some structure about the information we are trying to communicate, and this will make the user understand the system more easily. Data display format is the simplest form of visual structure.

Keyes (1993) says, "The structure of visual information is actually before the content is processed. By revealing a latent system that carries this content, the structure of visual information supports the reader's functions of navigation, directing, understanding, remembering, etc. This visual organization occurs unconsciously by the reader during the visual scanning process. Furthermore, this primary visual system strongly affects the reader in how to read, understand, interpret and remember the content" [15].

Prasad Bokil and Shilpa Ranade put a functional model of structural grid in visual communication (Fig. 8), which illustrates the basic aspects of the function of the grid: creation, correlation, and space management [24].

Creativity factor deals with the creation of any form. While the correlation factor is responsible for the structural correlation between elements and each other, the space factor is the process of ordering elements for space.

According to the model of Prasad Bokil and Shilpa, we can describe the functional model of visual structure for interface design in three tasks:

1. The process of designing and structuring elements, such as font structural grid, and the structural grid/axes for design icons, pictogram, and illustrations.
2. The processes of coordinating elements between them and each other, including the alignment and visual hierarchical and color coding.
3. The space distribution: starting from information architecture (as a conceptual arrangement), then the layout grid, and finally the responsive patterns (Fig. 9).

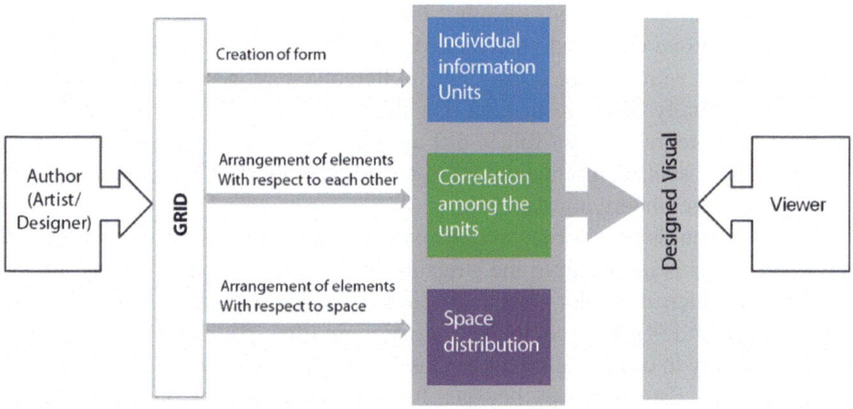

Fig. 8 The functional model of the grid in visual communication—Prasad Bokil and Shilpa Ranade, [24]

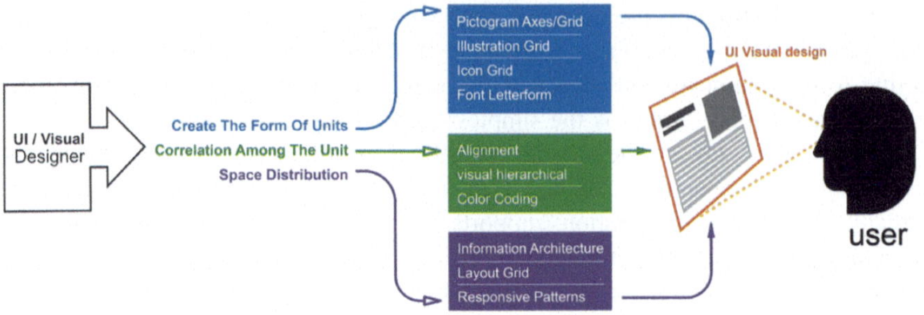

Fig. 9 The functional model of visual structure for user interface design

Color coding (The Hidden Structure). The color component can enhance or weaken the visual information structure. Its usefulness depends not only on color and its proportion but also where it is used. Some called it the hidden grid.

Color coding is one of the methods used in the structural correlation between elements and each other. The color is intended to group elements together, and is interpreted as similar regardless of spatial position, alignment, type of formal face, or information structure. The color coding plays a role in the work of visual assemblies of the graphic elements, even if they are spatially distant. It also contributes to the hierarchy in contrast, resulting in the separation of information that is more visually compelling than other less important information within the context [13, 15, 30].

Color coding refers to what are the categories under which information is displayed on interface. The color coding system allows for quickly identifying a relevant information category without having to read the contents or go into details for the first time. This allows focusing on this category of information, while the remaining information is excluded from the interest.

Color coding requirements. The basic requirement for color coding is to set the color function to be reflected, meaning that each set of specific data values are associated with one color and vice versa. Each color represents a fixed set of data values. In other words, color coding must differentiate between two different data values and visually distinguish them. On the other hand, visually similar colors mean that they represent data values that are close to some. Otherwise, perceptual differences in visual representation are not required to reflect the true size of the differences in the underlying data [5].

The effectiveness of color as an information signal depends on several factors: (1) Where color is used during the structure of the information, (2) What are the elements to which the color refers, (3) How does the distinction between color signals occur, and (4) What color characteristics are used. Although color coding is an excellent way to display information classification, only a few color codes can be viewed quickly. Estimates vary between five and ten symbols [15, 29].

In order to obtain an effective color coding, the color used must be relevant to the information presented and known to the user. In other words, he can interpret the meaning so that the person can recall the information he needs and the ease of excluding other information that is not needed. Otherwise, the use of color will hinder the performance and become a burden on user memory. The proper use of color requires analysis of user expectations and his experiences [8].

Structural Design for Icons and Imagery. There is no doubt that the structural design of the icons and pictogram has a strong role in unifying the visual character and identification of the elements. The designer's interest in creating the structural grid will undoubtedly help to create a unified logic for the construction of forms with the different possibilities and graphic solutions resulting from it.

The unity of the visual character of these elements is achieved through two things. The first of which are the grids and coordinates used in the design of the iconic elements, and the second, the percent of the mass of the space, that is, the ratio of the occupied space to the spaces formed for the shape [13].

Letterform. The designer should not neglect to make sure that the grid he uses to create the characters or his font meets the target for which the font was created, as appropriate for the nature and type of information that these characters will display, besides ensuring font clarity if used in small sizes or on low-resolution screens.

Layout Design Grid. We are in an interactive world where everything is constantly changing: content, display windows, screen, and so on; all change over time. The digital world is a dynamic world that evolves permanently from ever before, and our task as designers is to keep pace with that development. From here we try to explore the dynamics of network systems that display content in the highest degree of beauty and functionality.

The emergence of the World Wide Web and the evolution of online media has seen the migration of the page to the screen. Many of the skills and principles of layout design have directly shifted to these new media, although there are slight differences in some structural terms. In an internet environment, the basic functions of layout design are the layout itself. The design must be structured to obtain a specific response, whether for informational purposes, entertainment, or directing the reader.

There are many similarities between screen designs and printed page. However, the difference between the physical book and the web page is that the book is always divide into two parts: the right page and the left page, with a physical space between the pages, while the web page or screen designs in general is one entity: one page. Web page layout is therefore treated more as a panoramic [10].

The second fundamental difference between the electronic pages and the pages of the printed book is that the pages of the printed book are fixed and therefore the designer puts single possibility for the design grid; on the contrary in the web pages, for example, in the case of the web, the user can reduce the size of the page (window). Also, in the case of smart phone applications, the user can browse the application in the phone area, whether the width or length, and so on, which made the design process become complicated for the designer, and it has to rebuild the content according to a number of different sizes, with no harm to the quantity or importance of the content, in addition to ensuring that content is kept accessible to the reader in the same visual sequence (Hassan 2017).

Responsive pattern simply refers to a website layout that responds (scales itself) automatically to an array of devices (screen sizes) and resolutions. Basically, it is mobile and tablet-friendly. That way, whether you're using a personal laptop, a notebook computer, or you phone, you can access the same website on all of them without losing quality. The idea of responsive design is that the same website will look great while you're browsing the web on your phone and on your computer without sacrificing any of the quality or ease of navigation. It is important to stay up-to-date with the most popular screen sizes and resolutions when designing web and mobile sites.

There is a lot of tools that help them to test their response with their users like "responsivedesignchecker.com" which was initially created in 2012 as an exercise for Media Genesis developers to see if they could create a tool that could be used to test websites for mobile design.

4 Visual Perception

Visual perception is defined as "the process of organizing and interpreting the visual stimuli in the visual field, and processing them in the visual centers of the brain."

When we talk about the graphical user interface, there is no doubt that we are talking about a visual interactive environment in which the users interpret all of its elements, resulting in a kind of interaction between them and the interface system. The user relies on the sense of sight to interpret and understand that interface, that any defect in the process of visual perception of the elements of the interface of use, this will ultimately affect negatively on the interactive process. Hence, there is importance of talking about issues of visual perception for the design of the user interface, in the framework of improving the user experience and achieve more usability.

The Color. People have a cognitive limit to a number of visual cues that they can understand, manipulate, and use effectively before they become too distracting. Color can extend this limit by creating optical layers. However, if the color is used badly, it will have the opposite effect of reducing the cognitive limit. Color works to attract the eye, whether it occurs at the first or fifth level of the hierarchy.

Edward Tuft, 1990, in his book "Envisioning Information" (Tufte 1990), made an effective case for the color strategy in topographic maps, namely, the use of high-density colors on a floor of tints of saturated colors or on layers of unsaturated colors: (1) effective hierarchy, (2) independent information layers, and (3) distinguish between layers.

The hierarchy is by distinguishing the degrees of saturation and value. The small spaces have the highest saturation and have a strong vision like red. While the layers of information can be distinguished by changes in value (tints and quantities of dimmed white), there may be a different number of hues with the same value or degree of saturation as one layer. The distinction between layers is by changing the color or saturation of the single layer.

Icon Findability Testing. To gauge findability, icons must be shown in their native habitat—in the context of the full interface. In-context testing can help you determine if multiple icons appear too similar and users will have a difficult time distinguishing among them, or if the icon is hidden under a false floor or in an ad-rich area and is thus overlooked.

Time-to-locate tests are the best measurement of whether or not users can easily find an icon (or some other interface element) among the crowd of the full design. In these tests, participants must click or tap the UI element to achieve a given task. Measure how long it takes people to successfully select the correct icon, as well as the rate of first-click selections (i.e., how often their first click is on the right icon: wrong selections indicate that the icons are not suitably differentiable, while slow-but-correct selections are a discoverability issue [12].

Fonts. Michael Bernard and others studied the common types of fonts used on web pages—point 12 except Font Agency was used as a size of 14 because its letters are smaller—to take into account the differences in effective reading speed (reading time/accuracy), as well as recognizing the font clarity [19].

The test of clarity was performed by participants reading as fast as possible with 15 words changed (not reported). The original words in the section were replaced with similar words, e.g., Cake/Fake and the participant if he discovered it, he should read it aloud.

Effective reading grade or rate was used to determine the accuracy of the font legibility. In other words, legibility was tested by reading efficiency. The result was extracted by the ratio of the words that were replaced in the reading passage divided by the time consumed in reading, which was calculated by the stopwatch.

The result was not significant in the effect of font type on reading efficiency. This was not strange in the light of previous studies, which suggested that the font of this size (12) is fairly strong in its ability to convey meaning to the reader accurately.

As for the reading speed test, there were significant differences when studying the mean time for each font type. Regardless of the accuracy of the reading, when testing more, the Tahoma font was read much faster than the Corsiva font at a time difference of 40 s for two pages of text. Accordingly, different reading speeds, associated with different types of fonts, may have no real consequences for little text amount on the Internet—as long as fonts are used in their traditional state (Fig. 10).

Space (interface format). *Preattentive*. Neisser [22] presented the idea of a preattentive stage of visual processing (meaning vision before attention). Vision at that stage is not subject to any pressure. In the preattentive stage, everything can be manipulated simultaneously across the whole visual field. In visual information, we write the basic building blocks of the visualization process, called "Preattentive and Ease of Search Processing," in which we talk about the qualities that immediately capture our eyes when we look at any visual representation or representation. They can be seen in less than 10 ms, even before they make a conscious effort to observe their presence. Here is a list of preattentive features (Fig. 11) [31, 32].

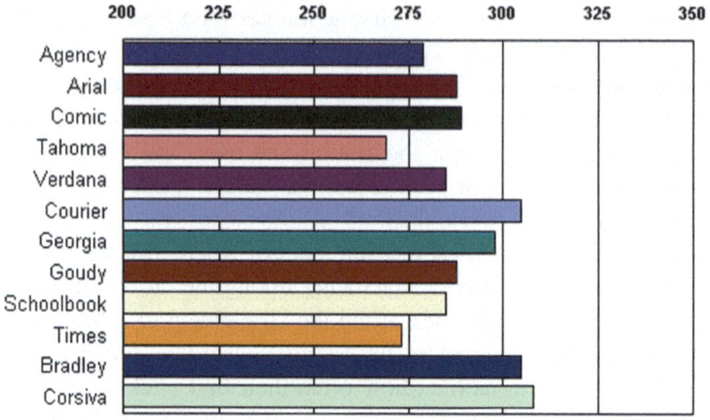

Fig. 10 The reading speed test [19]

Gestalt Theory of Visual Perception. Gestalt theory is one of the most important theories in the interpretation of visual perception issues. In general, Gestalt theory is based on how similar elements are grouped into groups to study the relationships between these groups and others, and study the relationship between the elements of a single group. The Gestalt theory is explained through six rules or laws, which are summarized below with a form to be developed ways to apply or use them through the user interface.

1. Approximate: This principle indicates that the elements or shapes as they come closer together tend to assemble in units, which are perceived as one. This is done by controlling the spaces between them (Fig. 12).

2. Similarities: Refers to similar or equal elements in shape, size, or color, tend to assemble together in units or formulas (Fig. 13).

3. Continuation: The concept of continuity indicates that regulation in the field of cognition tends to occur in such a way that the straight line continues for a straight line; the part of the circle continues in a round and so on (Fig. 14).

4. Closing: This law refers to incomplete or almost complete forms or subjects that tend to close or complete. They are perceived as complete or closed (Fig. 15).

5. Symmetry: Similar objects stand out as distinct formulas from other units within the realm of perception (Fig. 16).

6. Common Fate: The Gestalt law of common fate states that humans perceive visual elements that move in the same speed and/or direction as parts of a single stimulus. A common example of this is a flock of birds (Fig. 17).

Visual Scanning. Many eye tracking studies have focused on data collection for eye movement through internet, while participants are engaged in the task of searching for model information across web pages.

Russell studied eye tracking, and this study examines the contributions of eye tracking data to traditional usability testing procedures when using or interacting with a web page for the first time [25].

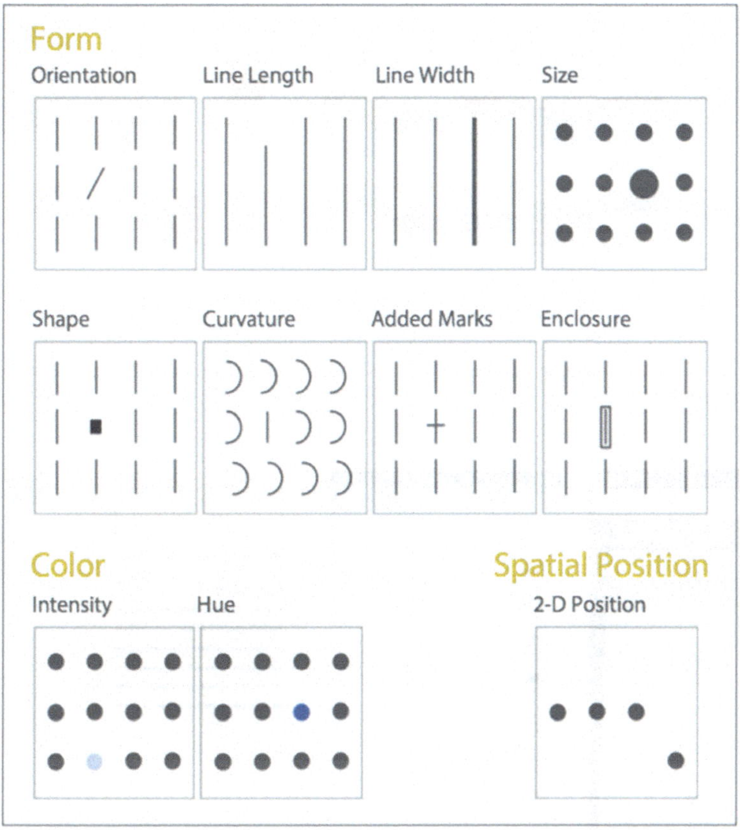

Fig. 11 Preattentive features, [31]

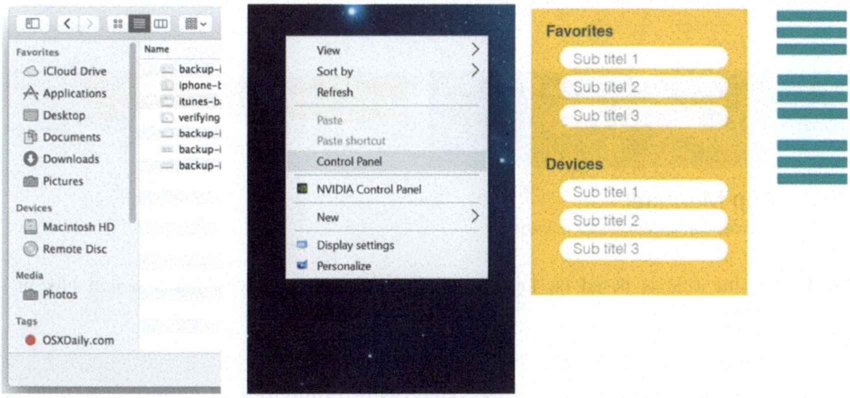

Fig. 12 Approximate principle can be seen in many interfaces, like the sidebar in mac systems, also in the right click menu in the windows systems

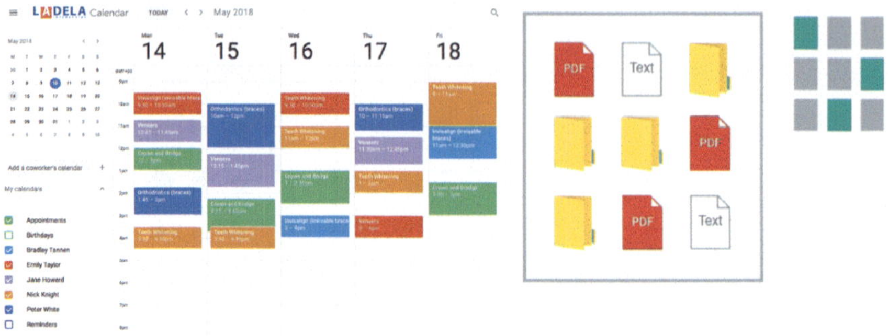

Fig. 13 Similarities principle can be seen in tasks segmentation in calendar, as the same color means the same task

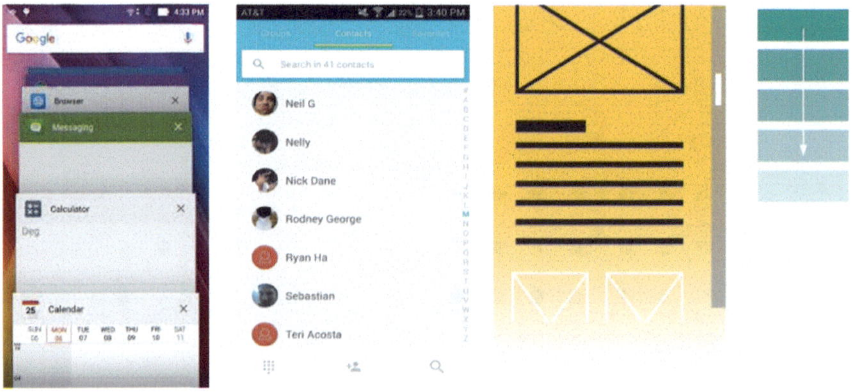

Fig. 14 The rule of continuity is clearly shown in the phone systems, where the user is browsing applications either in a vertical or horizontal direction. The user will continue in the same direction spontaneously without requiring an effort to think

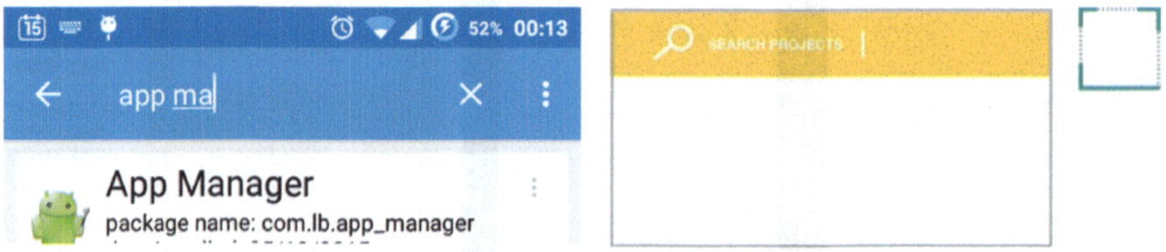

Fig. 15 Closing rule is based on logical thinking, for example expecting a search box in a navigation area

As part of the usability test, three sites were compared in a similar field. The user's eye movements during the initial login were recorded by users to visit each site. Eye tracking data was checked to gain additional knowledge about how the

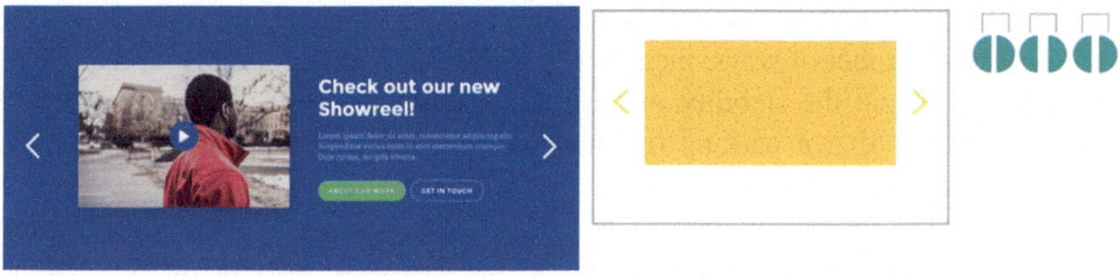

Fig. 16 The best practice for this rule is the slider in web pages, as we see the image, text and the two arrows as a one unit

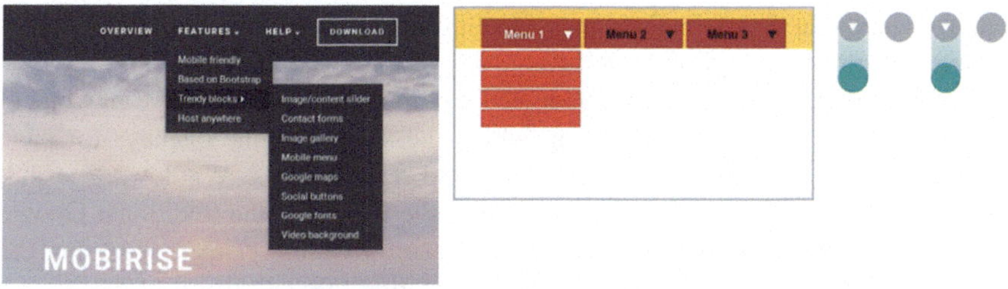

Fig. 17 We see words that slide from the list at the same speed and in the same direction as one object

user initially viewed the site, and what elements of the page attracted visual attention?

All three sites used in this study were e-commerce sites that specifically dealt with the sale of educational toys. During this process, respondents' responses were often a little more detailed than "is the site to play" and "it's aimed at parents," but this was expected. This would be just a way to determine whether users are able to recognize the purpose or domain of the site simply from the home page. However, the only data collected during this process is whether the design and information for the home page effectively communicates the site's objectives to the users. Adding eye tracking data allows designers to understand exactly what users are searching for to extract understanding and meaning from the site.

The results showed that the eye movement data completes users' verbal feedback in their reactions to the site. In particular, eye tracking data that received aspects of the site revealed more visual attention and in what order was seen.

You must investigate a number of different aspects of website design. For example, designers can investigate how users change their visual attention to different aspects of the page depending on: (1) the nature of the task; (2) repeated exposure to the site; and (3) the relative ranking of the page content.

The homepage of any site is often divided into sectors or areas (place of the logo —a place to search—a place for the navigation list, etc.) and then the data is scanned through the eye tracking in terms of installations recorded within these ranges or places, or area of interest.

The data that is concerned with eye fixation on the area of interest (AOI) can be compared in a number of ways, including: (1) The arrangement of area of interest in terms of the first in the stability of the eye or standing on it. (2) The number of records or times of eye parking on each scale. (3) Cumulative time for fixation recorded in the area of interest.

This provides a way to identify which aspects are viewed and cared for by the user. When he visit the home page of any site at first time, also the duration of the consideration, and the number of times viewed at the first meeting or interaction with the site.

One of the eye tracking output data is the representation of information reception areas of the screen, either by fixing the eye on it (number of stops fixations) or the length of time that the eye is fixed on it (dwell time), and this is represented by color coding (hotspots) through the interface.

C. Siu and B. Chaparro have conducted a comparative study of eye movement through two different types of layout, which the search engine displays for page results, namely, the grid system (reflecting the results of the Windows 8 search engine page) against the List menu (reflecting the results of the traditional Google search engine page) through two types of tasks (informatics versus navigational) to observe differences in visual patterns [see Formative (58-2), (59-2)] [4].

Users fixated first on the results located on upper left quadrants of the grid layout for the *info task* condition. Search results located on the bottom row and right column of the *grid* were attended to last. The *nav task* had the target location of the search task appeared on the 2nd, 4th, 5th, 6th, 7th, and 9th search results of the *grid* layout. Results indicated noticeable similarities to the *info task*. When performing a *nav task*, users fixated first on the upper left quadrants of the *grid* layout. Figure 4 shows the gaze orders and Fig. 5 shows an example of a heatmap generated from the gaze data (Figs. 18 and 19).

5 s test. Five second testing is a type of usability testing that enables you to assess how rapidly a message is communicated. This type of test offers both quantitative and qualitative feedback for optimizing a design. A 5 s test is done by displaying a participant a picture, after which the competitor answers questions based on his or her memory and design impression. Studies have discovered that users only take a few seconds before choosing whether to remain or leave the website [23].

Although the five second test is used in many ways, the main topics are:

- Do people understand the product or service?
- Do people feel they will receive a benefit from the page?
- Can people recall the company or product name?

These issues are essential because if a page transmits all this data rapidly and easily, the correct audience will be captured. This is a main factor in developing changes aimed at enhancing conversion and engagement.

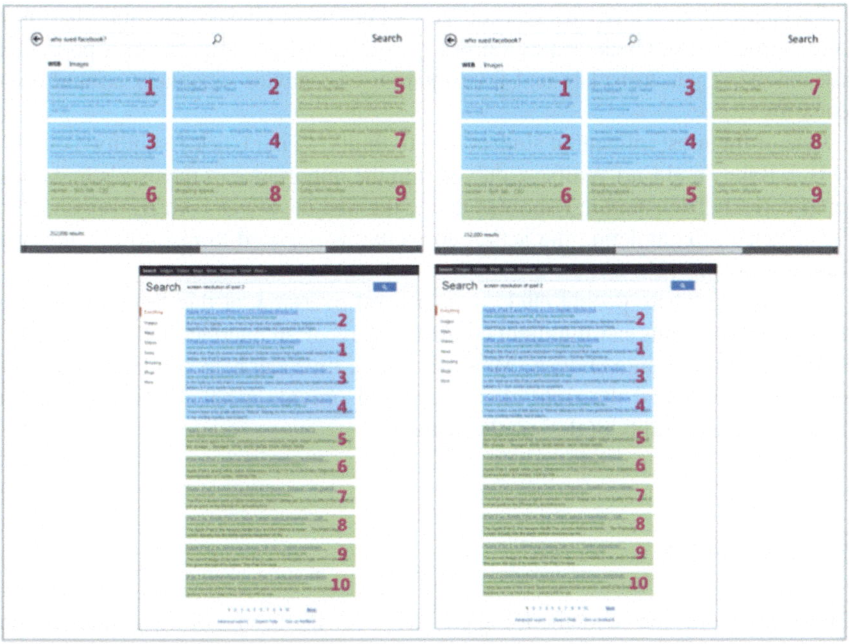

Fig. 18 Order of fixations of the info and navigation tasks for both layouts represented numerically [4]

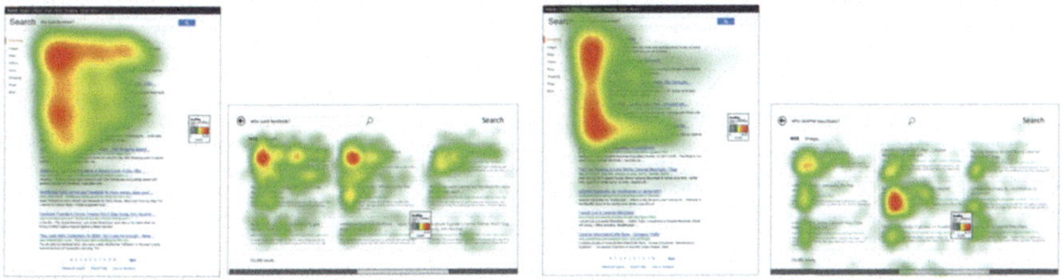

Fig. 19 Example of heatmaps generated from gaze patterns from individual participants performing an info task on both layouts

We can simply divide the results into one group of people who' got it and another group who did not. If you have more than 80% of participants in the first group, then your design is successful. If this is much lower, some modifications will probably have to be made.

The first click test is a method to measure the usability of a website, application or design by knowing how easily a specific task can be completed. The aim of the first click test is to verify that it is clear and easy to perform a given task by first clicking a user on an interface.

Packages of web analytics can tell you where users clicked, but not what they were trying to do. Click testing allows you to ask users to perform a particular task, allowing you to isolate and investigate user behavior separately around each scenario.

One of the most influential studies into usability, "First Click Usability Testing" by Bob Bailey and Cari Wolfson, delved into the importance of the user's first click being correct. Their findings showed that if the first click was correct, users had an 87% chance of completing the action correctly, as opposed to just 46% if the first click was wrong [2].

Visualizing where people click gives a good overview of the design. Clicks in unexpected locations can highlight confusing parts of an interface and are useful for informing future design choices. Measuring click time can help you figure out how easily users will find the right click place and provide a helpful benchmark for comparing the usability of design alternatives.

5 Conclusion

The research provides a methodological vision for the design of the graphical user interface, in addition to presenting the appropriate test proposals for each stage. This requires the recommendation of researchers interested in this area, in testing this methodology, and show its effectiveness on the experience of the end user.

The methodology is based on testing the basic elements of the GUI (color, Iconography, typography, and space) through three visual axes, namely: visual semiotics, visual structure, and visual perception.

References

1. Andersen, Peter Bøgh. 1992. *Computer semiotics*. Department of Information and Media Science, University of Aarhus, Niels Juelsgade.
2. Bailey, Bob. 2013. *FirstClick usability testing*. Web usability, Aug 10, http://webusability.com Accessed 20 May 2019.
3. Bradley, Steven. 2010. *Legibility and readability in typographic design*. https://vanseodesign.com, May 24, https://vanseodesign.com/web-design/legible-readable-typography/ Accessed 10 May 2019.
4. Chaparro, B.S. 2014. *The Software Usability Research Lab (SURL)*, Nov 15, http://usabilitynews.org/ Accessed 2 Nov 2016.
5. Christian Tominski, Georg Fuchs, Heidrun Schumann. 2008. Task-driven color coding. In *12th International Conference Information Visualisation. London*: IEEE, 373–380.
6. Darrodi, Maryam Mohammadzadeh. 2012. *Models of colour semiotics*. PhD Thesis, School of Design, The University of Leeds.
7. Dubay, William. 2004. *The principles of readability*. Impact Information, 126 E. 18th Street, #C204, Costa Mesa, CA 92627, (949) 631–3309.

8. Galitz, Wilbert O. (2002). *The essential guide to user interface design an introduction to GUI design principles and techniques*. Robert Elliott (ed.), vol. II. Wiley, Inc.
9. Gatsou, Chrysoula, Anastasios Politis, and Dimitrios Zevgolis. 2012. The importance of mobile interface icons on user interaction. *International Journal of Computer Science and Applications (Technomathematics Research Foundation)* 9 (3): 92–107.
10. Gavin Ambrose, Paul Harris. 2016. *Layout*. 2nd. AVA Publishing, 2005. Harley, Aurora. "Usability Testing of Icons." *Nielsen Norman Group*. Feb 7, https://www.nngroup.com. Accessed 15 May 2019.
11. Gray, W.S., Leary, N.E. 1935. *What makes a book readable?* Chicago, IL: University of Chicago Press.
12. Harley, Aurora. 2016. *Usability testing of icons*. Nielsen Norman Group. Feb 7, https://www.nngroup.com Accessed 15 May 2019.
13. Hassan, Ibrahim. 2017. *Graphical user interface design: between upgrading user experience and usability*. PhD Thesis, Graphic Department, Alexandria University, Alexandria.
14. Hassan, Ibrahim. 2018. The role of visual connotations and conceptual metaphors in enriching the typography design. In: *Fourth international conference of fine arts and community service (Visual arts between the problem of modernity and identity) Faculty of fine arts*. Luxor.
15. Keyes, Elizabeth. 1993. Typography, color, and information structure. *Technical Communication*, 638–654.
16. Klare, G.R. 1963. *The measurement of readability*. Ames, Lowa: Lowa State University Press.
17. Leeuwen, Van. 2005. Typographic meaning. *Visual Communication* 4 (2): 137–143.
18. Lemon, Mark. 2013. *Towards a typology of typographic signification*. Master Thesis, Department of Semiotics, University of Tartu, Tartu.
19. Michael Bernard, Barbara Chaparro, and R. Thomasson. 2000. Finding information on the web: does the amount of whitespace really matter? *The Software Usability Research Lab (SURL)*. Jan 8, http://usabilitynews.org. Accessed 2 Nov 2016.
20. Nadin, Mihai. 1990. Design and semiotics. In: *Semiotics in the individual sciences*, vol. 2, ed. Walter A. Koch 418–436. Bochum: Universitatsverlag Brockmeyer.
21. Nadin, Mihai 1988. Paradigm*, Interface design: a semiotic. *Semiotica*, April 3, 269–302.
22. Neisser, U. 1967. *Cognitive psychology*. New York: Appleton-Century-Crofts.
23. Perfetti, Christine. 2007. *5-Second tests: measuring your site's content pages*. UIE. Sept 11, www.uie.com. Accessed 29 May 2019.
24. Prasad Bokil, Shilpa Ranade. 2012. *Function -Behaviour- Structure Trpresentation Grids in Graphic design*. Design computing cognition (DCC'12). Texas USA, xx–yy.
25. Russell, Mark. 2005. *Software Usability Research Lab (SURL)*. Feb 13, http://usabilitynews.org/. Accessed 31 Oct 2016.
26. Shaikh, A.D. 2009. Know your typefaces! semantic differential presentation of 40 onscreen typefaces. The *Software Usability Research Lab (SURL)*. Oct 15, http://usabilitynews.org/. Accessed 3 Nov 2016.
27. Teenie Matlock, Bodo Winter. 2015. Experimental semantics. In: *The Oxford Handbook of Linguistic Analysis*, ed. Heiko Narrog Bernd Heine, 771–790. London: CPI Group (UK) Ltd, Croydon, Cro yy.
28. The Principles of Readability 2004.
29. Tufte, Edward R. 1990. *Envisioning information*. 4th Edition. Graphics Press.
30. Ulf Ahlstrom, Larry Arend. 2005. Color usability on air traffic control displays. In: *Human factors and ergonomics society 49th annual meeting*. USA: NASA Ames Research Center.
31. Ware, Colin. 2013. *Information visualization perception for design*, ed 3rd. Elsevier, Inc.
32. Wolfe, Jeremy M. 2000. *Visual attention*, vol. 2. Academic Press.

Steps for Using Green Information Technology in the New Administrative Capital City of Egypt

Hesham Mahmoud

Abstract Currently, most people feel worried about the environment. Around the world there is an increasing awareness that humanity is gradually pushing Earth towards extinction and annihilation. Everywhere, leaders and common citizens alike are getting more certain that improving the environment and reversing global warming is a matter of life and death. Conferences are held around the world, sometimes simultaneously, to research how to establish the best strategies and mechanisms for the preservation of the environment against all types of pollutions. All scientific norms establishing entities are now required to achieve that goal; each according to its jurisdiction. One of the major fields that are required to participate in the achievement of that goal is information technology. Information technology experts insist upon the necessity of using Green information technology techniques "Green computing" everywhere. Green computing should be used in all types of government works and works done by States. The aim of such use is to make computer hardware and software environment-friendly.

Keywords Green information technology · New administrative capital city
The strategy of egypt 2020/2030 · Proposition of application

1 Introduction

Generally, the environment of human beings represents their surroundings; namely, the framework they are living within that contains "soil, water, and air," and what could be classified under each of these three ingredients as to material components and life pulsating creatures, and the phenomena controlling the above as to weather, climate, wind, rain, gravity, magnetism, and other factors [1], and the reciprocal relationships between such factors. The concept of preservation of the environment

H. Mahmoud (✉)
Management Information Systems, Modern Academy for Computer Science and Management Technology in Maadi, Cairo, Egypt
e-mail: dr.heshammahmoud122@gmail.com

A. Z. Ghalwash et al. (eds.), *Internet of Things—Applications and Future*,
Lecture Notes in Networks and Systems 114,
https://doi.org/10.1007/978-981-15-3075-3_27

means all the procedures taken by public entities; whether affiliated to the state or independent entities, to mitigate the harmful effect of humans on the environment. There are many studies made in universities and specialized research centers to understand the environmental system and to define the effects of human activities that disequilibriate the environment and endanger the life of current and future generations. The aforesaid studies aim mainly at establishing, implementing, and rectifying appropriate measures for such disequilibrium.

We can simplify the ecological system through defining that system as "A group of exact reactions between living organisms that reside in a certain sector of nature and the surrounding media of that sector." If Homo sapiens intervene senselessly and thoughtlessly in that natural balance, they are bound inevitably to corrupt that system.

One of the examples of the rectifying procedures called for by information technology experts is the necessity of using green information technology in all sectors of the country such as government and private sectors alike. The use of green information technology should be particularly adopted in cities that have high environmental pollution.

First/The Research Problem: The Egyptian environment faces many challenges; especially the challenge of the rise in ratios of ecological pollution, especially in the city of Cairo, the current Capital city as a result of many and diverse factors. One of the factors adversely affecting the environment in Egypt is the use of traditional information technology; such as the use of computer sets harmful to the environment and to the health of humans.

These environmental challenges will increase in coming years with the high rise in the use of traditional computer sets in both the government and private enterprises. Such sets are much cheaper than new environmentally efficient ones and they are offered for exportation by many computer savvy nations. Those nations plan to renew their computer sets with new more ecologically friendly ones and get rid of old computers.

This is multiplied through the high rise in the number of the population of Egypt annually; a matter that is certain to immensely increase pollution and hazardous wastes in the current capital city of Cairo and make the problem of the existence of such pollution and wastes much more aggravated in the future than it is now.

Second/The Research Importance: The research suggests that the following should be considered by the Higher Management of the New Administrative Capital:

- Establishing procedures for the use of computers and all relevant gadgets and equipment (Provided for Governmental Entities and the Private Sector) for such equipment to be compatible with Green Information Technology.

  Such as

- Endorsing such technology for all governmental entities working in the new administrative capital city.

- The above should be done with the necessity of establishing procedures for follow up and to oblige all to commit to such procedures.
- Refraining from giving work licenses for companies that don't use components compatible with the aforesaid environmentally safe technology.
  - Enforcement of all instructions made for the aforesaid technology standards during operation.
- Use of techniques that conserves energy resulting from computer sets and relevant equipment.
- Furthermore, such energy efficient techniques should be utilized with very environmentally competent methods such as the use of Smart Computing.
- Disposing of electronic gadgets and relevant equipment such as old computer sets in a very safe and environment friendly ways and recycling discarded equipment and gadgets.
- Use of lap tops, LCD monitors, and compound sets of printer, photocopier, fax, and scanner all merged in the same set and this should be done with the goal of lowering electricity consumption kept in mind.

Third/Research Specialized Goals: This study aims to achieve the following

- Delineating the benefits of using green information technology.
- Achieving continuous improvement of information technology utilized in the new Egyptian administrative capital city.
- Contributing to making the ecological system in Egypt a clean one.
- Contributing to the implementation of the Egypt 2020/2030 strategy.
- Merging the Egyptian system with the world system.

Fourth/: Green Information Technology (Green Computing):
4/1 What is the Green Information Technology?

This definition means preserving the environment through the use of technologies and behaviors that are considered environment-friendly [2]. Green IT could be otherwise defined as the manufacturing and use of computers and mobile phones, and other relevant equipment, programs, and tools that work in a green way. Working in a green way, in turn, means taking into consideration the preservation of a clean environment free of unbeneficial technological wastes. Some also define Green IT as inclusive of green programming. There are four pivotal components for the application of green information technology that are

A. First Main Point of focus: Green Design

This Point of focus concerns designing computer systems and their relevant equipment such as servers and printers and other gear so that they would be safe for the environment and not hazardous to human health.

B. Second Point of focus: Green Manufacturing

This Point of focus is concerned with the manufacturing of computers and gadgets of computers such as mobile phones, tablets, and other gadgets in a way safe for the environment and producing items that don't harm the ecological system so that such production is done in a way that doesn't have an unsafe impact on the health of humans.

The following two figures illustrate some of the examples of material components of computer hardware and software agreeing with the aforesaid green technology concept: Green (mouse, flash drive, connectors, and rams) (Figs. 1 and 2):

C. **Third Point of focus: Green Use**

This Point of focus is concerned with the use of computers and their relevant hardware equipment such as to laptops, routers, and other gadgets in an environment friendly way and in a way that doesn't endanger health.

D. **Fourth Point of focus: Green Disposal**

This Point of focus tackles the recycling of computers and their relevant equipment such as tablets, printers, and other gadgets so that such recycling and disposal of technology hardware don't affect the environment negatively and such disposal would not be adverse to human health.

4/2 The main activities the four points of focus revolve around are

The four types of activities that are the main points of concentration concerning Green Technology are

- Server visualization.
- Environmental Sustainability.
- Eco-Labeling of IT Products.
- Data Center Design, Layout, and location [3].

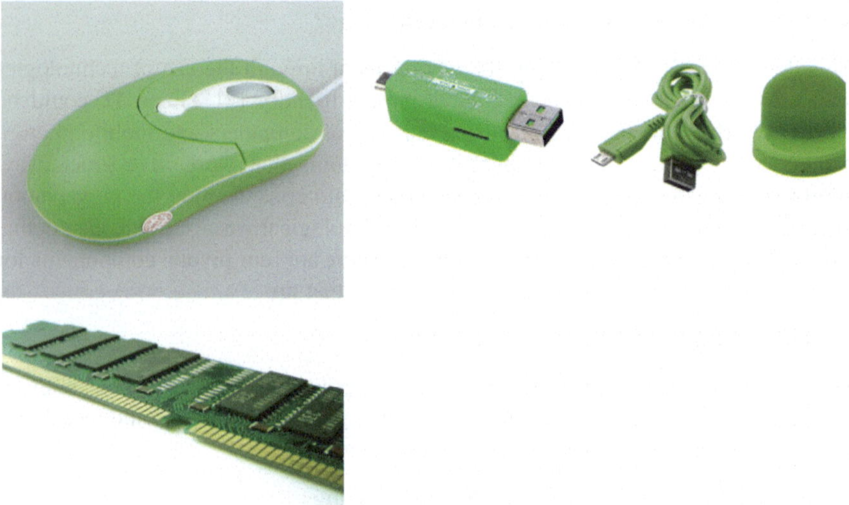

Fig. 1 Illustrates some of the computer hardware compatible with green technology. Green Printing Software

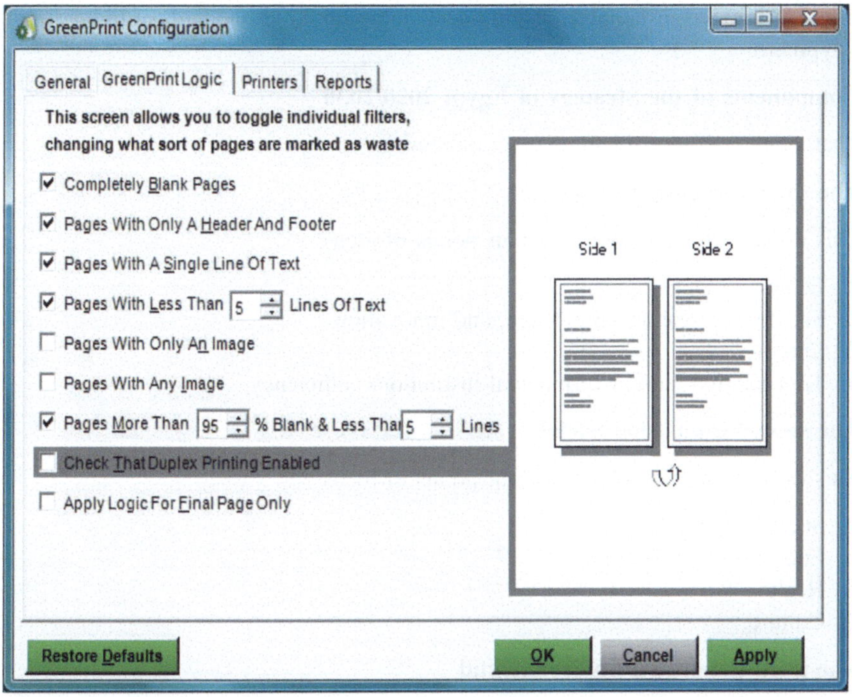

Fig. 2 Illustrates some of the examples of software compatible with Green Technology

- Energy Efficient Computing.
- Power Management.

Fifth/Egypt's New Administrative Capital City:

The new administrative capital of Egypt is a wide range project that was announced on 03rd of March, 2015, and it is currently under construction. It is located between the Greater Cairo Region and the Suez Canal Region, within a short distance of the regional circular road and the Cairo/Suez road. The new capital city will have within its area all the ministries, all the governmental authorities, a main recreational area, and an international airport. That city capital is designed to have many other activities and it will be initially situated on an area that equals 170 acres.

Sixth/The Strategy of Egypt 2020/2030:

The Vision of 2020/2030 will be a new Egypt that has a highly competitive, balanced and diverse economy that shall be driven by innovation and knowledge. The new Egypt will be based on justice, social integration, and contribution of all. That new intended version of Egypt will have a balanced and diverse ecological system that utilizes the genius of the geographic location of Egypt and its human resources. The strategy is based on the aforesaid two distinguished features of

Egypt to achieve sustainable development and to raise the standards of living for all Egyptians.

Components of the Strategy of Egypt 2020/2030

That strategy comprises three main dimensions

The First Dimension/Economic

This dimension is divided into four points of focus.

1. Economic Development.
2. Scientific research, knowledge, and innovation.
3. Energy.
4. Transparency and governmental institutions' efficiency.

The second dimension/Social

This dimension is divided into four points of focus

1. Social Justice.
2. Education and Training.
3. Health.
4. Culture.

The third dimension/Environmental

This dimension is divided into three points of focus

1. Environmental and social service.
2. Urban Development.
3. Higher Education.

First Point of Focus/The Environment and Social Service:

The vision of 2030 can be summed up in the idea that by 2030 the environmental dimension will have become an essential point of focus for all developmental and economic sectors in Egypt in a way that achieves the safety of natural resources and promotes the justice of their use and the best utilization of such resources and investing in them in a way that guarantees the rights of future generations in such resources, and works for the diversification of production facets and economic businesses and activities in a way that supports competitiveness, provision of new good jobs, elimination of poverty, and the achievement of social justice with the preservation of a clean, healthy, and safe environment for all people living in Egypt.

The Strategic Objectives of the aforesaid Points of Focus:

1. Achieving good Governance and maintaining the sustainability of natural resources for support of the economy, increasing competitiveness, and creation of new work opportunities.
2. Limiting pollution and integral management of wastes.

3. Maintenance of the equilibrium of ecological systems, biological diversity, and good governance of the above to achieve the sustainability of the aforesaid systems and diversity.
4. Egypt's fulfillment of its international and regional commitments towards ecological agreements and establishing mechanisms that guarantee that such commitments correspond closely with local policies.

Seventh/Contributions of this Study:

In the second decade of the twenty first century, all developed countries are competing for the use of advanced technology in their cities. The U.S. is one of those countries, and the U.S. proved that green information technology has a basic role in exterminating accumulated environmental problems in its cities. Green IT also helps in preserving the health of humans. Therefore, we suggest that Green technology should be used in Egypt's new administrative city; especially since that the city has not been opened yet, and therefore, there is ample time for the adoption of Green technologies in all facets of the new capital city.

The research suggests that the following should be considered by the Higher Management of the New Administrative Capital.

- Establishing procedures for the use of computers and all relevant gadgets and equipment (Provided for Governmental Entities and the Private Sector) for such equipment to be compatible with Green Information Technology.

  Such as

- Endorsing such technology for all governmental entities working in the new administrative capital city.
- The above should be done with the necessity of establishing procedures for follow up and to commit to such procedures.
- Refraining from giving work licenses to companies that don't use components compatible with the aforesaid environmentally safe technology.
  - Adherence to all instructions made for the aforesaid technology standards during operation.
- Use of techniques that conserves energy resulting from computer sets and relevant equipment.
- Furthermore, such energy efficient techniques should be utilized with very environmentally competent methods such as the use of Smart Computing.
- Disposing of electronic gadgets and relevant equipment such as old computer sets in a very safe and environment friendly ways and recycling discarded equipment and gadgets.
- Use of lap tops, LCD monitors, and compound sets of printer, photocopier, fax, and scanner all merged in the same set and this should be done with the goal of lowering electricity consumption kept in mind.

Eighth/Final Results:

Upon the foregoing, it is obvious that this research sheds light on many basic points relevant to Green IT and its reflection on the capability of implementing Green IT in Egypt, especially in the new administrative capital city. The points are as follows:

1. There is a definition of the concept of Green Information Technology and its types, tools, and how to implement it.
2. Green Information Technology could be used in the New Administrative Capital City in Egypt, and this could help prevent the occurrence of the same accumulative environmental problems prevalent in Cairo.
3. Green IT presents a unique opportunity for the Egyptian government to bury the wide gap between Egypt and advanced countries concerning the environment.
4. Such use of Green Technology could integrate Egypt within the global system and make Egypt increase its exports to other countries and that will make Egyptian trade merchandise more lucrative and attractive for advanced countries.

2 Conclusion

In 1972, the UN General Assembly announced 05th of June as the World Environment Day and this is on the anniversary of the UN Conference on the Human Environment, held in Stockholm in June 1972. On the same day, The UN Assembly sanctified the decision of the foundation of the UN Environment Program.

And for ten years; definitely since the UN Copenhagen Conference held during the period from 07th until the 18th of December, 2009 under the motto "Our planet is demanding out for us! Let us be United Nations in combating climate change,".

This motto shows the magnitude of environmental crimes committed against the universe and that motto sheds light on the subject of environmental changes and the resulting catastrophes suffered by the whole world since climate change came to constitute the main environmental challenge facing the world during that period.

And until the conference that will be held this year on 23rd of September, 2019 in the city of New York, U.S.A, and under the motto of "UN Climate Action," the UN kept exerting persistent efforts to conserve the environment against ecological change Currently.

All peoples of the world are now taking concrete measures to achieve environmental protection, and many conferences are held by world leaders for that purpose.

International companies are changing their methods of production so that all their products would be environment friendly. World renowned artists are contributing works urging nations to save the environment, and all are now demanded to contribute to help to protect the environment.

As to us, as specialists in information systems, we submit a paper to introduce Green Information technology to Egypt's New Administrative Capital City for the purpose of environment protection and to make the New City integrate with the new world environmental standards.

3 Future Work

1. The study suggests the use of Green Information Technology in the New Administrative Capital City of Egypt.
2. A further objective down the road comprises achieving the big Green shift, namely use of Green IT in other more traditional Egyptian cities.

References

1. Lynas, Mark. 2011. *The God Species: saving the planet in the age of humans*, National Geographic.
2. Murugesan, San and G.R. Gangadharan. 2012. *Harnessing Green IT*. Wiley.
3. Kenneth C. Laudon and Carol Guercio. 2018. *Management Information systems*. New Jersey: Prentice-Hall Inc.
4. Sites on the Internet about Green IT: http://www.moe.gov.bh/greenit.aspx, http://searchcio.techtarget.com/definition/green-IT-green-information-technology, https://ar.wikipedia.org.
5. Mahmoud, Hesham and Ahmed Attia. 2014. (New Quality Function Deployment Integrated Methodology for Design of Big Data E-Government System in Egypt), Big Data Conference, Harvard University, USA, Dec 14–16. https://www.iq.harvard.edu/files/iqss-harvard/files/conferenceprogram.pdf.
6. Mahmoud, Hesham. 2012. *Electronic Government*. Cairo: Administrative Professional Expertise Center–Bambak.
7. Mahmoud, Hesham. 2018. A proposal concerning satellite imagery within electronic government system in Egypt. In *Eighth international conference on ICT in our lives "Information Systems Serving the Community,"* Alexandria University, Egypt December, ISSN 2314-8942.
8. Mahmoud, Hesham. 2015. Transform the City of Sharm El Sheikh into an Electronic City–Application Model, Egyptian Computer Science "ECS" 2, 36–48. http://ecsjournal.org/JournalArticle.aspx?articleID=471.
9. Mahmoud, Hesham. 2016. Proposal for introducing NFC technology into the electronic government system in Egypt. *Egyptian Computer Science* "ECS" 1, 45–52. http://ecsjournal.org/JournalArticle.aspx?articleID=485.
10. Mahmoud, Hesham. 2016. A suggested novel application for the development of electronic government portal in Egypt. In *Sixth international conference on ICT in our lives "Information Systems Serving the Community,"* Alexandria University, Egypt. ISSN 2314-8942.
11. www.Egypt.gov.eg.
12. https://www.un.org/ar/climatechange/.
13. https://www.un.org/ar/climatechange/un-climate-summit-2019.shtml.

Load Balancing Enhanced Technique for Static Task Scheduling in Cloud Computing Environments

Ahmed H. El-Gamal, Reham R. Mostafa and Noha A. Hikal

Abstract Cloud computing has become an actual opportunity to supercomputing framework for creating proportional applications that associate enormous computational resources. Though, the difficulty acquired in constructing such parallel cloud mindful applications is higher than the ordinary parallel computing framework. It reports concerns for instance resource finding, heterogeneity, fault tolerance, and Tasks scheduling. Load balanced task scheduling in cloud computing is a vital issue, it is one of the NP-Completeness Problem, So Considered the focus of attention researchers in the cloud computing scope. The conventional Min-min and Max-min are a simple algorithm that creates a schedule that reduces the Makespan, regardless of load balancer, resource utilization, and resource usage effectively. To over-awed such restrictions, in this paper, a new load balance scheduling algorithm (SSLB) is designed and applied. This study focuses on static batch scheduling with the number of accessible resources and Received Tasks. Initially, it uses the Min-min or Max-min allocation map. SSLB utilizes the equal running time of the standard Min-min or Max-min. SSLB algorithm depends on the normal execution time rather than complete time as a determination premise. Experimental outcomes demonstrate the accessibility of load balance in cloud computing framework and overall reduce Makespan better than Min-min and original Max-min, also achieve of resource utilization and resource usage rate.

Keywords Cloud computing · Min-min algorithm · Max-min algorithm · Meta-Task scheduling · Load balancing · Resource utilization · Resource usage

---

A. H. El-Gamal (✉) · N. A. Hikal
IT Department, Faculty of Computers and Information Sciences, Mansoura University, Mansoura, Egypt
e-mail: eng.ahmed.sc@outlook.com

N. A. Hikal
e-mail: dr_nahikal@mans.edu.eg

R. R. Mostafa
IS Department, Faculty of Computers and Information Sciences, Mansoura University, Mansoura, Egypt
e-mail: reham_2006@mans.edu.eg

A. Z. Ghalwash et al. (eds.), *Internet of Things—Applications and Future*, Lecture Notes in Networks and Systems 114,
https://doi.org/10.1007/978-981-15-3075-3_28

1 Introduction

Numerous surveys record that Cloud Computing became one of the top ten technologies, Cloud computing presents a digital service delivery by the web through diverse packages which might be completed via PC frameworks in disseminated datacenters [1, 2]. Cloud system is getting more progressive currently. Cloud service providers are intended to offer services utilizing an enormous scale cloud framework with cost viability. Likewise, there are insufficient common enormous scaled applications, for example, social-networking–commerce, etc. These applications could assist in reducing the charges utilizing distributed computing. It offers infrastructure, platform, and programming that made accessible as membership-based services in a pay-as-you-go version to purchasers. Those services are known as

Software as a Service. SaaS is the supreme extensively recognized and broadly used a method of cloud computing. It offers every the cloudlet of a refined conventional application to numerous clienteles and frequently many clients, nonetheless over a Web program, not a "locally installed" application. Practically no code is working on the Users' local PC and the applications are regularly designed to accomplish particular cloudlets. SaaS removes client doubts about application servers, storage, application improvement, and associated, frequently concerns of IT.

Platform as a Service. PaaS offers virtual servers where customers can run present or new apps without having to worry about keeping OS, server hardware, load balancing, or computer capabilities. These sellers deliver APIs or development platforms to produce and operate cloud-based apps—e.g., via the Internet. Managed service providers that provide IT parts to observation devices and downstream apps such as e-mail virus scanning are included in this category on a recurrent basis.

Infrastructure as a Service. IaaS brings functions computing competency, normally as raw virtual servers, upon request that clients construct and accomplish. Cloud computing usually operates (but not always) grids or clusters or virtualized servers, networks, storage devices, and software in a multi-tenant architecture. IaaS is intended to enlarge or substitute the purposes of an entire datacenter. This reduces (time and cost) of the installation of capital equipment but does not reduce setup, integration, or administration costs and these activities need to be executed distantly.

In cloud computing, infrastructure, platform, and software are provided to the client as a service. As the number of clients increases, the task of allocating resources to each user each of their needs becomes hard. Here comes the notion of task scheduling. Task scheduling in cloud computing is a very operational and effectual concept as it allows the system to attain the greatest supply application and maximize throughput [3]. Many task scheduling algorithms have a hand in improved scheduling of tasks in the cloud and they afford optimization in the efficacy of numerous bounds, nevertheless, there is an opportunity to enhance [4]. Task scheduling is executed on the user side. Here the datacenter broker is the body where the scheduling is applied. Datacenter broker is the retailer between the client and the cloud service supplier. It accumulates all the statistics about the accessible sources, load on resources, and cost of communication, etc. Then as for the schedule, it assigns the cloudlets (tasks) to available Virtual Machines (VMs) [5]. A cloud computing environment consists of following main entities are Datacenter; which contain all hosts, hosts; which to host virtual machines (VM), Cloud Information Service which contain all information about hosts, virtual machine (VM) which run over hosts, tasks/cloudlet; which be execute on virtual machine (VM), Datacenter Broker; which to distribute all task for virtual machine according to specific calculation we will explain some it. Figure 1 displays the graphic demonstration of these objects and their working. A data center contains several hosts. Each host could have several VMs. Every host in datacenter takes the virtual machine on the base of their hardware specifications (bandwidth, memory, and processing power, etc.) [6]. Processing power is calculated in Million Instruction per Second (MIPS).

Cloud Information Service (CIS) is a source to supply cloud entities. Once a data center is created, it has to be registered on a CIS. When a user demands a service, cloudlet is (task) send to the datacenter broker that gathers the available data from the resources the CIS, and after that for the scheduling policy well-defined in the datacenter broker, it allocates the tasks for each VM [7]. In task scheduling, tasks are apportioned on two policies, whichever time-shared policy or Space shared policy [8]. In time-shared policy every VM has an alike amount of time, namely,

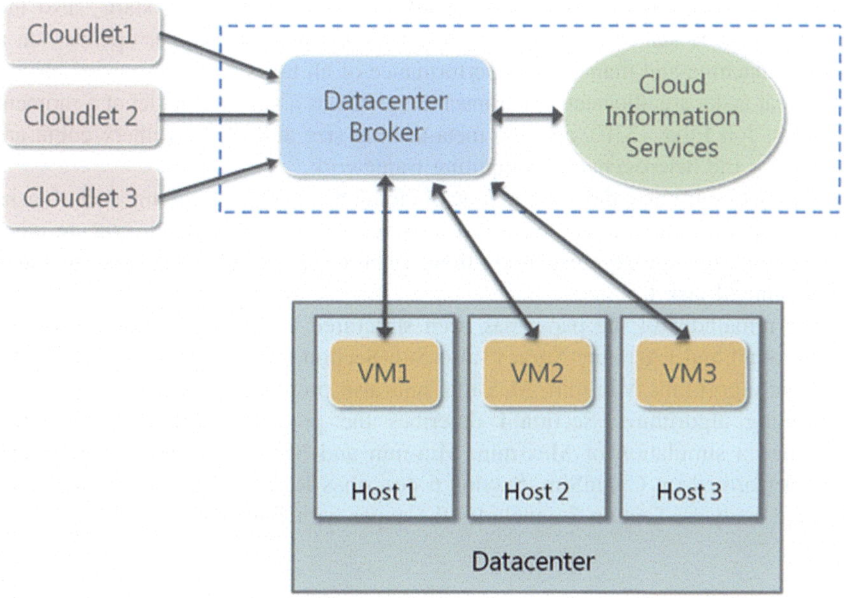

Fig. 1 Task allocation in cloud systems

tasks are performed in parallel. In space, the shared policy VM has an equal share of the processing element of the system [5].

2 Problem Definition

The problem of allocating tasks to resources is one of the main problems in large and small data centers where overloading only one resource and leaving others. This causes a stop to work on it and then stop the services provided by the resource almost permanently, which affects the efficiency of resources. So, must be found a solution to distribute loads appropriately, resource utilization, resource usage and at the same time taking into account the completion time of the task set, so the problem of allocating tasks to resources is one of the NP-completeness of the mapping issue.

Because of the NP-completeness of the mapping issue, the approaches developed are trying to find acceptable and cost-effective solutions in many special tasks. The suggested algorithms were created according to a series of assumptions in this research.

The applications to be performed consist of a set of individual tasks that are not interdependent, normally called meta-task, also the tasks have instruction set and associated data to be processed and the deadline or priority isn't considered, also the proposed algorithm considered a static scheduling algorithm. In turn, the expected execution time of the tasks on each machine in the computing environment. So, the matrix of execution time should be available before scheduling starts, also the process is to be statically mapped in batch mode, also the mapper operates on a distinct machine and manages the performance of all tasks on all resources, also in order that tasks are allocated, each machine executes a single cloudlet at a moment (Shortest Job First—SJF), also the meta-tasking size and the machinery count are known for the heterogeneous computing framework.

We try to minimize the waiting time of cloudlets through assigning a set of tasks having small execution time on a fast resource. On the other hand, execute other sets of tasks having large execution time concurrently on other resources to finish the submitted task list.

The remainder of the paper has been structured as follows: Sect. 3 describes Background and Literature Survey, also Sub Section (A) in 3 presents briefly two heuristic algorithms Min-min, Max-min and also presents literature for these task scheduling algorithms. Section 4 describes the proposed algorithm. Section 5 presented a simulation of Max-min, Min-min and SSLB algorithms which have been performed on CloudSim. Section 6 describes Results, Discussion, and comparative analysis. Section 7 concludes the paper with future scope.

3 Background and Literature Survey

This part introduces a description of some main terms used in task scheduling algorithms:

Meta-task (MT). It is a set of tasks, which will be mapped on available resources in cloud computing [9]. It can be shown as

$$\mathrm{MT} = \{T_1, T_2, T_3, \ldots, T_\mathrm{n}\}.$$

Makespan. It can be defined as the total time for executing a schedule (i.e., the number of cloudlets in meta-task) [10].

Minimum Execution Time (MET). Allocating a given cloudlet (T_i) to the best available resource (R_j) in order to reduce execution time for that particular task, without considering the machine's availability is known as MET [11].

Minimum Completion Time (MCT). Equation (1) refers to Assigning each given task to the available machine in order to complete the request as fast as viable, that could supply the quickest result, which means it considers the availability of the resource before assigning the task. Minimum completion time may be calculated by the addition of the execution time of undertaking task T_i, known as ET_{ij} and ready time or start time of resource R_j, known as r_j [12]. It can be represented as

$$\mathrm{MCT} = ET_{ij} + r_j \quad (1)$$

3.1 Task Scheduling Algorithms Based on Heuristic Approach

In the Cloud, a few cloudlets ought to be used to accomplish distinguishing intentions with the resources available. A scheduling algorithm is needed to perform an appropriate resource allocation map with the work scheduler. Different kinds of schedules are depending on other characteristics, for example, Static/Dynamic scheduling, Batch/Offline mode scheduling. Have been provided

Min-min Algorithm. This algorithm is based on the concept of marking a task having minimum completion time (MCT) first for execution on the resource, has the minimum time to complete (fastest resource). This algorithm contains two steps. First, the probable completion time of each task in the meta-task is calculated on every resource. Second, the cloudlet is designated and allocated to a consistent resource with the minimum expected completion time, and then the designated cloudlet is deleted from the meta-task. Replications of this method until all meta duties are mapped [13, 14]. The Min-min is a simple algorithm but it gives the fast result when the size of the task in meta-task is small as compared to large size task.

On the other hand, if huge size tasks overlay the number of smaller tasks, it gives short resource utilization and large Makespan because big size tasks should look forward to the completion of smaller tasks.

Max-min Algorithm. This algorithm overcomes the drawback of the Min-min algorithm (larger Makespan when numbers of large size tasks are greater than the small size task) [15]. The Max-min algorithm performs the exact steps as the Min-min algorithm but the main distinction comes in the second phase, where a task T_i is chosen which as the highest completion time instead of the lowest completion time as in Min-min and assigned to resource R_j, that gives the minimum completion time. Since it is named as the Max-min algorithm. This procedure is repeated until the meta-task gets empty or all the tasks are mapped [16, 17]. The main aim of the Max-min scheduling algorithm is to decrease the waiting time of large size cloudlets [18]. In this algorithm, small size tasks are concurrently performed with large size tasks, hence reducing the Makespan and supplying higher resource utilization.

Large numbers of algorithms for task scheduling are accessible to reduce the Makespan. All these algorithms attempt to discover resources to be assigned to the cloudlets that will reduce the total completion time of the cloudlets. Reducing the total completion time of the cloudlets does not mean reducing the real execution time of an individual cloudlet.

X. He et al. have introduced another algorithm taking into account the routine Min-min algorithm. The QoS guided Min-min algorithm, timetables assignments obliging high data transmission before the others. Subsequently, when the necessary bandwidth is very different for unique assignments, It gives preferred results over the Min-min algorithm. At whatever point the transmission capacity necessity of the majority of the task is very nearly the same [19].

K. Etminani et al. have a different algorithm using the algorithm Max-min and Min-min. One of these two algorithms is chosen according to the standard deviation of the ordinary time of performance for each of the assets. [13].

A. Afzal et al. have proposed another grid scheduling algorithm that minimizes the expense of the execution of work processes while guaranteeing that their related QoS requirements are fulfilled. The algorithm sees a network domain as a lining framework and timetables undertakings inside this framework. This algorithm is framework arranged and considers the execution cost. Thus, it is suitable for financial networks. Since the calculation is non-straight, as the span of the issue is gets huge the time it takes to get suitable planning gets to be long and unsuitable [20].

Mohammad I. Daoud and Nawwaf Kharma have proposed longest dynamic critical path (LDCP) calculation. The LDCP calculation has a period multifaceted nature of O $(m \times n)$ where m is processor number, and n is task number. In correlation, the time intricacy of the dynamic level scheduling (DLS) and heterogeneous earliest finish time (HEFT) calculations is O $(m \times n)$, individually, where e is the edge number. For dense DAGs with a comparative amount of edges to n, the time the many-sided quality of the HEFT calculation is O $(m \times n)$ [21].

T. Kokilavani et al. have proposed a Load Balanced Min-min (LBMM) algorithm. So, the LBMM algorithm decreases the Makespan and uses two stages to increase resource usage. The traditional Min-min algorithm is executed in the first phase and the tasks are rescheduled to use the unused resources effectively in the second phase. It uses the advantages of Max-min and Min-min algorithms and covers their disadvantages [22].

Lijun Cao, Xiyin Liu, Haiming Wang, and Zhongping Zhang (2014) OPT-Min-min suggested load balance programming algorithm. However, the results differ as the heavy load resources are defined differently. The OPT-Min-min algorithm has been intended to optimize the Min-min Scheduling Algorithm. The OPT-Min-min has been split into two phases. The Min-min algorithm has been used for preprogramming in stage one. Tasks on strongly charged resources are shifted to light charged resources in phase two. In OPT-Min-min, those resources are marked strongly loaded Make Min-min panels [23].

4 Proposed Algorithm

The Max-min assigns T_i to the R_j resource where the highest priority is for large tasks rather than for smaller ones. The Min-min algorithm assigns T_i tasks to the R_j resource where small tasks rather than large tasks have the greatest priority. The original Min-min and Max-min algorithms loss several main advantages as a load balancing for all submitted tasks in large cloud systems, between available resources in small cloud computing and small overall completion times. There are many objectives for the scheduling problem in the cloud computing environment. Our manipulated problem is concerned about minimizing the total waiting time for each task in the pending queue of tasks. In turn, we assign every task requires small execution time to the highest resource which can finish it correctly in the minimum completion time. However, large tasks are executed on the resources which complete them in higher completion time. The scheduling and assignment of the cloudlets have been done to execute the tasks concurrently for achieving the optimal minimum waiting time and higher resources utilization. Based on this, meta-tasks include homogeneous completion and execution tasks, we proposed a fundamental enhancement of static batch scheduling algorithms including Min-min and Max-min. It leads to an increase efficiency. The proposed enhancement increases the chance to execute tasks on resources simultaneously. Since the static batch algorithms are subject to load balance and small Makespan in a small distributed system, we focus on the static batch scheduling algorithm to derive the proposed algorithm that is called Static Scheduler Load Balancer (SSLB) algorithm. True, load balance improves efficiency in cloud systems but does not result in Makespan shortening. The algorithm calculates the expected completion time for each resource of submitted assignments. Then an initial scheduling map is achieved by using Min-min, Max-min, random, or other. Then the tasks are redistributed over the resources to guarantee load balance without increasing the Makespan. The algorithm focuses on reducing the overall Makespan which is the total completion time in a big distributed system, such as cloud system, performing duties simultaneously on accessible resources achieving load balance in a distributed system.

SSLB is a new scheduling algorithm classified as static scheduling algorithm. In which, we estimate the execution time for the submitted task before the scheduled start. From the estimated execution matrix, we can consider the Makespan that can be derived by other standard static scheduling algorithms such as Min-min and Max-min. The Makespan can be calculated as the total summation of the first minimum values of the execution matrix. We take the summation to the smallest count equals to the number of the submitted tasks. SSLB consumes allocation mechanism that can be a standard algorithm or random. So, SSLB can be considered an extension of the standard Min-min and Max-min if it uses Min-min or Max-min for getting the initial scheduling map. The initial map can be obtained in random, however, using a standard scheduling algorithm. This map is modified through subsequent steps for optimizing the final Makespan and increasing the load balance rate. SSLB redistributes the tasks which were associated with being executing on R_{j+1} to be associated with R_j. The redistribution considers the initial Makespan and the total number of tasks of R_j. SSLB scan the initial allocation map to reassign the tasks that can be executed on R_j instead of executed on R_{j+1} without increasing the last achieved Makespan. The redistribution of tasks is performed to increase the total number of tasks that are executed per resource and the total number were used from the resource set R.

Algorithm 1. Static Scheduler Load Balancer Algorithm

1) for all submitted tasks in meta-task; T_i
2) For all resources; R_j
3) $E_{ij} = \left|\left(\frac{T_i\,[MB]}{R_j\,[MBPS]} + \frac{T_i\,[MI]}{R_j\,[MIPS]}\right)\right|$
4) $E_k = sort(vector(E_{ij}))$
5) Set $makespan = \sum_{k=1}^{\|T\|} E_k$
6) Set $map =$ **schedule**(T, R)
7) For each R_j in all resources
8) for each T_i in $Tasks\ R_{j+1}$
9)   If r_j & $C_{ij} \leq makespan$ then:
10)   Remove T_i from $Tasks\ R_{j+1}$
11)   Add T_i to $Tasks\ R_j$
12)   Update r_j for selected R_j and R_{j+1}

The pseudo-code of the SSLB is shown in Algorithm 1. We denote the expected execution time matrix as E_{ij} that is defined as the required time by the resource R_j, to execute the task T_i, whereas r_j denotes the ready time of the resource R_j. Figure 2 is a flowchart of SSLB algorithm.

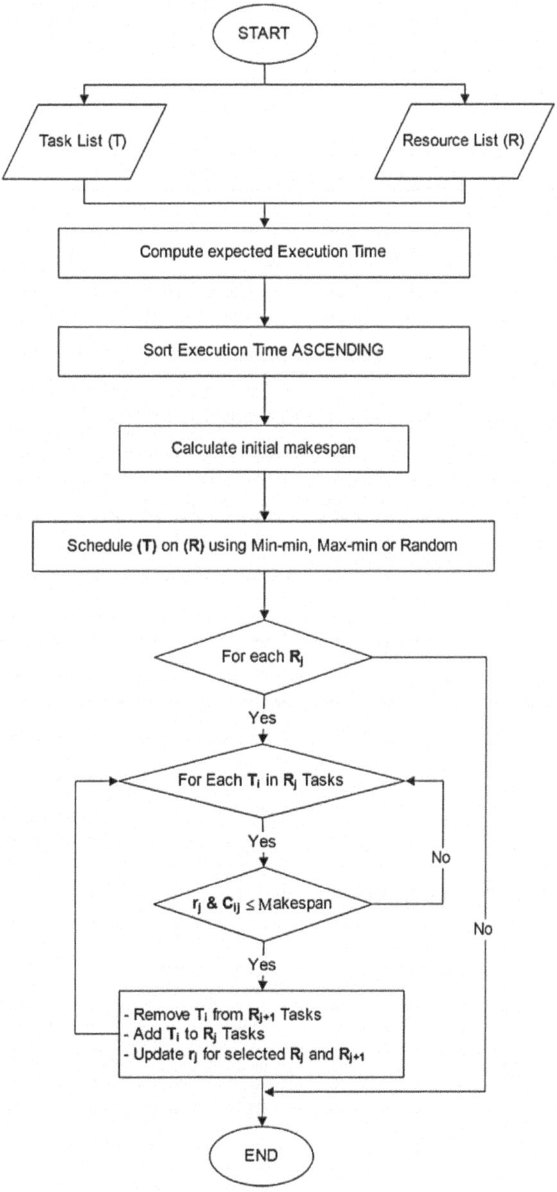

Fig. 2 Flowchart of SSLB algorithm using meta-task and resources set

In case, we initialize the allocation map using the Min-min or Max-min, the SSLB will have the same time complexity O (mn^2), similar to the original Min-min and Max-min, [13] where the number of resources is m and the number of tasks is n. Although having the same time of execution comparable to Max-min and Min-min, it generates better Makespan with more reliable scheduling schema. SSLB primarily promotes the load balance of accessible resources and allows the concurrent execution of submitted tasks with greater probability than the initial Min-min. The next chapter describes a straightforward example for exposing outcomes.

5 Demonstrative Example

Consider a cloud environment with two R_1 and R_2 resources and a meta-task group with four T_1, T_2, T_3, and T_4. The cloud scheduler should plan all assignments using the available R_1 and R_2 funds. The execution time of all duties is known prior to this issue. They can also be calculated if the number of instructions in each cloudlet and the computation rate of each resource is known. They are displayed (in a Second) in the table Expected Time to Compute (ETC). Table 1 reflects the amount of T_1 to T_4 instructions and information. Instructions volume is indicated in MI (Million Instructions) unit and data quantity is specified in MB (Megabytes).

The following Table 2 shows the processing speed and bandwidth of communication connections for all assets where processing speed is given in MIPS and bandwidth is specified in MBPS. Using the data given in Tables 1 and 2 to calculate the expected execution time of the tasks on each resource.

The following Table 3 represents the execution time of each task per available resource. In turn, as a result of having two resources for executing four tasks, we have eight values. For example, in order to execute task, T_1 on resource R_1, it will cost 18 s due to the following relation.

$$E_{11} = \left\lceil \left(\frac{88}{5} + \frac{20}{100} \right) \right\rceil = \lceil 17.8 \rceil = 18$$

In the same manner, the rest values of the execution time in the Table 3 are obtained.

Table 1 Tasks specifications

Task	Tasks specification	
	Instruction volume (MI)	Data volume (MB)
T_1	20	88
T_2	350	31
T_3	218	94
T_4	210	590

Table 2 Resources specifications

Resource	Resources specification	
	Processing speed (MIPS)	Bandwidth (MBPS)
R_1	100	5
R_2	300	15

Table 3 Execution time of the tasks on each resource

Task	Resources	
	R_1	R_2
T_1	18	6
T_2	10	4
T_3	21	7
T_4	121	40

Figure 3 represents the vector version, E_k of the estimated execution time matrix listed in Table 3. E represents the estimated execution time value from Table 3 while k is the subscript index of the value in the generated vector. The vector is sorted in ascending order. Every cell in the vector represents the execution time of a task on a resource. Here, we consider the minimal execution time whether the task is executed on whether resource. The vector is used to derive the expected Makespan that can be derived using the standard Min-min or Max-min. the expected Makespan can be derived using the constructed vector, E_k.

Figure 4 shows the allocation map of the assignments submitted to the available resources using the Min-min algorithm. Min-min choose the minimum completion time and so all tasks are scheduled to resource R_2 and resource R_1 remains idle. The Makespan produced by Min-min is 57 s. Figure 5 shows the allocation map of the submitted tasks on the available resources using the Max-min algorithm. In the same manner, Max-min produces Makespan equals to 57 sex. But the difference between Min-min and Max-min scheduling schema is that Min-min considers finishing small tasks using fast resources rather than executing large tasks on fast resources which is available in Max-min. SSLB algorithm generates an allocation map for the submitted tasks and available resources which is illustrated in Fig. 6. SSLB starts by initial allocation map and redistributes it in a manner that increases the load balance and optimizing the Makespan. SSLB generates less total completion time than Makespan generated by Min-min and Max-min. If we consider the initial schedule of SSLB is powered by Min-min algorithm. In turn, we will notice T_4 requires 40 s to be executed on R_2. While T_1 to T_3 consumes 49 s to be executed on R_1. So, SSLB iterates over the associated tasks of R_2 initially by Min-min for

$\boldsymbol{E_k}$	**4**	**6**	**7**	**10**	**18**	**21**	**40**	**121**

Fig. 3 Sorted vector of the execution time estimation for T on R

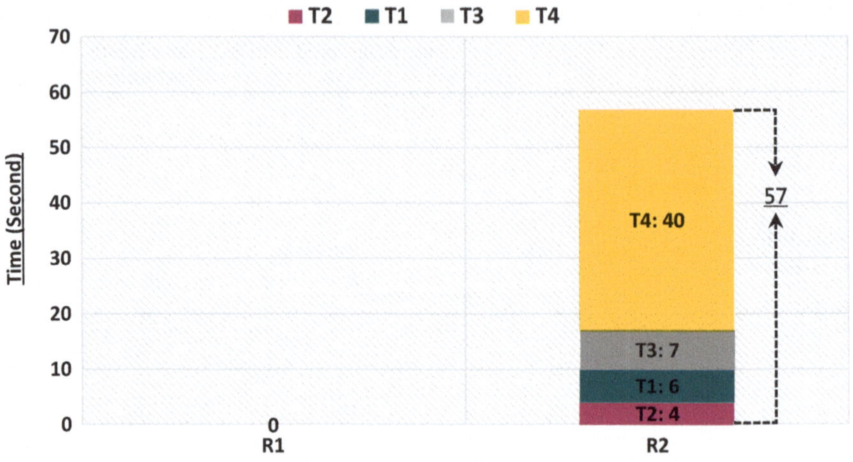

Fig. 4 Gantt chart for the standard Min-min algorithm

selecting task(s) for redistribution. The first iteration, SSLB moves T_2 from R_2 to be executed by R_1. Next iteration, SSLB reassign T_1 to be executed using R_1. In the third iteration, SSLB checks the ready time of R_1 and ensures the execution time of T_3 is less than last achieved Makespan; i.e., 57. Since the current ready time is 39 less than 57 and T_3 consumes 10 s on R_1, SSLB moves T_3 to be executed using R_1 instead of R_2. In the fourth iteration, SSLB failed to redistribute T_4 and reside assigned to be executed on R_2 with 40 s. In every iteration, SSLB updates ready time of the resources that were affected by redistributions. Hence, as a result of

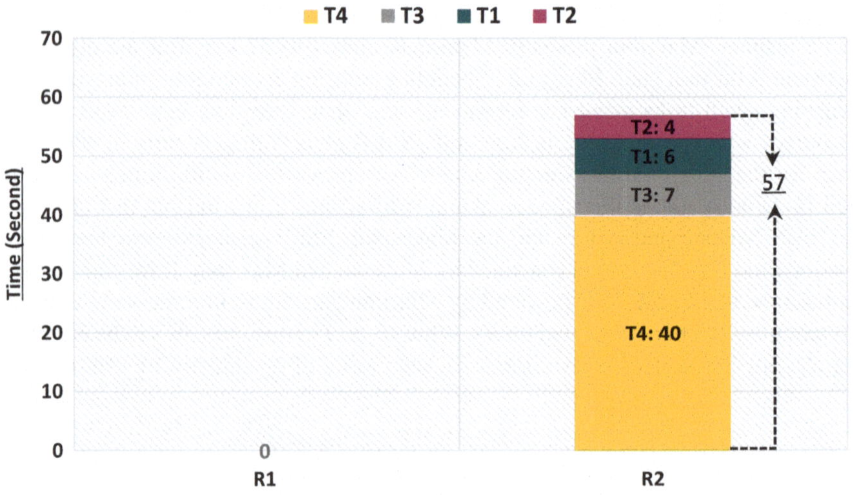

Fig. 5 Gantt chart for the standard Max-min algorithm

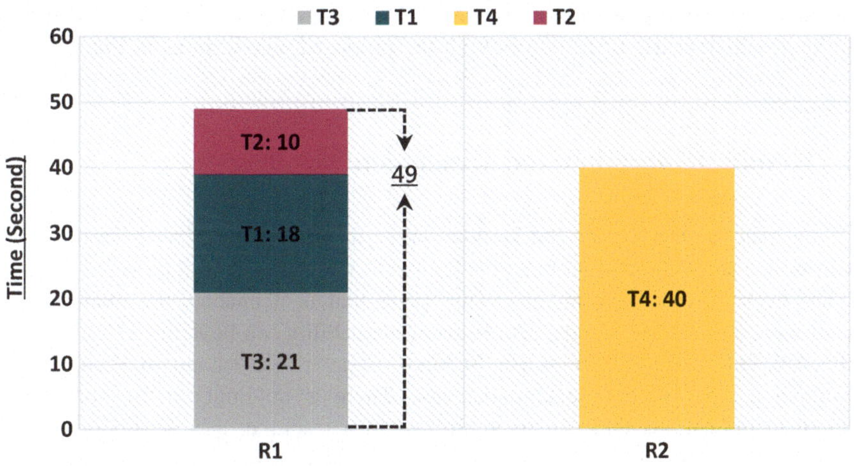

Fig. 6 Gantt chart for the standard SSLB algorithm

redistribution, the total Makespan is updated and become 49 instead of 57. Besides, the total number of used resources is increased from one resource to be two resources. SSLB algorithm generates mapping schemes with better total Makespan. Besides, the distribution of the tasks increases the load balance efficiently.

In Fig. 7 we will observe through comparison Appear the differences between Makespan in the previous algorithms, According to SSLB algorithm, the resources were used with the highest efficiency in mapping the submitted tasks based. In the

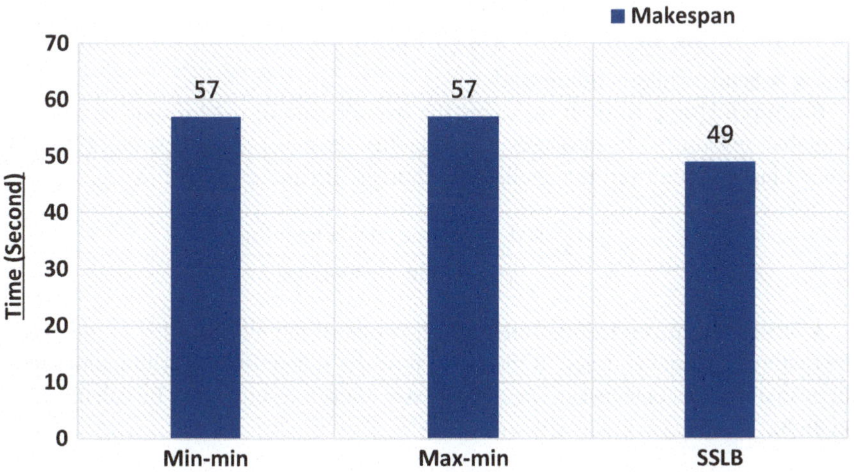

Fig. 7 Comparison of the Makespan of SSLB verse Min-min and Max-min

end, there is a clear difference in the rate of Makespan, the distribution of loads and the more efficient use of resources with the proposed algorithm as in Fig. 6.

6 Results Evaluation and Discussion

To evaluate and compare SSLB with other algorithms such as Max-min and Min-min, a simulation environment is known as CloudSim toolkit powered by Java 8 has been used. CloudSim has some good that is it enables the modeling of heterogeneous resource kinds, also Resource capability can be defined (in the form of MIPS (Million Instructions per Second) as per SPEC (Standard Performance Evaluation Corporation) benchmark), also There is no limit on the number of cloudlet applications that can be submitted to a resource, also It supports static simulation in addition to dynamic schedulers, and It was used in many types of research to assess outcomes such as [13, 20 and 21].

6.1 Performance Metrics

There are distinct performance metrics for assessing these algorithms to show the preeminence of SSLB compared to other algorithms. Some of these metrics are presented here.

Makespan. Makespan is a measure of the heterogeneous computing systems throughput, such as a cloud. It can be calculated as the following relationship:

$$\text{Makespan} = {}^{\max(\text{CT}_i)}_{t_i \in \text{MT}} \tag{2}$$

which is better while is minimized

Resource Usage Rate. It measures the overall ratio of used resources from the available resources. It measures the quantity level of used resources for the scheduling schema from the available resources. It can be defined as

$$R_u = \frac{\|\text{Used Resources in scheduling}\|}{\|\text{Available Resources}\|} \tag{3}$$

Average Resource Utilization Rate. It is one of the metrics used in measuring the quality of resource usage. It measures how the scheduling algorithm employees the resource to execute tasks. The average use of each resource can be calculated by relation:

$$ru_j = \frac{\sum_i (te_i - ts_i)}{T} \tag{4}$$

where t_i has been executed on m_j

Average Resource Utilization of Total Resources (ARUR). It is considered as the average of resource utilization rate for all resources available in scheduling problem. It is calculated using the following relationship:

$$ru = \frac{\sum_{j=1}^{m} ru_j}{m} \tag{5}$$

where ru within the range **0** to **1** and $\boldsymbol{m}$ is the total number of resources.

Load Balancing Level. It means square deviation of ru is described as

$$d = \sqrt{\frac{\sum_{j=i}^{m} (ru - ru_j)^2}{m}} \tag{6}$$

And the relative division of d over ru that determines the amount of load balancing is

$$\beta = \left(1 - \frac{d}{ru}\right) * 100\% \tag{7}$$

The highest and most effective amount of load balancing is obtained if d is equal to zero and β equals 1. Some scheduling algorithms will have better efficiency if d is near to 0 and β is closed to 1. It's the other metric used in.

Table 4 shows the various accessible resources of problem samples used for assessment [6]. In which each resource is defined using Million Instructions Per Seconds (MIPS) and Megabytes Per Seconds. MIPS represents the computation speed of the available resource whereas, MBPS specifies the bandwidth transfer rate.

Table 4 Problem Samples Specification of resources [6]

Problem of sample	Resource	MIPS	MBBS
P_1	R_1	50	100
	R_2	100	5
P_2	R_1	150	300
	R_2	300	15
P_2	R_1	300	300
	R_2	30	15
P_4	R_1	100	5
	R_2	300	15

Table 5 represents the different tasks submitted for each problem sample in meta-tasks. Each task is characterized using Million Instructions (MI) and Megabytes (MB). MI represents the total amount of instructions which is required by the task to be completed. MB denoted the total amount of data that is used to complete the task correctly. Table 5 represents four different problem sets that were used before in many different scheduling research studies in recent years. Based on the given details in Tables 4 and 5, we calculate the Makespan of each problem sample using distinct scheduling algorithms; Min-min, Max-min, and SSLB in Table 6.

Table 6 compares SSLB against the standard Min-min and Max-min using the stated four problem sets due to Makespan. In which, SSLB optimizes the Makespan rather than Min-min and Max-min. Initially, SSLB allocates tasks over resources under restriction of the expected Makespan. SSLB redistributes the tasks over the resources. The redistribution focuses on moving tasks into other available resources consuming less than or equal to the Makespan and increasing the resource usage, resource utilization, and load balance.

Figure 8 illustrates the Makespan comparison in a bar chart for figuring out the enhancement in the Makespan that is achieved by SSLB. We will observe through comparison the differences between Makespan in the previous algorithms, According to SSLB algorithm, the resources were used with the highest efficiency in mapping the submitted tasks based. In the end, there is a clear difference in the rate of Makespan, the distribution of loads and the more efficient use of resources with the proposed algorithm as in Fig. 6.

Table 7 measures the total number of the used resources in the final allocation map. Since SSLB is designed for increasing the load balance between the resources,

Table 5 Problem samples meta-tasks specification [6]

Problem of sample	Task	MI	MB
P_1	T_1	128	44
	T_2	69	62
	T_3	218	94
	T_4	21	59
P_2	T_1	256	88
	T_2	35	31
	T_3	327	96
	T_4	210	590
P_3	T_1	20	88
	T_2	350	31
	T_3	207	100
	T_4	21	50
P_4	T_1	20	88
	T_2	350	31
	T_3	218	94
	T_4	210	590

Table 6 Comparison of Makespan in seconds of problem samples using SSLB verse Min-min and Max-min

Problem of sample	Min-min	Max-min	SSLB
P_1	11	11	10
P_2	9	9	8
P_3	5	5	4
P_4	57	57	49

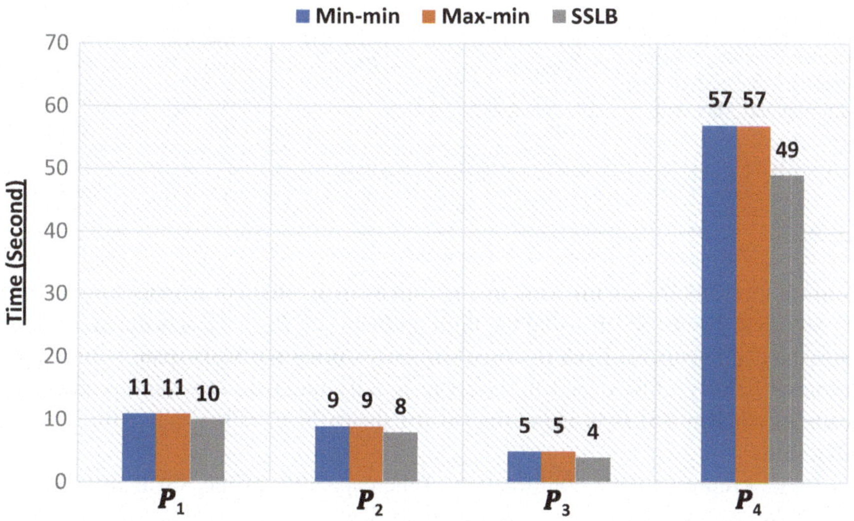

Fig. 8 Comparison evaluation of SSLB verse Min-min and Max-min according to Makespan

SSLB achieved higher rate in using a number of resources of the available resources. Also, resource usage is a quantity indicator that is affected positively to a higher number of using resources which is the main objective of SSLB. In turn, Table 7 concludes, SSLB is fully resourcing usages since it concerns using the total number of available resources Using the fourth performance metric stated in relation (4).

Table 8 compares ARUR. ARUR is a qualitative indicator which measures the ratio of employing the resources during the Makespan duration. ARUR measures the average resource usage rate of all available resources. The details of Table 8

Table 7 Comparison of resource usage ratio of problem samples using SSLB verse Min-min and Max-min

Problem of sample	Min-min (%)	Max-min (%)	SSLB (%)
P_1	50	50	100
P_2	50	50	100
P_3	50	50	100
P_4	50	50	100

Table 8 Comparison of ARUR of problem samples using SSLB verse Min-min and Max-min

Problem of sample	Min-min (%)	Max-min (%)	SSLB (%)
P_1	50	50	90
P_2	50	50	87.5
P_3	50	50	100
P_4	50	50	90.8

Table 9 Comparison of load balance level of problem samples using SSLB verse Min-min and Max-min

Problem of sample	Min-min (%)	Max-min (%)	SSLB (%)
P_1	0	0	88.9
P_2	0	0	85.71
P_3	0	0	100
P_4	0	0	89.9

confirm that SSLB uses the available resources in an efficient manner ranges from 87.8% up to 100%. SSLB using the third problem set, P_3, have scheduled the tasks over the two resources. Each resource was busy during the Makespan. In turn, each resource is fully utilized and the average is fully utilized. The same problem set leads to confirm SSLB is increasing load balance level rather than Min-min and Max-min.

Table 9 represents the confirmation of the capability of the SSLB in guarantee load balance level in the final allocation map. The capability ranges from 85.7% up to 100%. Min-min and Max-min failed to generate allocation map having resource usage rates so that, they failed to make a sense in load balance level. It is obvious that the SSLB algorithm can be considered as an efficient scheduling algorithm that can optimize Makespan, load balance, increase resource usage and has complexity O (mn^2).

Based on the results, our proposed SSLB always optimizes the Makespan and is smaller than the original Min-min and Max-min. SSLB scheduling also provides competitive task implementation using accessible resources and load balance in cloud computing.

7 Conclusions and Future Work

Min-min and Max-min algorithms are commonly applicable in small scale distributed systems. When the number of small tasks is less than the number of the large tasks in a meta-task, the Min-min algorithm schedules tasks, in which the Makespan of the system relatively depends on how many, executing small tasks concurrently with the large one. If can't execute tasks concurrently, Makespan becomes large. To overcome such limitations of the static batch algorithm, a new load balance scheduling algorithm, SSLB, is designed and implemented. Initially, it uses the Min-min or Max-min allocation map. This study concerned the static batch scheduling problem with the number of available resources and the submitted tasks. SSLB consumes the same running time of standard Min-min or Max-min. The study can be further extended by applying the proposed algorithm on actual cloud computing environment and considering many other factors such as scalability, availability, stability, and others. Also, in the future, we can improve the presented algorithm to consider priority and cost.

References

1. Bitam, S. 2012. Bees life algorithm for job scheduling in cloud computing. In *Proceedings of the third international conference on communications and information technology*, 186–191.
2. Armbrust, M., and A. Griffith. 2009. A berkley view of cloud computing. In *UCB/EECS, EECS department, UCB*.
3. Devipriya, S., and C. Ramesh. 2013. Improved Max-min heuristic model for task scheduling in the cloud. In *2013 international conference on green computing, communication and conservation of energy (ICGCE)*, 883–888.
4. Wu, X., M. Deng, R. Zhang, B. Zeng, and S. Zhou. 2013. A task scheduling algorithm based on QoS-driven in cloud computing. *Procedia Computer Science* 17: 1162–1169.
5. Priyadarsini, R.J., and L. Arockiam. 2014. Performance evaluation of min-min and max-min algorithms for job scheduling in the federated cloud. *International Journal of Computers and Applications* 99 (18): 47–54.
6. Patel, G., R. Mehta, and U. Bhoi. 2015. Enhanced load balanced min-min algorithm for static meta-task scheduling in cloud computing. *Procedia Computer Science* 57: 545–553.
7. Kaur, N., K. Kaur. 2015. Improved max-min scheduling algorithm. In *IOSR Journal of Computer Engineering (IOSR-JCE)* 17 (3): 42–49.
8. Kumar, S., and A. Mishra. 2015. Application of Min-min and Max-min algorithm for task scheduling in cloud environment under time shared and space shared vm models. *International Journal of Computing Academic Research (IJCAR)* 4 (6): 182–190.
9. Buchheim, C., and J. Kurtz. 2016. Min-max-min robustness: a new approach to combinatorial optimization under uncertainty based on multiple solutions. *Electronic Notes in Discrete Mathematics* 52: 45–52.
10. Sharma, N., S. Tyagi, and S. Atri. 2017. A survey on heuristic approach for task scheduling in cloud computing. *International Journal of Advanced Research in Computer Science* 8 (3).
11. Amalarethinam, D.G., and S. Kavitha. 2017. Priority-based performance improved algorithm for meta-task scheduling in a cloud environment. In *2017 2nd international conference on computing and communications technologies (ICCCT)*, 69–73.
12. Chen, H., F. Wang, N. Helian, and G. Akanmu. 2013. User-priority guided Min-Min scheduling algorithm for load balancing in cloud computing. In *2013 national conference on parallel computing technologies (PARCOMPTECH)*, 1–8.
13. Etminani, K., and M. Naghibzadeh. 2007. A min-min max-min selective algorithm for grid task scheduling. In *2007 3rd IEEE/IFIP international conference in central Asia on internet*, 1–7.
14. Anousha, S., and M. Ahmadi. 2013. An improved Min-Min task scheduling algorithm in grid computing. In *International conference on grid and pervasive computing*, 103–113.
15. Bhoi, U., P.N. Ramanuj, and others. 2013. Enhanced max-min task scheduling algorithm in cloud computing. *International Journal of Application or Innovation in Engineering and Management (IJAIEM)* 2 (4): 259–264.

16. Ming, G., and H. Li. 2012. An improved algorithm based on max-min for cloud task scheduling. In *Recent advances in computer science and information engineering*, 217–223. Springer.
17. Sharma, G., and P. Banga. 2013. Task aware switcher scheduling for batch mode mapping in a computational grid environment. *International Journal of Advanced Research in Computer Science and Software Engineering* 3.
18. Kanani, B., and B. Maniyar. 2015. Review on max-min task scheduling algorithm for cloud computing. *Journal of Emerging Technologies and Innovative Research* 2 (3).
19. He, X., X. Sun, and G. Von Laszewski. 2003. QoS guided min-min heuristic for grid task scheduling. *Journal of Computer Science and Technology* 18 (4): 442–451.
20. Afzal, A., A.S. McGough, and J. Darlington. 2008. Capacity planning and scheduling in Grid computing environments. *Future Generation Computer Systems* 24 (5): 404–414.
21. Daoud, M.I., and N. Kharma. 2008. A high-performance algorithm for static task scheduling in heterogeneous distributed computing systems. *Journal of Parallel and Distributed Computing* 68 (4): 399–409.
22. Kokilavani, T., D.D.G. Amalarethinam, and others. 2011. Load balanced min-min algorithm for static meta-task scheduling in grid computing. *International Journal of Computer Applications* 20 (2): 43–49.
23. Cao, L., X. Liu, H. Wang, and Z. Zhang. 2014. OPT-min-min scheduling algorithm of grid resources. *JSW* 9 (7): 1868–1875.

Comparative Study of Big Data Heterogeneity Solutions

Heba M. Sabri, Ahmad M. Gamal El-Din, Abeer A. Amer and M. B. Senousy

Abstract Recently, increasingly huge volumes of data are generated from a variety of sources. Existing data processing technologies are not suitable to cope with the huge amounts of generated data. Big data means enormous amounts of data, such large that it is difficult to collect, store, manage, analyze, predict, visualize, and model the data, which cannot be processed by conventional tools. Heterogeneity is one of the major features of big data and heterogeneous data result in problems in data integration and big data analytics. Many algorithms were defined in the analysis of large dataset. This paper addresses some different algorithms that handle the problem of heterogeneity and compare between them which can help the researchers to choose the best one. The objective of this paper is to perform a comparative analysis between four algorithms on the basis of different factors.

Keywords Big data · Data heterogeneity · Heterogeneity algorithm

1 Introduction

Heterogeneous data includes various types which are often generated from Internet of Things (IoT) such as image, audio, and video. These types of data are usually unstructured and semi-structured data, so there are challenges of dealing with heterogeneous data. The following are displaying the big data heterogeneity problem, big data heterogeneity types, and big data heterogeneity degrees.

H. M. Sabri (✉) · A. M. Gamal El-Din · A. A. Amer · M. B. Senousy
Sadat Academy for Management Sciences, Cairo, Egypt
e-mail: heba.ict@gmail.com

A. M. Gamal El-Din
e-mail: ahmedamado91@gmail.com

A. A. Amer
e-mail: abamer_2000@yahoo.com

M. B. Senousy
e-mail: badr_senousy_arcoit@yahoo.com

A. Z. Ghalwash et al. (eds.), *Internet of Things—Applications and Future*,
Lecture Notes in Networks and Systems 114,
https://doi.org/10.1007/978-981-15-3075-3_29

1.1 Big Data Heterogeneity Problem

One of the most significant problems that is faced by business organizations and researchers is the heterogeneity of big data. Generally, heterogeneity is a characteristic of all types of systems [1]. According to the point of view of data warehousing, data heterogeneity is defined as "data coming from diverse data sources and provided to the user with a unified interface" [2]. From another perspective, it is defined as "any data with high variability of data types and formats" [3]. So, from the previous definitions, big data heterogeneity can be defined as "any massive data gathered from diverse data sources with high variety of data types and formats."

The data created by users are heterogeneous in nature while data analysis algorithms presume homogeneous data for improved processing and analysis. In addition to that, it is impractical to transform all unstructured data to structured data and, therefore, data must be initially organized before analysis [4].

Data usually comes in four basic types such as structured, semi-structured, unstructured, and quasi-structured data. The structured data is well organized and can be easily accessed, reserved, restored, and analyzed. Semi-structured data is a type of data with no rigid data model, e.g., XML, RSS. Unstructured data cannot be stored in a conventional database, unlike the structured data. It is estimated to be 80–90% of the world data. Quasi-structured data is almost identical to the unstructured data. It is inconsistent data formats which can be structured by doing extra effort, tools, techniques, and time [5].

A huge set of heterogeneous data contains different types of information such as image, audio, and video files. Also, it takes account of Internet search indexing, genetic information-related datasets, medical data, social network-oriented data, and log records that have been collected by sensor nodes. The previously mentioned different types of data are generally unstructured and semi-structured data and it is unviable to process and store the data using the current relational databases [6]. Besides that, heterogeneous data, like streaming multimedia data, is constantly generated at a high pace and from time to time has to be handled in a real-time fashion [7].

The solution of the big data heterogeneity problem is to build a unified data model [8].

1.2 Big Data Heterogeneity Types

Heterogeneous data are any data with high variability of data types and formats. They are possibly ambiguous and low quality due to missing values and high data redundancy. Because of the variety of data acquisition devices, the acquired data are also different in types with heterogeneity. Second, they are at a large scale. The following are the types of data heterogeneity and shown in Fig. 1:

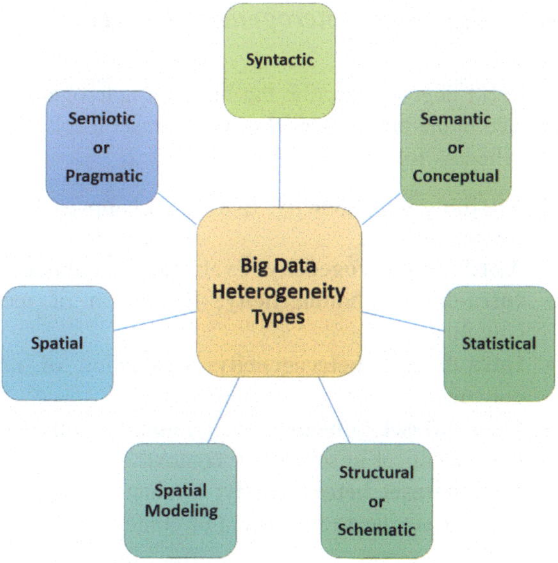

Fig. 1 Big data heterogeneity types

- **Syntactic heterogeneity**: It denotes variety in data types, file formats, data encoding, data model, and similar [9].
- **Semantic heterogeneity**: It is also known as conceptual heterogeneity or logical mismatch and denotes variations in meanings and interpretations [9].
- **Statistical heterogeneity**: It denotes variations in statistical features among the various parts of an overall dataset [9].
- **Structural or schematic heterogeneity**: It denotes identical data model (e.g., relational) while the schemas are different, including the type of an attribute (for example, °F versus °C) [10, 15].
- **Spatial modeling heterogeneity**: In spatial modeling heterogeneity, identical geographical phenomena may be modeled differently, for example, street as a line in one dataset, and as a polygon in another dataset [10].
- **Spatial heterogeneity**: A distinct feature of spatial data is that data allocation is not regularly the same in the whole study area [11].
- **Lexical heterogeneity**: It happens when required rows have identical structure through (various) databases, but various representations are utilized to denote the same real-world object [2].
- **Terminological heterogeneity**: It refers to differences in names when denoting the same entities from diverse data sources [3, 16].
- **Semiotic heterogeneity**: It is also known as pragmatic heterogeneity and denotes various interpretation of entities by people [3].

1.3 Big Data Heterogeneity Degrees

Different degrees describe the big data heterogeneity depend on whether the difference in the attribute level of that dataset levels. The degrees ofheterogeneity are classified as follows and shown in Fig. 2:

- **Attribute homogeneity**: Evaluation happens at column level for the same data formatting.
- **Attribute heterogeneity**: Evaluation of columns with formatting variabilities.
- **Intra-dataset homogeneity**: Evaluation of merging several identical data columns.
- **Intra-dataset heterogeneity**: Evaluation of merging several various data columns.
- **Inter-dataset homogeneity**: Evaluation considering several identical datasets (e.g., identical in schema or context).
- **Inter-dataset heterogeneity**: Evaluation considering several various datasets (e.g., dissimilar to in schema or context) [12].

The rest of the paper is structured as follows: Sect. 2 after introduction presents the related work, Sect. 3 presents the comparison of the different techniques of handling heterogeneity, Sect. 4 discusses the comparison of the algorithms, and Sect. 5 provides the concluding remarks.

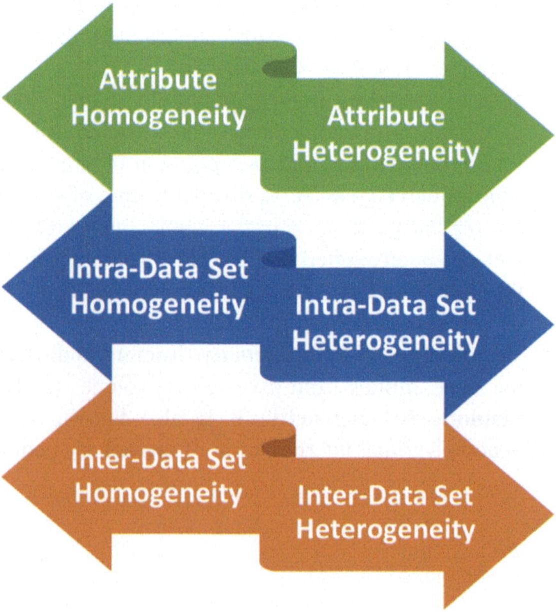

Fig. 2 Big data heterogeneity degrees

2 Related Work

There are a number of significant research manuscripts that discuss handling the heterogeneity of data, and the study of Gaffar et al. presented a modern conception through adding structure to highly unstructured heterogeneous data which improved the whole process of reserving and restoring the data. Human–Computer Interaction (HCI) patterns were used as a case study to introduce their proof of concept. The system depended on a primary XML database and the process helped feeding the patterns into the database and rewrote them in a semantical reusable format. The achieved results were creation of a process that assists in the accommodation and distribution of patterns, and the implementation of a system that when applying a process will aid beginner users get patterns in an organized manner and capable to utilize them effectively in their design artifacts [13].

Also, the study of Kumar et al. aimed to structure the unstructured data in a structured form by applying context and usage patterns of data items to elicit separate segments of data and identify relationships among the elicited data. The converted data and relationships put in a structured schema can aid to assure good performance. Moreover, a graph clustering algorithm, Markov's algorithm was used to cluster relevant data items. Two modules were built that are interfaced to the current Hadoop application. The data context analysis module is for analyzing the data context, while the data usage pattern analysis module is for analyzing the data usage patterns. The result of the two modules was merged to structure the data. A group of clusters was obtained according to the outcomes of the usage pattern analysis module. The result proved that the approach was capable to structure data which is making the data can use by the user [14].

In addition, the study of Kang et al. constructed a big data model which relied on ontology through using semantic web technology, and also proposed an ontology-based semantic model and ontology-based key/value storage model. The data is reserved in a key/value model which could be managed instantly by HBase NoSQL databases that is simple for data updating dynamically and fulfill the requirements of intense parallel data processing in a big data environment. The results were construction of a key/value storage model on the basis of ontology to resolve the issue of heterogeneous data storage and establishing an initial model system on the basis of HBase implementation and verification [8].

Furthermore, the study of Sindhu and Hegde proposed a framework for processing different big data which can be in structured, semi-structured, and unstructured formats, which is an efficient algorithm for managing data heterogeneity in three basic structures. In other words, transformation of centralized structured data to distributed structured data, unstructured to a structured data, and semi-structured to a structured data. The proposed approach uses the idea of text mining for transforming the semi-structured and unstructured data into structured data. The efficacy of the proposed framework was evaluated using processing time basically as it can eminently present the total time that could be used working on huge data. The achieved results from the study conclude that proposed framework is completely able to perform competently the transformation of centralized to distributed structured data in shorter processing time. The main contribution of the proposed research is that the computation of the proposed algorithms is so competent, and shorter processing time is consumed during the process of data transformation in HBase [6].

3 Comparison of Algorithms that Handle Heterogeneity

Here the researchers have compared the four clustering algorithms on the basis of different factors associated with them (Table 1).

4 Discussion and Results

Gaffar et al. created a process by using HCI patterns to assist in the accommodation and distribution of patterns, and the implementation of a system that when applying a process will aid beginner users to get patterns in an organized manner and capable to utilize them effectively in their design artifacts as the system depends on a primary XML database and the process helps in feeding the patterns into the database and to rewrite them in a semantical reusable format.

Kumar et al. use data context and usage patterns, in which the user requirement for big data examination is excluded for unstructured data which in turn saves a lot of time to the user. The outcomes proved that the approach is capable of structuring data, and therefore makes the data better usable by the user.

Table 1 Comparison between different techniques of handling heterogeneity

Research title	Structuring heterogeneous big data for scalability and accuracy [13]	Efficient structuring of data in big data [14]	Research on construction methods of big data semantic model [8]	A framework to handle data heterogeneity contextual to medical big data [6]
Authors	Ashraf Gaffar, Eman Monir Darwish and Abdessamad Tridane	Ashwin Kumar T. K., Hong Liu, and Johnson P. Thomas	Li Kang, Li Yi and Liu Dong	Sindhu C. S. and Nagaratna P. Hegde
Research year	2014	2014	2014	2015
Publisher	IJDIWC	IEEE	WCE	IEEE
Research area	General	Social media	Commerce	Health Care

(continued)

Table 1 (continued)

Algorithms	HCI (human–computer interaction) patterns	Context patterns, usage patterns and Markov's clustering algorithm	Construction of big data model relied on ontology by using semantic web technology, and proposed an ontology-based semantic model and ontology-based key/value storage model	A proposed framework for processing different big data which can be in structured, semi-structured, and unstructured. An efficient algorithm for managing data heterogeneity in three basic structures. In other words, transformation of centralized structured data to distributed structured data, unstructured to a structured data, and semi-structured to a structured data. The proposed approach uses the idea of text mining for transforming the semi-structured and unstructured data into structured data
Inputs	Images	Text	Text	Text
Research title	Structuring heterogeneous big data for scalability and accuracy [13]	Efficient structuring of data in big data [14]	Research on construction methods of big data semantic model [8]	A framework to handle data heterogeneity contextual to medical big data [6]
Tools	ORACLE RDBMS with support to XMLDB, ZOPE and Apache-Xindice XMLDB	Hadoop	Hadoop and H-Base.10 working nodes, each node of the system is basically configured as a dual-core CPU@3.2 GHz, 4G RAM, 500G hard disk space	Hadoop and H-Base
Future work	–	Expand the approach to fulfill other needs of big data like velocity	Solve the mapping issue in integration from existing database systems to big data management system	–

Kang et al. proposed a method in which the achieved outcomes were constructing a key/value storage model which relied on ontology to resolve the issue of heterogeneous data storage and establishes an initial model system on the basis of HBase implementation and verification.

Sindhu and Hegde proposed a method in which the efficacy of the proposed framework was evaluated using processing time basically as it can eminently present the total time that could be used to working on huge data. The proposed algorithms are computationally efficient and take short processing time.

The most common software tools used were Hadoop and HBase, as Hadoop is from the most well-known tools used in big data processing and HBase is the Hadoop database that works on the top of Hadoop framework and Hadoop Distributed File System (HDFS).

5 Conclusion

Due to increase in the amount of data in different fields, it becomes difficult to handle the data, to find associations, patterns, and to analyze the large datasets. In this paper, we discussed different algorithms in different research areas such as social media, health care, and commerce.

The main goal of our paper is to make a survey of various algorithms dealing with the problem of heterogeneity in big data that handles a massive amount of data from different sources and improves overall performance of systems.

References

1. Jirkovský, V., M. Obitko, and V. Mařík. 2016. Understanding data heterogeneity in the context of cyber-physical systems integration. *IEEE Transactions on Industrial Informatics* 13 (2): 660–667.
2. Cuzzocrea, A. 2015. *Record linkage in data warehousing. Encyclopedia of information science and technology*, 3rd ed, 1958–1967. IGI Global.
3. Wang, L. 2017. Heterogeneous data and big data analytics. *Automatic Control and Information Sciences* 3 (1): 8–15.
4. Desai, P.V. 2018. A survey on big data applications and challenges. In *2018 second international conference on inventive communication and computational technologies (ICICCT)*. Coimbatore.
5. Manikandan, G., and S. Abirami. 2017. Big data layers and analytics: a survey. In *3rd international conference on computer & communication technologies (IC3T 2016)*. Vijayawada, Andhra Pradesh.
6. Sindhu, C.S., and N.P. Hegde. 2015. A framework to handle data heterogeneity contextual to medical big data. In *2015 IEEE international conference on computational intelligence and computing research (ICCIC)*. Madurai.
7. Zhang, Y., J. Ren, J. Liu, C. Xu, H. Guo, and Y. Liu. 2017. A survey on emerging computing paradigms for big data. *Chinese Journal of Electronics* 26 (1): 1–12.

8. Kang, L., L. Yi, and L. Dong. 2014. Research on construction methods of big data semantic model. In *Proceedings of the world congress on engineering (WCE 2014)*. London.
9. Heureux, A.L., K. Grolinger, H.F. Elyamany, and M.A. Capretz. 2017. Machine learning with big data: challenges and approaches. *IEEE Access* 5: 7776–7797.
10. Balasubramani, B.S., and I.F. Cruz. 2019. *Spatial data integration. Encyclopedia of big data technologies*, 1541. Springer.
11. Jiang, Z., and S. Shekhar. 2017. *Overview of earth imagery classification. Spatial big data science classification techniques for earth observation imagery*. Springer.
12. Micic, N., D. Neagu, F. Campean, and E.H. Zadeh. 2017. Towards a data quality framework for heterogeneous data. In *2017 IEEE International Conference on Internet of Things (iThings) and IEEE Green Computing and Communications (GreenCom) and IEEE Cyber, Physical and Social Computing (CPSCom) and IEEE Smart Data (SmartData)*. Exeter.
13. Gaffar, A., E.M. Darwish, and A. Tridane. 2014. Structuring heterogeneous big data for scalability and accuracy. *International Journal of Digital Information and Wireless Communications (IJDIWC)* 4 (1): 10–23.
14. Kumar, T.K.A., H. Liu, and J.P. Thomas. 2014. Efficient structuring of data in big data. In *2014 international conference on data science & engineering (ICDSE)*. Kochi.
15. Salam, M.A., H.M. Zawbaa, E. Emary, K.K.A. Ghany, and B. Parv. 2016. A hybrid dragonfly algorithm with extreme learning machine for prediction. In *2016 international symposium on innovations in intelligent systems and applications (INISTA)*. Romania.
16. Anwar, A.S., K.K.A. Ghany, and H. Elmahdy. 2015. Human ear recognition using SIFT features. In *3rd world conference on complex systems*. Morocco.

Comparative Study: Different Techniques to Detect Depression Using Social Media

Nourane Mahdy, Dalia A. Magdi, Ahmed Dahroug and Mohammed Abo Rizka

Abstract Social media is one of the most influential means of communication in the present day. Youths and young adults use these sites to communicate with their peers, share information, reinvent their personalities, and showcase their social lives. Nowadays, not only teenagers but tens of millions of individuals use internet for the most part of their daily activities and share information, files, pictures, and video, as well as creating blogs to communicate their thoughts, personal experiences, and social ideals [1]. Therefore, the potential of social media of predicting and detecting, even the prior onset of major depressive disorder in online personas, is very high. In this paper, we discuss how the employment of different machine learning methods and techniques helps in detecting depression on social media [2].

Keywords Survey · Social media · Depression · Detecting depression · Comparative study

1 Introduction

Depression is a common and serious medical illness that negatively affects how you feel, the way you think, and how you act, it causes feelings of sadness and/or a loss of interest in activities once enjoyed. It causes severe symptoms that affect how you feel, think, and handle daily activities, such as sleeping, eating, or working, it can lead to a variety of emotional and physical problems and can decrease a person's

N. Mahdy (✉) · A. Dahroug · M. A. Rizka
Technology & Maritime Transport, Arab Academy for Science, Cairo, Egypt
e-mail: nourane-mahdy@hotmail.com

D. A. Magdi
Information System Department, French University in Egypt, Cairo, Egypt

D. A. Magdi
Computer and Information System Department, Sadat Academy for Management Sciences, Cairo, Egypt

A. Z. Ghalwash et al. (eds.), *Internet of Things—Applications and Future*, Lecture Notes in Networks and Systems 114,
https://doi.org/10.1007/978-981-15-3075-3_30

ability to function at work and home. To be diagnosed with depression, the symptoms must be present for at least two weeks (The American Psychiatric Association (APA)).

According to the Centers for Disease Control and Prevention (CDC), 7.6% of people over the age of 12 have depression in any 2-week period., and according to the World Health Organization (WHO), depression is the most common illness worldwide and the leading cause of disability. They estimate that 350 million people are affected by depression, globally. With the era changing, we, including depressed users, almost cannot survive without social media. Researchers thereby started to analyze the online behaviors of depressed users and develop computational models to predict the emergence of depression, even in cases where social media users are not yet aware that their health has changed [3].

These researches had found that Twitter covers a wide range of topics, and that Twitter microblogs are effective tools for analysis of mental health attitudes and can be a replacement or a complement for the traditional survey methods depending on the specifics of the research question [4]. By applying machine learning to the psychology area for detecting depressed users in social network services and utilizing vocabulary and man-made rules to calculate the depression inclination of each micro-blog [5].

2 Social Media

Social media by definition are websites and apps that allow consumers to generate and share content or engage in social networking. Social media has grown exponentially since 2004 and has not yet reached its peak of popularity [6]. There is no denying that platforms for social media are now a significant source of news and data. Each platform has a distinctive audience with popular social media platforms increasing in terms of size (Oberlo, 2019). The worldwide use of social media is increasing. It is undoubtedly one of the most common internet operations engaged by customers. The 2019 social media statistics indicate that there are 3.2 billion users globally of social media, and this amount is only increasing. That is roughly 42% of the present population (Emarsys, 2019). Emarketer has broken down the use of social media and the findings are exciting, to say the least, by generation. In order to break it down, 90.4% of Millennials, 77.5% of Generation X, and 48.2% of Baby Boomers are active users of social media (Emarketer, 2019).

One of the reasons for this elevated use of social media is that mobile user possibilities are constantly improving which makes access to social media easier by the day, regardless of where you are, especially that most social media networks are accessible as mobile applications or optimized for mobile browsing, making it simpler for users on the go to access their favorite sites. We are all slowly becoming addicts to social media in this day and age. Whether it's scrolling down our limitless Facebook feeds, or posting the ideal brunch picture on Instagram before eating, social media has become unavoidable. These social media statistics show that an average of 2 h and 22 min is spent on social networks and messaging per individual per day (Globalwebindex, 2018)

Brands are riding the social media marketing wave. 73% of marketers think that their social media advertising attempts were "slightly efficient" or "very efficient" for their company (Buffer, 2019). In their advertising approach, brands continue to include social media—and for all the correct reasons. Whether it's marketing influencer or story advertisements, they're attempting everything [7]. Social media enables brands to access affordable marketing, communicate with their audience, and create brand loyalty. But the precise effect of social media is hard to assess, as each social media platform measures activity differently.

3 Depression

Depression is more than a sensation of sadness. From time to time, everyone feels angry or unmotivated, but depression is more severe. It is a mood disorder with extended emotions of sorrow and loss of interest in everyday activities. If for a period of at least two weeks these symptoms persist, it is classified as a depressive episode. It is a significant suicide risk factor. The profound sorrow and desperation that accompanies depression can make suicide feel like the only way to escape pain (psycom, 2019).

Depression affects people, regardless of their background, from all walks of life. It can also impact individuals of all ages. There are several possible causes of depression. They can range from biological to circumstantial such as family history, childhood traumas, and medical conditions [8].

Unfortunately, there is still a stigma surrounding mental health problems, and some individuals see illnesses like depression as a weakness. But mental health problems are not always preventable since anyone can develop certain physical health problems [9].

Gaining a deeper understanding of depression can assist in starting the recovery journey. Taking time to know more about the causes and symptoms of depression will help you considerably when considering therapy techniques.

4 Depression and Social Media

Social networks have become widely-used and popular mediums for information dissemination, as well as facilitators of social interactions, and individuals are progressively utilizing social media platforms, such as Twitter and Facebook, to impart their considerations and insights with their contacts [10]. The postings on these destinations are made in a naturalistic setting throughout the daily activities and happenings, these contributions and activities provide valuable insight into individual behavior, experiences, opinions, and interests.

Social media, therefore, offers a means to capture behavioral characteristics appropriate to the thinking, mood, interaction, activities, and socialization of an individual [11]. The emotion and language used in posts on social media may show emotions of worthlessness, guilt, helplessness, and self-hatred that characterize major depression [12]. In addition, sufferers of depression often withdraw from social and activity circumstances, such changes in activity could be significant with changes in social media activity. We are pursuing the hypothesis that changes in language, activity, and social ties can be used together to build statistical models to detect and even predict MDD in a fine-grained way, including ways to complement and extend traditional diagnostic methods [13].

Detection of mental illness in social media can be regarded as a complex task, primarily because of the complicated nature of mental disorders [14]. This study area has begun to develop in the latest years with the ongoing rise in popularity of social media platforms that have become an essential component of the lives of people.

This close relationship between social media platforms and their users has resulted in these platforms reflecting on many levels the private lives of consumers. Researchers are provided with a wealth of data about one's lives in such a setting. In addition to the level of difficulty in defining mental illnesses through social media platforms, it was not commonly accepted to adopt supervised machine learning methods such as deep neural networks due to the problems in acquiring adequate quantities of annotated training data.

For these reasons, we are attempting to define the most efficient profound neural network architecture among a few chosen architectures that have been effectively used in the processing of natural language assignments. With restricted unstructured text information obtained from the Twitter social media platform, the selected architectures are used to identify users with indications of mental illness (depression in our situation) [15].

5 Related Studies

Machine learning is a trending concept in which the program learns to classify or make predictions based on patterns found in the training dataset, because of its enormous ability to model almost any data and make predictions at a superhuman pace, machine learning found itself employed in every single place possible [16]. That is why it plays a fundamental role in extracting correlation patterns between depression and a variety of user's data captured from social media.

in this section, we present different machine learning frameworks that have been proposed to accomplish the given tasks.

Author	Topic	Aim	Method	Result
Yu-Tseng Wang	A neural network approach to early risk detection of depression and anorexia on the social media text	An approach to early risk detection of depression and anorexia on social media in CLEF eRisk2018	Combine TF-IDF information and convolutional neural networks (CNNs)	Their models achieve ERDE 5 of 10.81%, ERDE 50 of 9.22%, and F-score of 0.37 in depression detection and ERDE 5 of 13.65%, ERDE 50 of 11.14%, and F-score of 0.67 in anorexia detection
Zunaira Jamil	Monitoring tweets for depression to detect at-risk users	The goal of this thesis is to apply Natural Language Processing and Machine Learning techniques to build a system that given a set of tweets from a user1 can identify at-risk tweets and hence at-risk users. This leads to a secondary goal of the thesis, which is to identify relevant tweets for analysis from a large amount of Twitter data	Support Vector Machine (SVM)	A trained user-level classifier Support Vector Machine (SVM) that can detect at-risk users with a recall of 0.8750 and a precision of 0.7778. A tweet-level classifier uses SVM and performs with a recall of 0.8020 and a precision of 0.1237
Moin Nadeem et al.	Identifying depression on twitter	Establish a method by which recognition of depression through analysis of large-scale records of user's linguistic history in social media and posits a new methodology for constructing a classifier by treating social as a text-classification problem, rather than a behavioral one on social media platforms	Crowdsourcing and BOW	86% classification accuracy the Naïve Bayes' algorithm yielded an A-grade ROC AUC score of 0.94, a precision score of 0.82, and an 86% accuracy; a Bag-of-Words approach was determined to be a useful feature set, and they determined bigrams to present no significant advantage over a unigram-based approach

(continued)

(continued)

Author	Topic	Aim	Method	Result
Md. Rafiqul Islam et al.	Depression detection from social network data using machine learning techniques	Analyze social network data for the user's feelings and sentiments to investigate their moods and attitudes when they are communicating via these online tools	Decision tree, k-Nearest Neighbor, Support Vector Machine and LIWC	After analyzing 7146 depressive indicatives Facebook comments to identify the most influential time. They got 54.77% depressive indicative Facebook users communicate with their friends from midnight to midday and 45.22% from midday to midnight. To prove that all of the classifiers results are almost between 60 and 80%
Munmun De Choudhury et al.	Predicting depression via social media	Build an SVM classifier that can predict, ahead of the reported onset of depression of an individual, his/her likelihood of depression	Crowdsourcing and Support Vector Machine	The classifier yielded promising results with 70% classification accuracy
Ahmed Hussein et al.	Deep learning for depression detection of twitter users	They present a system that identifies users at the risk of depression from their social media posts	Convolutional Neural Networks (CNNs)	CNN-based models perform better than RNN-based models
Sayanta Paul et al.	Early detection of signs of anorexia and depression over social media using effective machine learning frameworks	The paper presents different machine learning techniques and analyze their performance for early risk prediction of anorexia or depression. The aim of this challenge is to detect signs of such diseases from the posts or comments of individuals over social media	Bag-of-words ada boost, random forest, logistic regression and support vector machine meta map Glove and Fast text word embeddings	

(continued)

(continued)

<table>
<tr><th>Author</th><th>Topic</th><th>Aim</th><th>Method</th><th>Result</th></tr>
<tr><td></td><td></td><td></td><td></td><td>
<table>
<tr><th>Methods</th><th>$ERDE_5$</th><th>$ERDE_{50}$</th><th>Fmeasure</th><th>Precision</th><th>Recall</th></tr>
<tr><td>RKMVERIA (SVM using BOW)</td><td>12.17%</td><td>8.63%</td><td>0.67</td><td>0.82</td><td>0.56</td></tr>
<tr><td>RKMVERIB (LR using BOW)</td><td>12.93%</td><td>12.31%</td><td>0.46</td><td>0.81</td><td>0.32</td></tr>
<tr><td>RKMVERIC (RF using BOW)</td><td>12.85%</td><td>12.85%</td><td>0.25</td><td>0.86</td><td>0.15</td></tr>
<tr><td>RKMVERID (RNN using GloVe)</td><td>12.89%</td><td>12.89%</td><td>0.31</td><td>0.80</td><td>0.20</td></tr>
<tr><td>RKMVERIE (AdaBoost using BOW)</td><td>12.93%</td><td>12.31%</td><td>0.46</td><td>0.81</td><td>0.32</td></tr>
</table>
</td></tr>
<tr><td>JT Wolohan et al.</td><td>Detecting linguistic traces of depression in topic-restricted text: attending to self-stigmatized depression with NLP</td><td>This paper examines whether a machine learning approach based on linguistic features can be used to detect depression in Reddit users when they are not talking about depression, as</td><td>LIWC-ontology-Linear Support Vector Machines TF-IDF weighted combinations of word and character n-grams</td><td>T^2 intervals about LIWC index scores and two classification tasks are consistent with this belief. There appear to be substantial differences in depressed users' language when they are explicitly</td></tr>
</table>

(continued)

(continued)

Author	Topic	Aim	Method	Result
		would be the case with those wary of depression stigma		discussing depression and when depression-related data is withheld the best performing model combined word- and character n-grams with LIWC features
Michael M. Tadesse et al.	Detection of depression-related posts in Reddit social media forum	Examine Reddit users' posts to detect any factors that may reveal the depression attitudes of relevant online users	LIWC dictionary, LDA topics, and N-gram features. Logistic Regression, Support Vector Machine, Random Forest, Adaptive Boosting and Multilayer Perceptron classifier	The proposed method can significantly improve performance accuracy. The best single feature is bigram with the Support Vector Machine (SVM) classifier to detect depression with 80% accuracy and 0.80 F1 scores. The strength and effectiveness of the combined features (LIWC + LDA + bigram) are most successfully demonstrated with the Multilayer Perceptron (MLP) classifier resulting in the top performance for depression detection reaching 91% accuracy and 0.93 F1 scores
Hayda Almeida	Detecting early risk of depression from social media user-generated content	This paper presents the systems developed by the UQAM team for the CLEF eRisk Pilot Task 2017. The goal was to predict as early as possible the risk of mental health issues from	n-grams, dictionary words, selected Part-Of-Speech (POS), and user posting frequency. N-gram features were extracted as of Bag-Of-Words (BOW), bigrams, and trigrams	

(continued)

(continued)

Author	Topic	Aim	Method	Result
		user-generated content in social media		
Maxim Stankevich et al.	Feature engineering for depression detection in social media	This paper considers different feature sets for depression detection task among Reddit users by text messages processing and evaluates the applicability of stylometric and	(BoW), tf-idf, Svm and n-gram	SVM model with the tf-idf + stl + mrph feature set achieves the best F1-score (63.36%)

	ERDE 5	ERDE 50	F1	P	R
UQAMA	14.03%	12.29%	**0.53**	0.48	**0.60**
UQAMB	13.78%	12.78%	0.48	0.49	0.46
UQAMC	13.58%	12.83%	0.42	0.50	0.37
UQAMD	**13.23%**	**11.98%**	0.38	**0.64**	0.27
UQAME	13.68%	12.68%	0.39	0.45	0.35

Table 5. Performance results on the eRisk test set

(continued)

(continued)

Author	Topic	Aim	Method	Result
		morphological features, also perform a comparison of our results with the CLEF/eRisk 2017 task report		
Hong-Han Shuai et al.	Mining online social data for detecting social network mental disorders	A machine learning framework for detecting SNMDs, namely Social Network Mental Disorder Detection (SNMDD)	SNMD-based Tensor Model (STM)	The accuracy and AUC of STM are 89.7% and 0.926. That STM is able to derive more precise and accurate latent features than Tucker to achieve the best performance in SNMDD

6 Comparison of Conclusion

Comparative analysis using different machine learning techniques such as CNN, SVM, and language processing and detecting techniques such as LIWC, BOW, TF-IDF, and crowdsourcing.

The comparative analysis of the different machine learning techniques indicates that SVM with the LIWC model is the best model for knowledge extraction and classification according to Michael M. Tadesse et al. Paper with 91% accuracy.

Hence this survey provides different papers working with different possible techniques that helps detecting depression from Twitter, it helps automatically select important and significant knowledge from a huge amount of data, to predict as early as possible with the high accuracy the risk of depression or mental illness, and raise the awareness of such a problem.

7 Conclusion

Depression is a significant contributor to the worldwide disease burden. Traditionally, by referring to the criteria of clinical depression, doctors diagnose depressed individuals face to face. However, in the early phases of depression, more than 70% of patients would not consult doctors, which leads to further deterioration of their conditions. Meanwhile, individuals are increasingly relying on social media to reveal feelings and share their daily life, leveraging social media to help identify mental and physical illnesses. Inspired by these, our job seeks to detect depression in a timely manner through the collection of social media data.

References

1. Ibrahim, A.A., and R.G., Idris. 2017. Psychometric properties of social media and academic influence scale (SMAIS) using exploratory factor analysis. *International Journal of Information and Education Technology* 7(1): 15
2. Nadeem, M. 2016. Identifying depression on Twitter. arXiv:1607.07384.
3. Shen, G., J., Jia, L., Nie, F., Feng, C., Zhang, T., Hu, T.S., Chua, and W., Zhu. 2017. Depression detection via harvesting social media: a multimodal dictionary learning solution. In *IJCAI* 3838–3844.
4. Zaydman, M. 2017. Tweeting about mental health. Doctoral Dissertation, Pardee Rand Graduate School.
5. Wang, X., C., Zhang, Y., Ji, L., Sun, L., Wu, and Z., Bao. 2013. A depression detection model based on sentiment analysis in micro-blog social network. In *Pacific-Asia conference on knowledge discovery and data mining* 201–213. Berlin: Springer.
6. Chan-Olmsted, S.M., M., Cho, and S., Lee. 2013. User perceptions of social media: A comparative study of perceived characteristics and user profiles by social media. *Online Journal of Communication and Media Technologies* 3(4): 149–178.

7. Glucksman, M. 2017. The rise of social media influencer marketing on lifestyle branding: A case study of Lucie Fink. *Elon Journal of Undergraduate Research in Communications* 8(2): 77–87.
8. Wirback, T. 2018. Depression among adolescents and young adults: Social and gender differences.
9. Barbara, Hocking, and Paul, Morgan. 2013. A life without stigma: A SANE report. Victoria: SANE Australia.
10. Amedie, J. 2015. The impact of social media on society.
11. De Choudhury, M., M., Gamon, S., Counts, and E., Horvitz. 2013. Predicting depression via social media. In *Seventh international AAAI conference on weblogs and social media.*
12. De Choudhury, M. 2013. Role of social media in tackling challenges in mental health. In *Proceedings of the 2nd international workshop on Socially-aware multimedia* 49–52. New York: ACM.
13. Guntuku, S.C., D.B., Yaden, M.L., Kern, L.H., Ungar, and J.C., Eichstaedt. 2017. Detecting depression and mental illness on social media: an integrative review. *Current Opinion in Behavioral Sciences* 18: 43–49.
14. Seabrook, E.M., M.L., Kern, and N.S., Rickard. 2016. Social networking sites, depression, and anxiety: A systematic review. *JMIR mental health* 3(4): e50.
15. Orabi, A.H., P., Buddhitha, M.H., Orabi, and D., Inkpen. 2018. Deep learning for depression detection of twitter users. In *Proceedings of the Fifth Workshop on Computational Linguistics and Clinical Psychology: From Keyboard to Clinic* 88–97.
16. Shrestha, K. 2018. Machine learning for depression diagnosis using twitter data.